COMPREHENSIVE AGRICULTURE RESOURCE BOOK
For college and high schools in Eastern and Southern Africa
Full colour version

BY

AUGUSTINE AFULLO AND BENJAMIN DANGA

CO-PUBLISHED BY
1. WAMRA TWECHNOPRISES,
P.O. BOX 36665-00200,
NAIROBI, KENYA
+254722690956
Cell: +254-722690956

2. CREATE-SPACE, USA

Printed by
AMAZON, USA

All Photographs used in this book have been produced and processed by the authors except where acknowledged.

First published 2014
Set in Sylfaen 10/12

CITATION:
AFULLO A AND DANGA B (2014) COMPREHENSIVE AGRICULTURE RECOURSE BOOK. Full colour Version. WAMRA TECHOPRISES, NAIROBI, KENYA AND CREATESPACE ONLINE EDITION

ISBN: 9966-7205-6-1

DEDICATION
This book is dedicated to all its readers, who are committed to eradicating food shortage and hunger in the world, especially Sub-Saharan Africa, as well as to Governments which make it their business to make food security a priority in their conunties.
To be one of those the book is decidicated to, Sign if you believe in this cause.
Sign……………………………………………………………………..
Name……………………………………………………………………..

PUBLISHED BY
WAMRA TECHNOPRISES,
P.O. BOX 36665-00200,
NAIROBI, KENYA.
Tel: +254-722690956/ 738410345
wamra_technoprises_org@yahoo.com
First published 2014
Set in Sylfaen 10/12
ISBN: 9966-7205-6-1

About The Authors

Professor Augustine Afullo is an environmental health consultant, manager and trainer. He holds a doctoral (PhD) degree in Environmental science from Commonwealth Open University, British Virgin Islands (UK), a Master of Science degree in water and environmental management (MSc-WEM) from WEDC's Loughborough University, a Master of Philosophy degree in Environmental Health (MPhil- Environmental health) from Moi University, Eldoret, Kenya and a Bachelor of Science degree in Agriculture (BSc Agric) from the University of Nairobi, CAVS, Kabete, Kenya. Augustine has professional post-graduate qualifications in health education, environmental science, education and professional certificate qualifications in occupational safety and health (OSH) (chartered institute of environmental health (CIEH-UK)), research methods (RSA) and landscaping, ornamentals and lawn management (BCA, Botswana). Professor Afullo is a former Assistant professor in the North Central College, Naperville, Illinois, USA. He is an alumnus of the Fulbright Scholar in Residence (SIR) and commonwealth association alumni. He is also an associate member of the African Waste and Environment Management Centre (AWEMAC), a National Environment and Management (NEMA-Kenya) registered and licensed lead expert in environmental assessment, a registered trainer in the chartered institute of environmental health (CIEH-UK), a former member of the chartered institute of water and environmental management (UK), and the African studies association (ASA). Augustine has over 20 years of teaching, capacity building, training, lecturing and research experience in environmental management, water, sanitation, and hygiene (WASH) in Eastern, Southern Africa and the United States of America. He has done many environmental impact assessments, monitoring and evaluation (M&E) tasks as the lead consultant, and supervised at least forty post graduate (advanced diploma, Masters and doctoral) students. Dr Afullo has published widely in refereed environment, occupational health, sanitation, hygiene, water, food and nutritional security and public health journals. Augustine is currently a reviewer of a number of academic journals as well as a member of the advisory board of the world research journal of environment and waste management.

Dr Benjamin Danga is a Soil scientist, trained agricultural educationist and a consultant in Food security, and Integrated Soil fertility management. He holds a doctoral (PhD) degree in Soil Science from Egerton University (with Israel), a Master of Philosophy degree in Soil Science from Moi University, Eldoret, Kenya and a Bachelor of Science degree in Agricultural education and extension (AgEd) from the Egerton University, Kenya. Benjamin has over 20 years of teaching and teacher training, capacity building, lecturing and research experience in food security sustainable production systems, soil and water quality, water supply, sanitation and hygiene in disadvantaged communities especially in Arid and semi-arid areas and schools. He has won at least five academic research grants, and did at least twenty monitoring and evaluation (M&E) assignments as the consultant. In addition, Ben has supervised over ten post graduate students. Dr Danga has published widely in refereed agronomy, soil fertility, food security, water use efficiency, sanitation, hygiene, and water journals.

Table of Contents

MODULE 1: GENERAL AGRICULTURE

1.1 TRENDS OF AGRICULTURE
1.2 ENVIRONMENTAL INFLUENCES
1.3 PRINCIPLES OF LAND USE
1.4 AGRICULTURAL TECHNOLOGY
- DEVELOPMENTAL STAGES
- IMPACT ON THE ENVIRONMENT
- TEMPERATURE AND EVAPORATION

MODULE 1: GENERAL AGRICULTURE

1.1 INTRODUCTION

Definition of Agriculture: The word “Agriculture” (Pur, Temo thuo, Kilimo) is derived from two Latin words, Ager – meaning land or field; & Cultura – meaning cultivation. Therefore the term agriculture means cultivation of land for food production. Thus agriculture or farming is generally understood to mean the production of crop plants through cultivating the soil and the rearing of livestock. The foods can be from plants or animals, and some of the plants may not be directly edible-but are instead produced for other purposes e.g. industrial or as cash crops.

Agriculture is also understood to mean the science and art of producing crops and livestock for economic purposes. On the other hand, a concise definition of agriculture is given as the science and art of cultivating the soil, production, feeding and management of crops and livestock, preparation and processing of their products and by-products for the use of man. It also involves the disposal of these products through marketing. Thus, the small-scale industries connected with the processing of agricultural products also come under agriculture.

Agricultural production involves use of natural resources of the earth for the production of food, fodder, Natural fibers, lumber, energy, medicine etc. The Agricultural resources used for agricultural production include;-

- Agricultural land
- Agricultural labor
- Water
- Agricultural machinery
- Fertilizers

Land is the most important resource, and the primary aim of agriculture is to cause the land to produce more abundantly, and at the same time, to protect it from deterioration and misuse. It is synonymous with farming–the production of food, fodder and other industrial materials.

Agriculture as art, science and business of crop production;

Agriculture is defined as the art, the science and the business of producing crops and the livestock for economic purposes.

As an art: it embraces knowledge of the way to perform the operations of the farm in a skillful manner. The skill is categorized as;

Physical skill: It involves the ability and capacity to carry out the operation in an efficient way for e.g., handling of farm implements, animals etc., sowing of seeds, fertilizer and pesticides application etc.

Mental skill: The farmer is able to take a decision based on experience, such as (i) time and method of ploughing, (ii) selection of crop and cropping system to suit soil and climate, (iii) adopting improved farm practices etc.

In case of students, agriculture as an art means involvement in practical e.g. budding, grafting, planting, planting, thinning, pruning, emasticulation, cultivation etc

As a science: It utilizes all modern technologies developed on scientific principles such as crop improvement/breeding, crop production, crop protection, economics etc., to maximize the yield and profit. For example, new crops and varieties developed by hybridization, transgenic crop varieties resistant to pests and diseases, hybrids in each crop, high fertilizer responsive varieties, water management, herbicides to control weeds, use of bio-control agents to combat pest and diseases etc.

As the business: As long as agriculture is the way of life of the rural population, production is ultimately bound to consumption. But agriculture as a business aims at maximum net return through the management of land, labour, water and capital, employing the knowledge of various sciences for production of food, feed, fibre and fuel. In recent years, agriculture is commercialized to run as a business through mechanization

1.2 WHY IS AGRICULTURE IMPORTANT

Agriculture is important to man because it provides us with a reliable source of food, clothing, and shelter. In East Africa, nearly 80% of the population depends on agriculture for a living. Besides, agriculture contributes 40% of the region's GDP (IFPRI, 2004), thus it remains a key driver of the Eastern Africa's economies.Many developing countries in sub-Saharan Africa have agriculture-based economies. The agricultural sector plays a significant role in the economy in four key respects:

(i) It provides a basis for subsistence of the majority of the population. About 80% of the population living in rural areas derives their livelihoods mainly from agricultural activities.

(ii) It is the largest employer in the economy, accounting for 60% of the total employment

(iii) It provides, directly and indirectly, the primary source of foreign exchange earnings, with horticultural products, tea and coffee being leading earners.

(iv) It is central to food security.

Agriculture also has a high multiplier effect, which means that agricultural investment can generate high economic and social returns and enhance economic diversification as well as social development

1.3SCOPE AND BRANCHES OF AGRICULTURE

Agriculture is an applied science which makes use of the knowledge of the basic sciences such as Biology, Chemistry, Physics and Geography. It is therefore a subject discipline with a wide range of activities. Its main branches therefore comprise the following specialized disciplines.

Agronomy – crops and soil sciences

The word agronomy has been derived from the Greek work "agros" meaning fields and "nomes" meaning management. It is the specialized branch of the agricultural science dealing with crop production and soil management. It is the study of the relationship between crops and soils.While soil science involves the study of soil components as they relate to crop production, crop science is concerned with the physiological development and production of crops for the use of man.

Animal Science

It studies the scientific methods of animal breeding, feeding, management and disease control for obtaining the best economic returns from animals (livestock) and their various products and also the processing of products derived therefore. Animal breeding deals with the genetic improvement of animals by selection, hybridization etc while animal nutrition is the study of feeds, ingestion, digestion and assimilation of the food materials by the animals. It includes Livestock production and management, which is the science that deals with the production, rearing management of the animals that are kept on the farm (farm animals) other than poultry. They include dairy and beef cattle, sheep, goats, pigs etc

Horticulture

It is that branch of agricultural science which relates to the raising of gardens or orchards including the growing of vegetables, fruits, flowers and ornamental shrubs and trees. Horticulture can be further divided into (1) Pomology or fruit cultivation; (2) Olericulture or vegetative cultivation (3) Floriculture or cultivation of flowers and ornamental plants; and (4) Fruit and vegetable preservation.

Agricultural Engineering

This is the study that deals with the farm machinery and mechanization. Includes study and application of agricultural Farm Machinery, Farm Structures, Irrigation and Drainage Engineering. It is the science dealing with the application of knowledge, techniques and disciplines of various fields of engineering to the solution of problems arising in the field of agriculture and rural living, with the object of reducing labour, improving agricultural productivity and raising the standard of living of farmers and increasing the over-all earning per worker in the field Agricultural Engineers are inventors of agricultural implements, they design fabricate and maintain various agricultural implements and equipment used within the agricultural sector.

Agricultural Economics

It is the branch of economics that is concerned with the farm management and production. It includes Agribusiness, Agricultural Marketing, Rural Community Development.

Pisciculture (Fisheries) and Aqua farming

This refers to the production and management of fish and their products. The branch includes aquaculture which is the raising of plants or animals as fish or shell-fish, in a sea, lake, river, or any other body of water.

Apiculture (Beekeeping, apiology) - It is the science and art of raising honey bees, beekeeping and bee-culture.

Sericulture - It is the science that deals with the raising of silk cocoons, the basic material for the production of silk.

Agricultural Biology - This is another discipline related to crop as being affected by pests and diseases and their control for increased agricultural production. It therefore deals with plant pathology (diseases and disorders of plants), entomology (insect pests and their management) and nematology (study of nematodes in relation to crop plants).
Forestry - This relates to the cultivation of economic trees for the use of man.

Agricultural Extension - Devise and employ various methods of effectively transmitting agricultural innovations to the farmers. The agricultural extension worker is mainly concerned with the behaviour of the rural population and their attitudes toward changes. They help to build up local farmers groups and organizations so that they can benefit from extension programmes. Agricultural extension therefore provides the indispensable elements that farmers need to improve their agricultural productivity.

Agroecology – deals with Agroecosystem analysis, Biodiversity, Climate change and agriculture, Ecology & Ecosystem.

Wild Life - This is the discipline that deals directly with the production and management of wildlife.

Thus agriculture is related to almost all subjects, and borrows a lot of ideas from biology, chemistry, economics, engineering, mathematics, geography, physics, among others. Indeed, it is a life science and a general agriculturist can specialize later in literally any field of choice later in life.

1.4 TRENDS OF AGRICULTURE

Agriculture as an art and science developed over millions of years from the most rudimentary methods such as gathering and hunting, to the use of sticks, and subsequently to the most advanced and sophisticated stage of use of machinery. This subtopic will be discussed under (i) Developmental stages (ii) Impact on the environment

1.4.1Developmental Stages In Agriculture:

The various stages in agricultural development are:

- Hunting and gathering
- Domestication
- Shifting cultivation
- Settled agriculture (subsistence)
- Settled agriculture (Commercial farming

(a) Hunting & Gathering:

Ancient man used to survive by collecting roots, leaves, barks, fruits and seeds of plants from the bushes and forest. This is called gathering. But wild animals were also sought for, chased, trapped and killed, then slaughtered as a source of meat and skin. This is called hunting. Hunters could travel on foot for long distances. The forests and bushes were very rich in plant and animal materials for gathering and hunting. With time, the human population started increasing, and the food resources also started reducing. There was thus more competition for the limited resources and often-caused conflicts among the hunters and gatherers.

Before the invention of agriculture, all humans probably obtained the food they needed for survival through hunting for animals, fishing, or gathering. Hunters and gatherers lived in small groups. i.e. hunter gatherers settled in small communities in the Fertile Crescent, a narrow band of land arcing across the Near East. The men hunted game or fished, and the women collected fruits, berries, nuts, and roots. This division of labor sounds like a stereotype but is based on evidence from archaeology and anthropology. The group traveled frequently, establishing new home bases or camps. The direction and frequency of migration depended on the movement of game and the seasonal growth of plants at various locations. New evidence suggests that people cultivated rye over 13,000 years ago, while continuing to hunt animals and gather wild plants in the Fertile Crescent. Hunter-gatherers turn to agriculture only as a last resort when population pressure forces them to acquire more food on available space.

Figure 1: The Early Man – as (a) hunters and (b) gatherers. Source: From Mazoyer, M., 2006

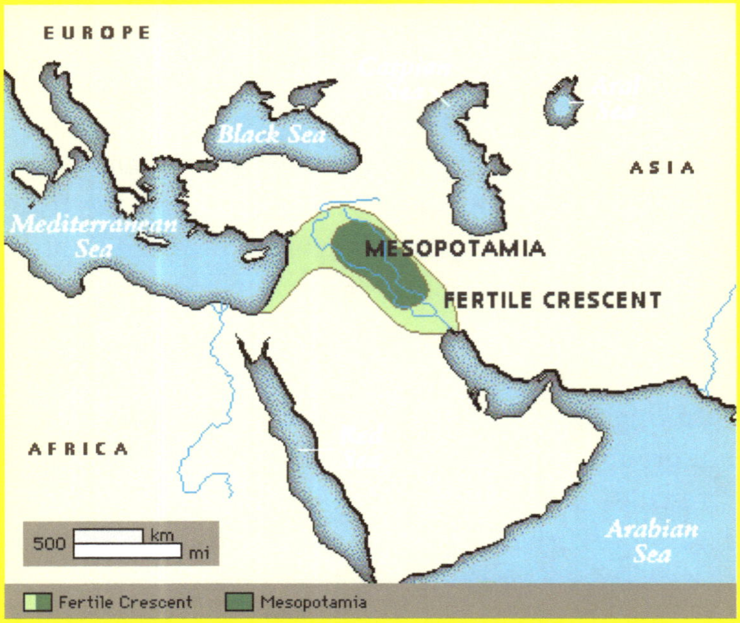

Figure 2: Map showing location of the Fertile Crescent. Source: From Mazoyer, M., 2006.

Today, contemporary hunting and gathering societies are isolated groups living on the periphery of world settlement, but they provide insight into human customs that prevailed in prehistoric times, before the invention of agriculture. It is estimated that only a quarter-million people, or less than 0.005% of the world's population, perhaps still survive by hunting and gathering.

Impact on the environment:

Hunting and gathering initially had minimal effects on the environment. However, with the population growing, there was, and has been a tendency of over-exploitation of the animal and plants resources used as food, medicinal, ornamental. Excessive illegal hunting, now called poaching, has led to extinction of some vital plant and animal species.

(b) Domestication

On the other hand, man had learnt that the seeds of fruits collected from the wild could germinate and grow right in their homesteads. These plants seemed to grow better in fertile soils. This encouraged man to start introducing the same to their homesteads. Thus, within a short time, cereals, pulses and other crops were soon being cultivated in homes/ gardens. Some were even moved long distances across continents by the very early adventurers such as Europeans and Arabs coming to Africa. They brought new seeds and plant materials with them, and they also carried some back to their homes. The earliest crops to be cultivated (sown and reaped) were wheat, barley, sorghum and millet". With this movement also came some of the worst weeds ever seen.

Similarly, man started developing a close relationship with some friendly and docile animals that served as sources of his food. These included goats, sheep, cattle, pigs, chicken, etc. He took them home and used them as source of eggs, milk or meat. However, the first animals to be kept were oxen and goats. Man tamed these animals and herded them in groups in his settlement. Since then, they have become increasingly dependent on man for existence. "The process of man introducing and growing plants and rearing animals brought from the forest / wild into homes so that they are within his control is called domestication. These livestock resources have been tamed more and more overtime by selection and culling, and their yields have also tremendously increased. The same method led to taming of domestic pets such as dogs, cats etc. The process continues, with some people having managed even to domesticate some snakes, buffaloes, etc.

Summary of Timeline of the Inventions of Agriculture

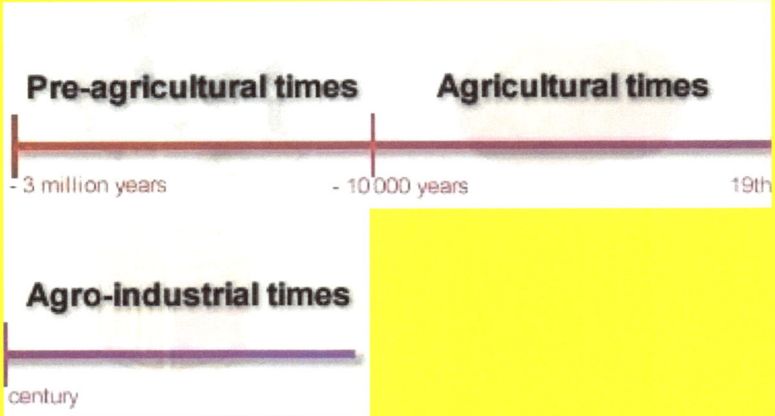

- 9500BC - Earliest evidence for domesticated wheat.
- 8000 BC - Evidence for cattle herding
- 7000 BC - Cultivation of barley; animals are domesticated
- 6500 BC - Cattle domestication in Turkey
- 6000 – 5000BC - Irrigation aids farming
- 5500 BC - Sumerians start organized agriculture
- 5400 BC - Archaeological proof for domestication of chicken
- 5000BC - Africa grows rice, sorghum
- 4000BC - Ploughs make an appearance in Mesopotamia
- 4000BC - Horse is domesticated in Ukraine
- 3000 BC – Turmeric is harvested at Indus Valley
- 3000 – 2700 - Maize is domesticated in Americas
- 2737BC - Tea is discovered – in china
- 2000 BC - First windmills in Babylon
- 1000 – 600 BC - Sugar processing in India
- 500 BC - Row cultivation in China
- 200 – 100 BC - Multi-tube seed drill invented in China
- 700 – 1300 AD - Arab Agriculture Revolution
- 1000 AD - Coffee originates in Arabia
- 1492 - Columbian exchange changes agriculture
- 1599 – First practical greenhouse is created
- 1658 - Ranch operation reaches the U.S
- 1700 – 1850 - British Agricultural Revolution
- Mar 14, 1794 - Cotton gin is invented
- 1800 – 1900 - Chemical fertilizer began to be used

- 1837 - John Deere invents steel plough
- 1860 – 1870 - Hay cultivation changes
- 1866 - Gregory Mendel describes Mendelian inheritance
- 1879 - Milking machine replaces hand milking
- 1892 - First practical gasoline-powered tractor
- 1900 – 1950 – Birth of industrial agriculture
- 1930 – 1935 – First aerial photos for agriculture
- 1939 – 1970 - DDT becomes a rage
- 1944 - Green Revolution begins in Mexico
- 1972 - Organic movement starts taking roots
- 1996 - Commercial cultivation of genetically modified plants

Environmental effects of domestication:
Populations of animals that were heavily preyed on by others have increased due to protection from man. Also, an ecological imbalance in some cases occurred due to sudden withdrawal of important participants from the food chain- e.g. goats' removal from bushes left their preys and predators unattended. This change in population of plants and/ or animals resulting from withdrawal of a member of a food chain of which they are a part is an example of an ecological imbalance. If the effects are extreme, it can lead to ecological catastrophe.

(c) SHIFTING CULTIVATION:
Man started farming but soon realised that crop yields continued to fall if produced in the same field repeatedly for a long time. Thus he took to shifting cultivation. This means clearing, cultivating and planting crops in one field for some years then move to a fresh site when the old site is exhausted. As an exhausted field was left to rest, up to 10 years or more could pass before it was used again. Thus the farmer kept shifting his farming activities from one part of his field to the other. He mostly used slash and burn methods. At this stage, only crude tools and equipment such as metals, sharpened sticks, etc were used, and food was produced purely for family consumption. The production of food enough for only the family is called subsistence farming. Later, metal hoes and axes, and more recently, ox ploughs were introduced. This increasing level of sophistication in farming also increased efficiency and therefore productivity (yields).

Environmental Impacts of shifting cultivation:
Bushes are cleared, leading to loss of vital plant and animal resources- especially those considered as weeds.Often, fire is used to burn what has been slashed. This at times causes ecological imbalances in the fields as fire would eliminate completely those plants without enough underground reserves since they have no chance of regenerating.

(d) SETTLED AGRICULTURE.
When population started increasing, there was less and less land for continued movement and shifting. The farmers therefore started settling down. This increased their capacity to produce more food than the family needs. At the same time, the farmer also realized that there were often items that he did not have and was not able to produce which the neighbours had. The crop and animal (food) resources were in such times used as a medium of trade, with the food items being exchanged with the goods, which the farmer lacked. This was called barter trade. However, some people were not able to produce food for themselves and had to buy. Settled agriculture encouraged more and more production to meet this challenge caused by demand. Soon, some farmers started specializing in producing specific food items. This increased their farming efficiency. It also encouraged land consolidation in areas where land had been subdivided into small uneconomical units. With larger pieces of land, mechanized farming is most efficiently done.

Intensive agriculture using sack gardens, common where land is scarce, or in urban farming Right: Multi-storey garden

Environmental Impacts of settled agriculture

There is over-exploitation of resources around the homesteads, leading to land degradation, soil erosion and related hazards. Bushes and forests are cleared to pave ways for settlement; this at times lead to extinction of some rare plant and animal species. There are land squabbles as dishonest neighbours push boundaries.

Livestock-crop conflict ensues.

1.5 AGRICULTURAL TECHNOLOGY

Definition: - Agricultural technology is the use of methods, ideas and skills that make crop and livestock production easier and more efficient. Some technologies (ideas) are modern, and others traditional. Examples of agricultural technologies include

- Use of machines to cultivate, harrow, harvest and process farm produce (i.e. mechanization). Examples are tractors, combine harvestors, cultivators, planters etc.
- Use of pesticides such as fungicides, nematicides, herbicides etc which reduce farm losses; Examples are dithane M 45, Furadan, 2,4-D, DDT, etc

- Use of organic fertilizers (manures) that enrich the soil and make the crops grow better, and maintain the soil structure, rendering agriculture sustainable. E.g. kraal manure, compost manure, poultry manure etc.
- Use of inorganic fertilizers which add specific nutrients to soils, thereby making them available to crops- e.g. Urea, Super-phosphates, CAN / LAN, 2:3:2, etc
- Supplementing natural water through irrigation

1.6 Farming systems as a technology transfer approach.

Farming systems include mixed farming, crop rotation, monoculture, large-scale commercial farming, etc. All, however, began with hunting and gathering. Different systems were and are still used at different times, and often gave way for others depending on the new problems and ideas thought to be able to help people cope with these problems (technologies). At each farming stage, there was a problem for which a solution had to be sought. Through this, different farming systems have been adopted, each with its own weaknesses, which led to discovery of the new ideas and thus necessitating a change of system. Thus technologies (ideas) have been gradually transferred from one stage and farming system to another.

At hunting and gathering stage, people were walking long distances to get food. This became a problem as population grew. Thus they resorted to domestication. This involved selection, planting (using varied spacing which was improved with time and experience). Domesticated animals and plants (crops) were selected for superiority, while others were slaughtered / or used as food (culled). This formed the basis of controlled breeding of crops and livestock. Best bulls were used as sires, and others were not given a chance (e.g. bull exchange programme in villages). Related animals were not allowed to breed, thus reducing / minimizing inbreeding.

As small-scale farmers, first settled farmers realized there were increasingly lower yields as the same crop and field was used. They thus started using mixed cropping and shifting cultivation. However, land became scarce thus they could not shift any more. They resorted more to mixed farming as a way of minimizing total losses, as well as using biological pest repellants interplanted with other crops. This formed the basis of biological pest control. They also resorted to crop rotation. With more people being unable to produce, demand for food for sale increased. Thus some farmers resorted to large-scale farming. To accomplish this well, they had to mechanize so as to be more efficient. Thus the idea of use of improved tools and equipment in agriculture came into being.

To properly manage large-scale farming, people resorted to monoculture. This was to ease the use of e.g. combine harvesters, which could not be used well in mixed farm systems. The monoculture brought with it a lot or risks- including possibility of total crop loss due to diseases and pests. This led to the discovery and use of pesticides (herbicides, acaricides, nematicides, bactericides, insecticides etc). With time, the demand for food increased once more and land was becoming less and less- with no extra space for extensive farming. This led to an increased use of inorganic fertilizers, which were able to release their nutrients quicker, and facilitate faster growth of crops, and give higher yields. Again, it was realized that still with all these inputs (fertilizers, pesticides etc) it was not possible to produce more that a given level for each crop. People thus resorted to looking for and using high yielding crop and livestock varieties and breeds. Selection became more intense. This led to the production and use of the hybrid seeds. Once in a while, it was possible to run at a complete agricultural loss due to drought. People thus resorted to irrigation. But with water becoming scarce, more efficient use of irrigation water has been reached by getting more refined irrigation methods.

MODULE 2: THE PLANT AND PLANT GROWTH

2.0 INTRODUCTION
2.1 Genetic Factor / Crop Variety
2.2 Environmental Factors
2.3 PRINCIPLES OF PLANT GROWTH
Growth
Movement (tropisms)
Reproduction
Respiration
Excretion
Sensitivity
Nutrition
2.4.1FUNCTIONS OF PARTS OF A PLANT
ROOTS
STEMS
THE PLANT LEAF
FLOWERS
2.5FOOD STORAGE IN PLANTS

MODULE 2: THE PLANT AND PLANT GROWTH

2.0 INTRODUCTION

Plant growth is defined as the progressive development of an organism. It is expressed in terms of dry weight, height, or diameter. Plant growth is affected by genetic factors and environmental factors. The environment components are climate, edaphic (soils), pests / diseases etc. Knowledge of these factors and how they affect plant growth is important because we can manipulate them to our advantage- to maximize the returns from farming.

2.1 Genetic Factor / Crop Variety

Crop species and varieties that are genetically superior, or capable of making maximum use of the supply of available nutrients should be selected. Planting of new high yielding crop varieties and hybrids results in large increases in crop yields. However, the high crop yields produced with new hybrids and crop varieties will require more plant nutrients. This is because under poor / low fertility conditions, the higher yielding varieties cannot develop their full yield potential. In fertile soils, the higher yielding varieties will deplete the soils more rapidly leading to declining yields if supplemental nutrients are not supplied. Like soil nutrient supplementation, water supply, pests control etc must be kept at optimum to ensure the maintenance of the productivity of high yielding varieties. Hence it is important to ensure optimum management of all factors influencing crop yield. This is further explained by the limiting factor concept.

2.2 Environmental Factors

The Environment is the aggregate of all the external conditions and influences affecting the life and development of crop plants. The major factors include:

Temperature
Moisture supply
Light / radiant energy
Composition of the atmosphere / Air
Soil physical characteristics eg texture, structure & composition of soil air
General soil fertility / supply of mineral nutrient elements
Soil reaction / pH / acidity
Biotic factors – weeds, pests & diseases of crops
Absence of growth-restricting substance

Soil management e.g. Tillage, compaction, drainage, fertilizer placement
Crop management practices e.g. crop rotation, plant population, herbicides

2.3 PRINCIPLES OF PLANT GROWTH

A plant is a form of life. It is only over a period of time when changes in a plant appear. Plants exhibit the following properties of living organisms: Growth, Movement (tropisms), Reproduction, Respiration, Excretion, Sensitivity and Nutrition

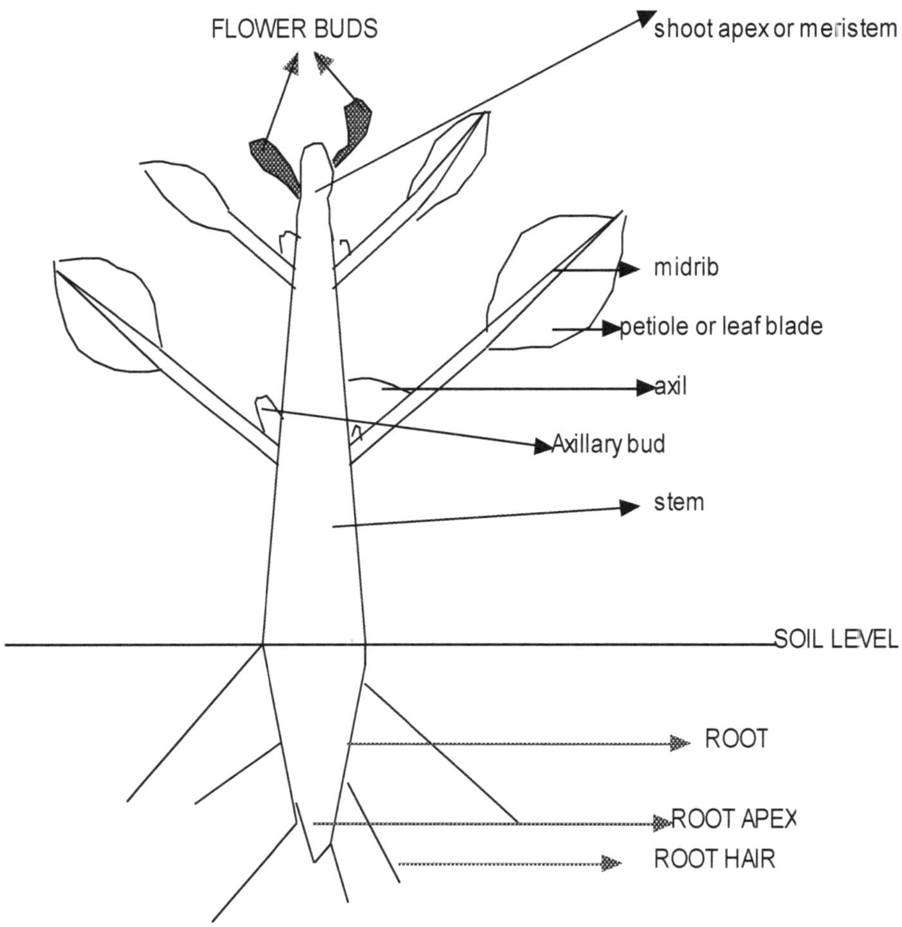

Figure 14: Parts of a Plant

A plant is made up of the root and the shoot. The root is the part below the ground, while the shoot is the part above the ground. The shoot is further made up of the stem, leaves, branches, flowers (and fruits). Between the stem and a leaf is an axil. Branches often grow from the axillary buds within the axils. An illustration is shown below.

2.4.1FUNCTIONS OF PARTS OF A PLANT

ROOTS

Hold the plant firmly in the soil. They absorb water and dissolved substances (minerals) from the soil. Some act as storage organs for food, while some also help retain moisture within the plant, especially the succulent roots which contain a lot of water. Most food storage roots are also succulent, i.e., they also store water

STEMS

They hold the shoot in place. They expose the leaves, branches and flowers to the light and air. They also transport dissolved minerals in solution from the roots to the rest of the plant,

THE PLANT LEAF

These spread out to expose a large surface area to the sunlight. They absorb this light energy and Use it for photosynthesis. They release excess heat from the plant via transpiration, thus cooling the plant. They transport manufactured foods from the leaves to the roots for use and / or storage.. Green stems also carry out photosynthesis.

FLOWERS

These carry the male and female organs necessary for the production of seeds. They are carried high on the plants so that pollinating agents can reach them. The flowers are also sources of food to many insects which draw nectar. The insects end up pollinating the plants. Through this dependence, some plants have always depended on some insects for survival and vice versa, thereby exhibiting a situation called co-evolution. This means the two have evolved together for generations.

2.5FOOD STORAGE IN PLANTS

After photosynthesis, food from the plant leaves are translocated to different parts for either storage or use. Those translocated for storage may be taken to any one or a combination of the following:

1. Stem: stem tubers (potato), sugarcane, and other trees. The starch may be stored in the collenchyma and parenchyma tissues within the cortex, while cellulose may be stored as wood within the vascular bundle;

2. Leaves: cactus, bulb onions, Opuntia (the plants just behind the Lion house office), the Aloe etc.

3. Fruits: mango, banana, pineapple, tamarind as the fleshy/juicy component, or cellulose may be deposited as protective material around the seeds, as in the case of cotton.

4. Roots: carrot, root tubers (e.g. sweet potato, cassava) etc

5. Seeds e.g. cereals (maize, sorghum, millet (see figure below), wheat, barley, oat) and pulse seeds e.g. beans, peas grams etc, and oil crops e.g. groundnuts, simsim etc. Once these are stored in the above locations, they may be used in growth of new / young plants in both vegetative propagation and in sexual reproduction.

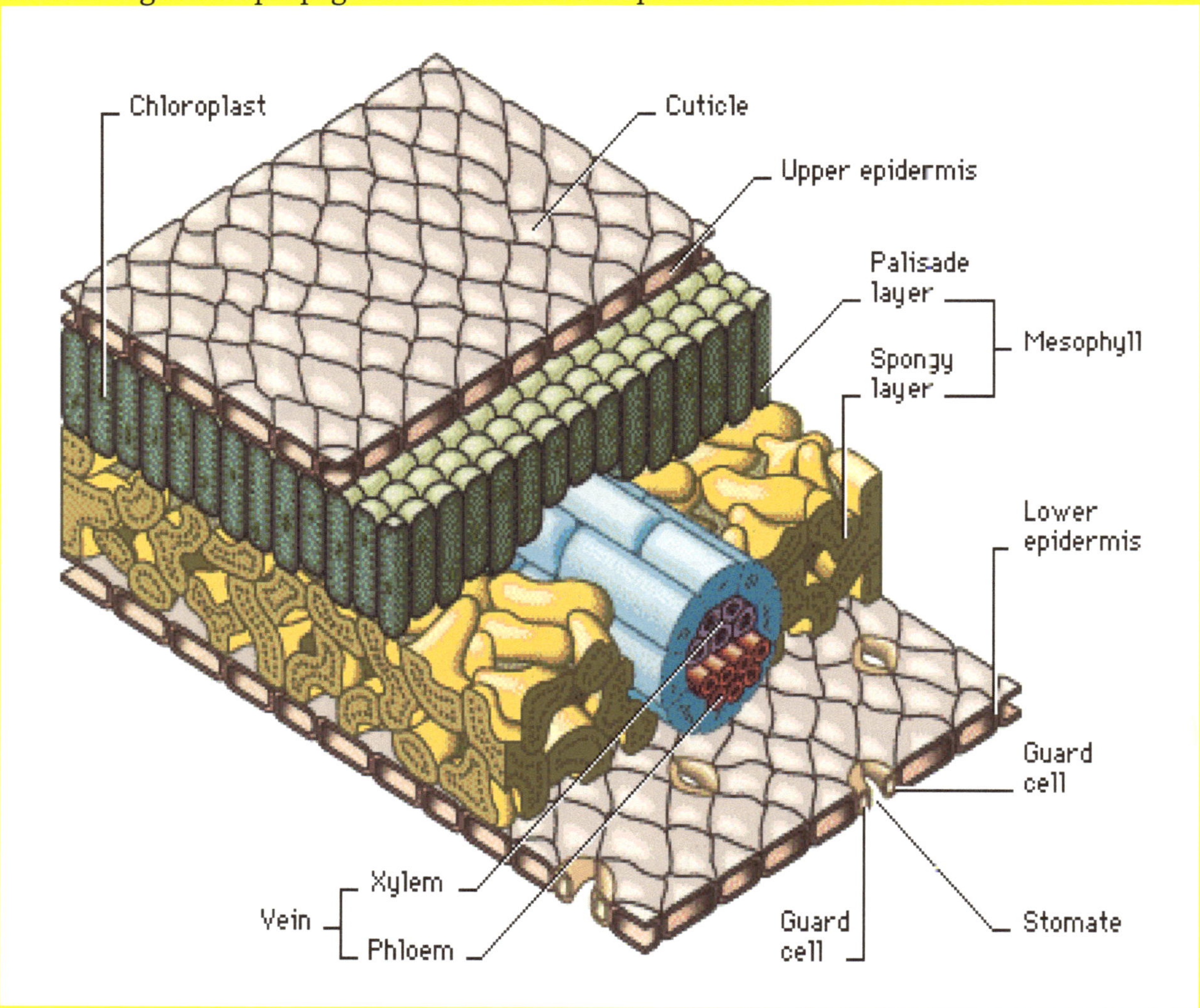

Figure 15: A cross-section of a plant leaf (source: Wikipedia)

Water is required for photosynthesis, dissolving of minerals and manufactured foods, as well as cooling of the plant. Oxygen is needed for respiration to give energy which is then used in other plant internal processes such as photosynthesis. Carbon-dioxide is a reactant in photosynthesis. Other requirements are minerals salts and other external stimuli such as light, support media etc. Support media can be soil or anything else. In hydroponics, plants grow without the soil as a medium. They are put in water with balanced nutrients. Among the minerals are major, minor and micronutrients. In a later chapter, some of the minerals and their functions in plants are described. The nutrients can come from organic fertilizers (manures) or inorganic fertilizers.

TRANSVERSE SECTION OF A LEAF

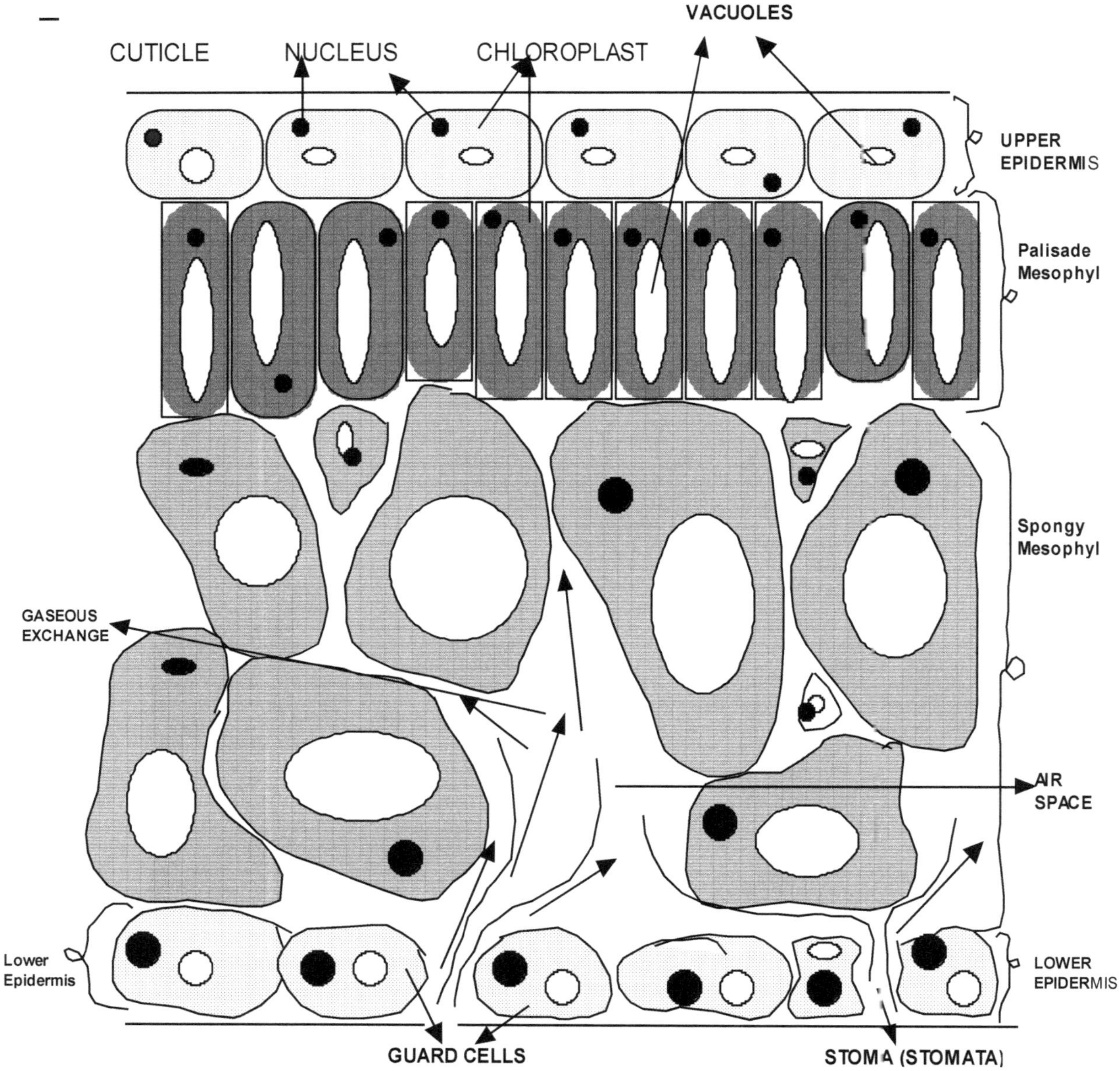

Figure 16: The internal structure of a leaf.

MODULE 3: ENVIRONMENTAL FACTORS INFLUENCING AGRICULTURE

3.12 Evaporation
3.12 Evaporation

MODULE 3: ENVIRONMENTAL FACTORS INFLUENCING AGRICULTURE

3.0 Introduction

The environment is the total sum of all conditions and circumstances inside and outside an individual. They can be classified as biotic and abiotic factors. Biotic factors are associated with living, while abiotic factors are associated with the non-living. Abiotic factors include climate and soil (edaphic), while the biotic factors include pests, disease pathogens, parasites, vectors, weeds and predators. The environmental factors discussed here include: temperature, evaporation, wind, humidity, rainfall, frost, photoperiodism and effects of day length.

3.1ABIOTIC FACTORS

3.11 TEMPERATURE AND FROST

Temperature is the degree of hotness or coldness of a place or object. It is measured in degrees Celsius or centigrade (oC), or the Fahrenheit (F) using Six's Maximum and minimum thermometer.

General effects of temperature:

Generally, rates of reactions increase with temperatures until a peak (optimal temperature) is reached beyond which the rates drastically decline, sometimes falling to zero. For man, this optimal temperature is 36.90C. For livestock, it varies with species. Some of the optimal body temperatures of some livestock are given below:

Table 1: Optimal body temperatures for different livestock

Livestock	Cattle	Sheep & goat	Pig	Chicken	Horse	Dog
Optimal body temperature 0C	38.5-39.5	38.5-40.5	30.0-39.5	40.5-43.0	37.5-38.5	38.0-39.5

Extremely high atmospheric temperatures would therefore make the animals restless, and reduce their appetite and production. Low temperatures inactivate enzymes. Enzymes are biological catalysts that change (increase or decrease) rate of reactions within the body (metabolisms). They become inactive in low temperatures, but are denatured in high temperatures. Enzymes are proteinous in nature, and tend to lose their shape and they become ineffective. In this state, they are said to be denatured. This is what happens to an egg which quickly solidifies when heated up. The temperature-reaction rate relationship is as shown below.

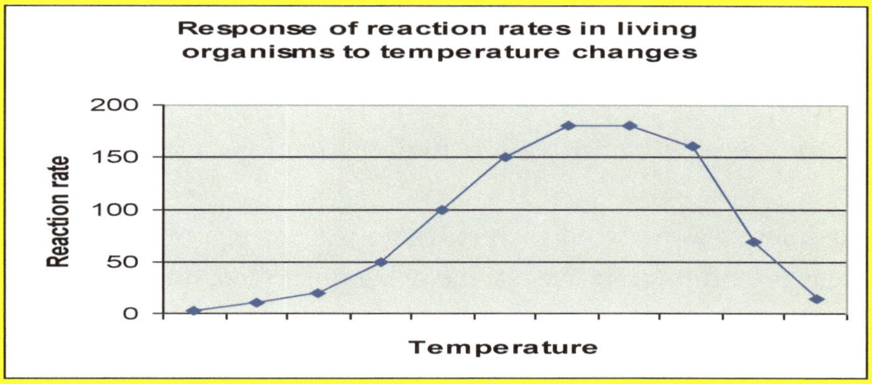

Figure 2: The response of growth rate to temperature changes

VERY LOW TEMPERATURES AND FROST

Effects of very low temperature on Agriculture (Called frost when it is extreme)

- Poor or no germination
- Very slow growth and maturity of crops
- Cold- sensitive crops often die
- Fluids may freeze inside cells and not only become unavailable to plants, but also destroy cells. This is called frost. Some crops are too sensitive to frost, including maize, beans, coffee etc
- Young and delicate animals may die from pneumonia
- Increased appetite in livestock
- Sometimes some diseases may increase e.g. a number of fungal diseases in sorghum which increase as July approaches in Eastern Africa- leading to seeds with black fungal spores (called yugni).

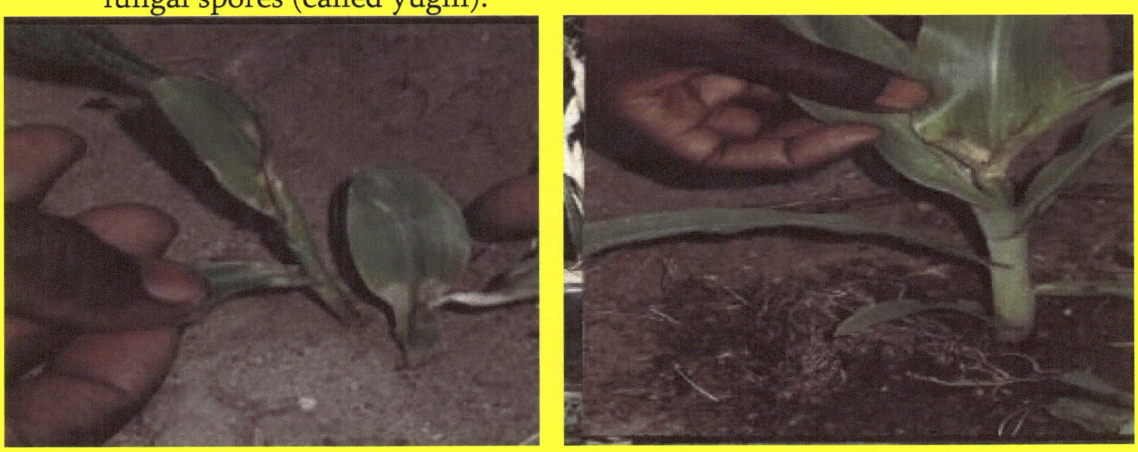

Figure 3: Maize crop affected by frost in Botswana during the may-July mild winter.

How to cope with very low temperatures.

- Build good livestock house
- Provide delicate livestock with some means of heating
- Plant cold- tolerant crops;
- Use green house to produce cold- sensitive but high value crops such as flowers and vegetable seedlings.
- Time production schedules so as to avoid such extreme seasons e.g. plant sorghum early so that July reaches after grains/seeds have been formed, and hopefully filled.

The effects of snow on agriculture.

However, when its snow, falling over a period, and standing on the surface over a couple of days, then the concept of outdoor agriculture is completely forgotten. It can only happen under foced systems such as indoor greenhouses.

Plants Uncathed by extremely cold winter weather

This is how 'tall' the snow can be during winter

When the snow falls, it has its pressure on the leves of plants, which quickly lose their capacity to sirvive, and either dry up or fall off. This transition is exhibited by changing leaf colours, and eventually giving in and drying / falling off the plants. Most leaves fall during the fall season (September to December). Come January, most plants have no leaves, and stay in the domant form for 3 months, with temperatures much below 00C (under 32F). In the Americas, this coincides with the winter season. By April, the temperatures start to improve upwards, and the buds form, later leaves, abd agriculture is possible again. This is the spring season.

The fall season: Sept- Dec Season in USA

The Early winter season (Dec-January) in the USA

The late winter season (Feb-March) in the USA

Snow capped tree (resistant plant, with leaves) in the middle of winter

Snow capped shrub able to cope with extreme winter tempearatures Early 'spring', leaves start being green, plants resurrect

Left: end of snow..Effect of snow on grass Right; Early spring; the grass springinging

VERY HIGH TEMPERATURES: EFFECTS ON AGRICULTURE

- Increases evaporation and enhances drying up of soils and water reservoirs;
- Heat stress to livestock, thus depressing appetite and growth;
- Some cold- loving crops cannot grow at all, and become stunted;
- May cause colouration and / or premature ripening of fruits;
- Increased transpiration and wilting of crops and livestock feeds;
- May cause cracking of some soils (especially the black cotton soils / vertisols) thereby interfering with growth and expansion of roots of crops

Figure 4: Maize crop affected (stunted) by very high temperatures

How to cope with high temperatures

- Mulch crops to reduce water loss from the soil surface i.e. evaporation;
- Plant heat/drought resistant and tolerant crops;
- Use greenhouse and / or shading to establish delicate crops e.g. seedlings (see figure below);
- Build a strong livestock house with a solid wall on the side facing the sun at its strongest time of the year;
- Water livestock more regularly; it is better ad lib water supply;
- Irrigate crops more frequently (may have dangers of leaching and soil erosion);
- Preserve feeds during times of plenty so as to use during extreme weather conditions when livestock are kept indoors;
- Keep livestock breeds that are heat tolerant.
- Tap solar energy so that it can be used to manipulate and modify the local environment to suit production. This is useful since the next world war is likely to be over energy. So alternative energy sources should be researched and sought.

Figure 5: The green cover used to reduce the amount of light to which crops are exposed.

Various forms of a green house

HIGH TEMPERATURES IN AFRICA

Tropical environments exhibit high temperatures reaching even 38oC. The lowest at times reach 15oC at night in some places. This is partly attributable to the high tropical and equatorial insolation. This makes Africa and the tropical world generally very hot, resulting in lots of heat stress to plants and animals. These figures are shown below:

Table 2: Insolation by region (Source: Obara and Ogonda, 1990)

	Region	Insolation calories/cm2/day
1	Temperate 50o-60oS/N	40 – 450
2	Semi arid tropics	200 – 450
3	Equatorial 8oN-8oS	400 – 450
4	Subtropical 30 - 40oS/N	750

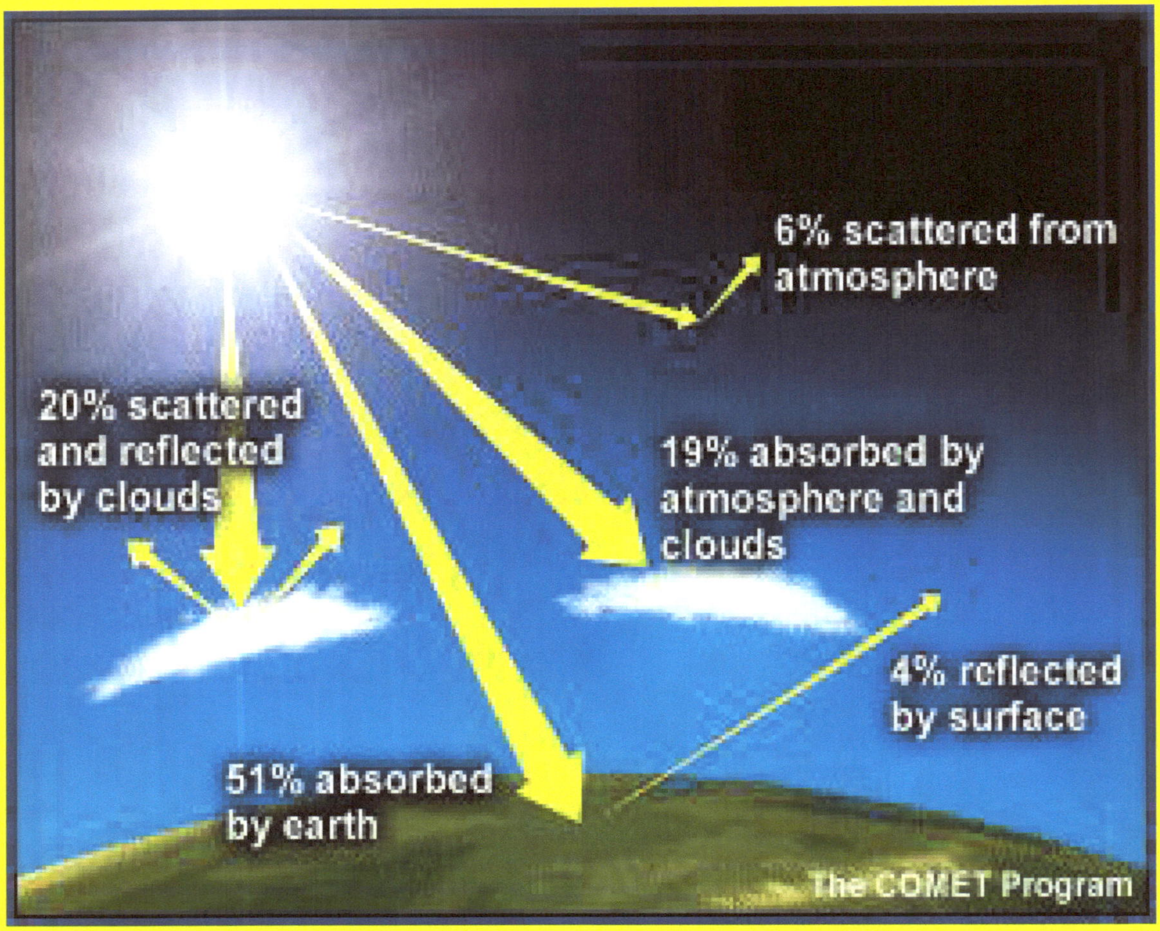

The fate of solar radiations: plants help to fix it for all living organisms (Source: Wikipedia)

The insolation of the tropical region is about 2 million Kcal/km2/yr (8400 MJ). This makes the temperatures of this region always high throughout the year. The high temperatures cause hot climates, which increase the evaporation and transpiration losses from the surface and plants respectively. This loss therefore worsens the moisture availability in the tropics. Most of the time, there is a net moisture deficit, rendering only stress resistant or tolerant plants and animals to be able to perform well. In Botswana- in Southern Africa, there has been a record 520C recorded. The common daily maximum temperature is 32-420C between August and February every year. Yet, solar power is not used by more than 1% of all energy requirements. This is the gold mine Africa should strive to tap and utilize. In the temperate counties, however, having sunshine does not necessarily translate to heat, as can be seen in this snow-filled ground on a sunny day.

Snow capped ground surface on a sunny day

3.12 EVAPORATION

Evaporation is the change of the state of water from liquid to gaseous state. This means the vapour resulting from the process simply disappears, and thus the water is lost. Evaporation may take place from the soil surface or from a water-body such as a dam, lake, river etc. Loss of water by a plant through its leaves is called transpiration. It may be increased by wing, high temperatures, increased surface area e.g. by ploughing etc. Intensive cultivation, e.g. many soil operations e.g. ploughing, harrowing, discing etc break soil clods, increase soil surface area and generally increase evaporation. This can be reduced by minimum or zero tillage. It can be measured using an evaporation pan.

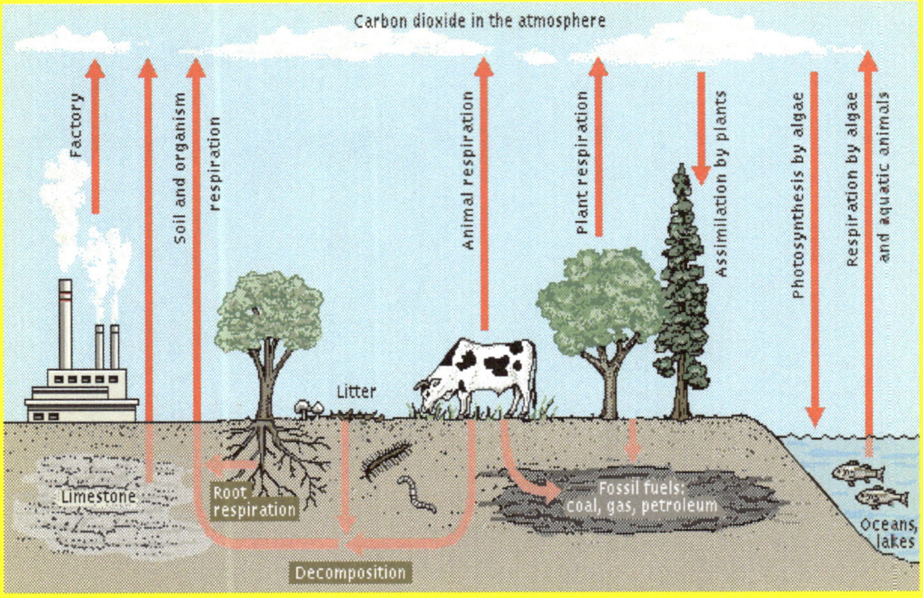

BOX 2: THE CARBON CYCLE Source: Wikipeadia

Positive effects of evaporation

- Is a vital player in the water cycle leads to water vapour which may condense and fall back as rainfall.
- Increases the amount of water in the atmosphere (humidity), and thus reduce transpiration loss.
- Leads to quick relief from water logging or flooding;
- Evaporation from the surface of animal bodies has a cooling and comforting effect.

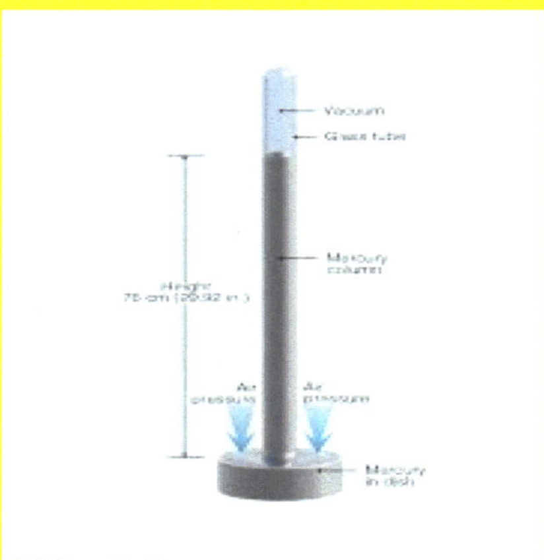
Mercurial barometer

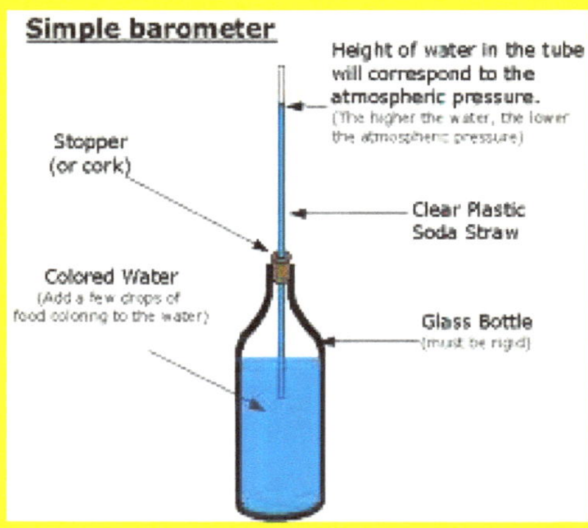

Simple barometer

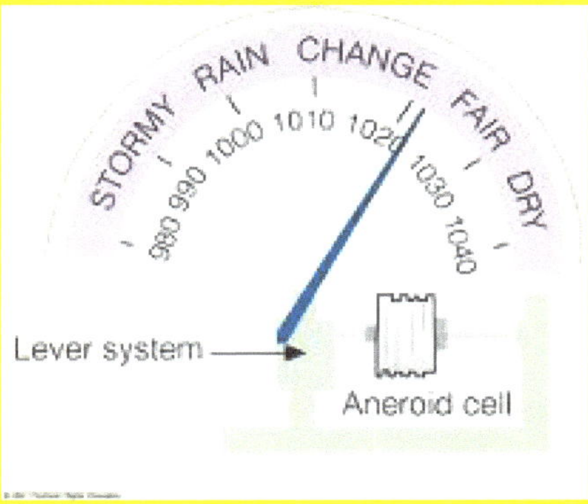

Figure 6: Instruments for measuring environmental factors affecting agriculture (Air pressure); Left: Aneroid barometer; Right: Pocket barometer

Negative effects of evaporation

- Plants may wilt since the soil becomes drier. They may die later;
- Reduced crop and livestock yields resulting from
- moisture stress/deficiency;
- Reduced feed since less plants will grow;
- May lead to high soil temperatures;

How to cope with the negative effects of evaporation

- Mulch crops to reduce evaporation;
- Plant drought resistant crops;
- Irrigate;
- Use shade trees to reduce moisture stress.

3.13 WIND

This is moving air (or air in motion). Its speed is measured using a cup anemometer, while the direction is measured using a wind vane.

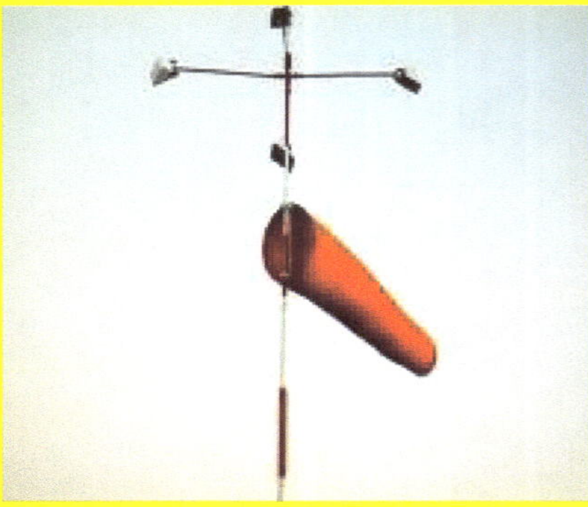

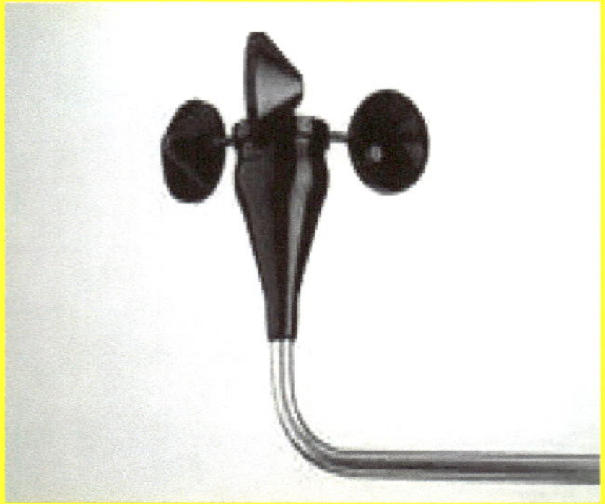

Wind anemonmeter

Positive effects of wind:

- Causes pollination in wind pollinated flowers, thereby making it possible to produce viable seeds;
- Causes seed dispersal in wind dispersed seeds, thereby reducing crowding and competition among parent plant and its offsprings;
- Has a cooling effect on plants and animals when it is very hot;
- May carry away poisons from a congested and contaminated atmosphere e.g. air pollutant (problem is, this pollutant will cause havoc elsewhere-so it is simply a problem postponed or transferred);
- Causes quick drying up of waterlogged soils and flooded lands.
- Negative effects of wind
- Causes falling over of crops (lodging), especially tall varieties;
- Can destroy farm structures;
- Increases water loss from the plant (crop) and from the soil (evapo-transpiration);
- Can cause soil erosion;
- Cause restlessness in livestock.

How to cope with these negative effects of wind:

- Mulch to reduce wilting of crops;
- Construct strong wind breaks such as a line of trees;
- Make strong livestock houses;
- Plant high value-wind prone crops in a greenhouse.

3.14 RELATIVE HUMIDITY (RH)

Humidity is the amount of moisture in the atmosphere at a given temperature. The humidity may increase until no more water can even evaporate and change into vapour; instead, some vapour may even start condensing back into liquid water. The level at which no more water can be accepted by the atmosphere because there is enough is called the saturation level. Relative Humidity is the ratio of humidity at a given temperature to the saturation level at the same temperature, i.e.

RH = 100x the amount of water vapour in the atmosphere at a given temperature /
The amount of water vapour in the atmosphere at saturation point at the same temperature

This is expressed as percentage. It is measured using a hygrometer.

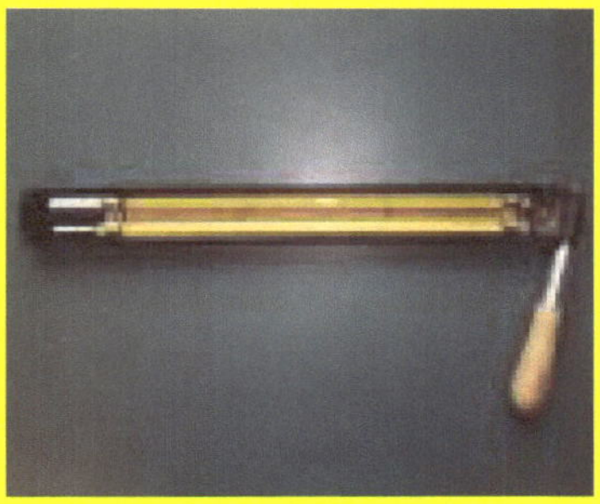

Various forms of Psychrometer

Very high RH leads to:

- Low evaporation;
- Low transpiration;
- Heat stress in livestock, since the rate of evaporation of sweat is greatly suppressed;

- Increased incidences of fungal diseases in crops.

Coping with effects of high relative humidity:

- Spray crops regularly using fungicides;
- Plant crops adapted to high RH e.g. cashew nuts, coconuts and other coastal crops;

Agricultural Effects of Low RH:

- Increased evaporation and thus desiccation in livestock (especially Bos Taurus);
- Fast drying up of dams;
- Increased transpiration in crops thus causing wilting, low yields and death.

Figure 7: Effects of very low rainfall AND / OR low relative humidity AND / OR high temperatures on crops

How to cope with effects of low RH

- Use greenhouse / Establish delicate high value crops in a greenhouse;
- Have a well-ventilated livestock house.
- Shade;
- Mulch;
- Plant drought tolerate crops/ crops with thick cuticle which reduce transpiration;
- Irrigate regularly;
- Water livestock regularly (ad libitum).

3.15 LIGHT

Light is a form of energy, which enables us to see. The light comes from the sun (Solar energy) in form of rays. As these move towards the ground level, they get dispersed, absorbed and even reflected. Some surfaces are able to cause considerable dispersal of solar rays into its individual units (radiations)- some being visible to man (thus called visible part of the spectrum), and others invisible (e.g. ultra-violet and infrared). It is the visible part that we are concerned with here. Most of the light comes form the sun; the amount of sunshine reaching the ground can be measured using a sunshine recorder.

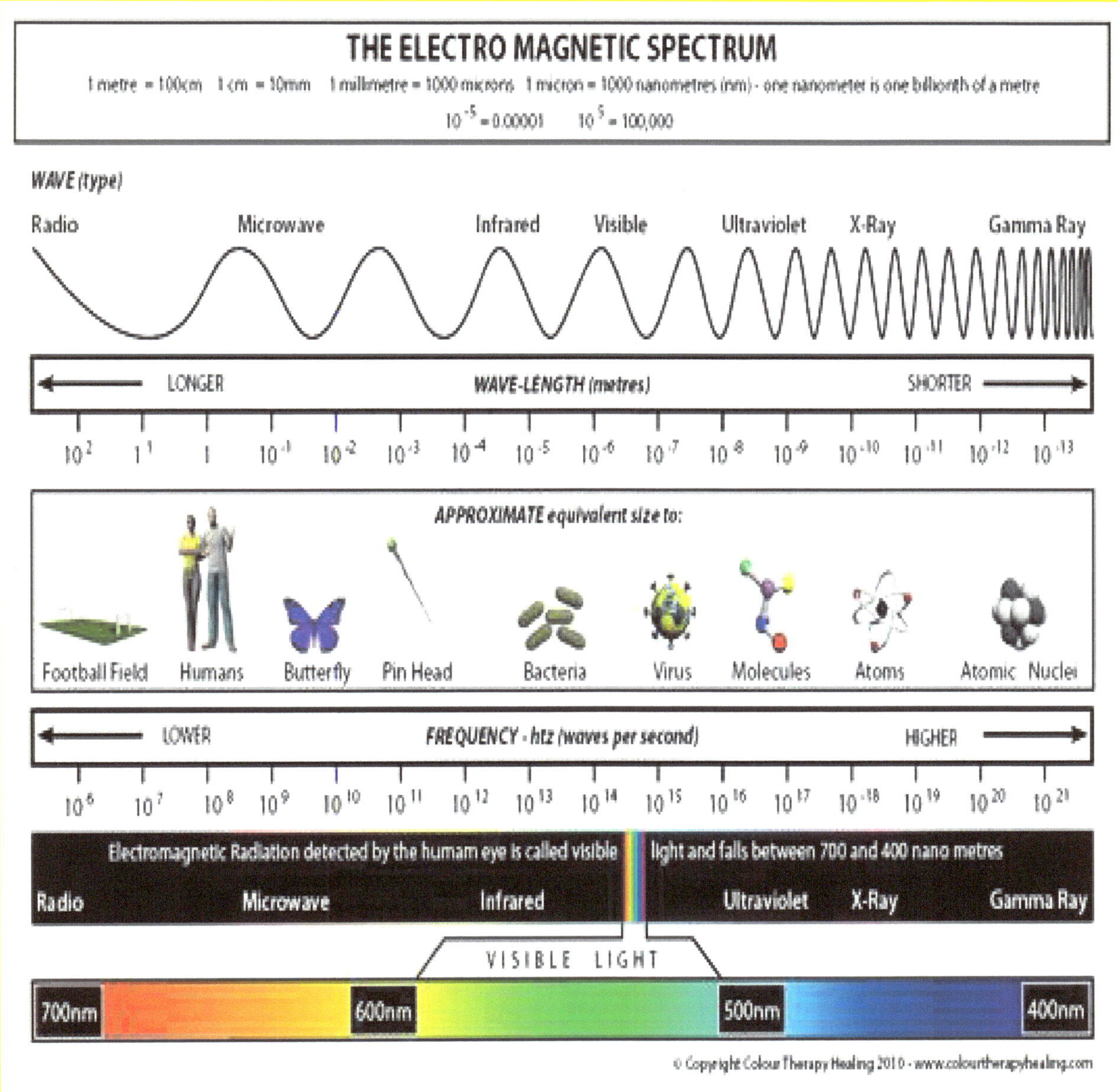

Figure 9: The solar radiation (Electro-magnetic spectrum (EMS))

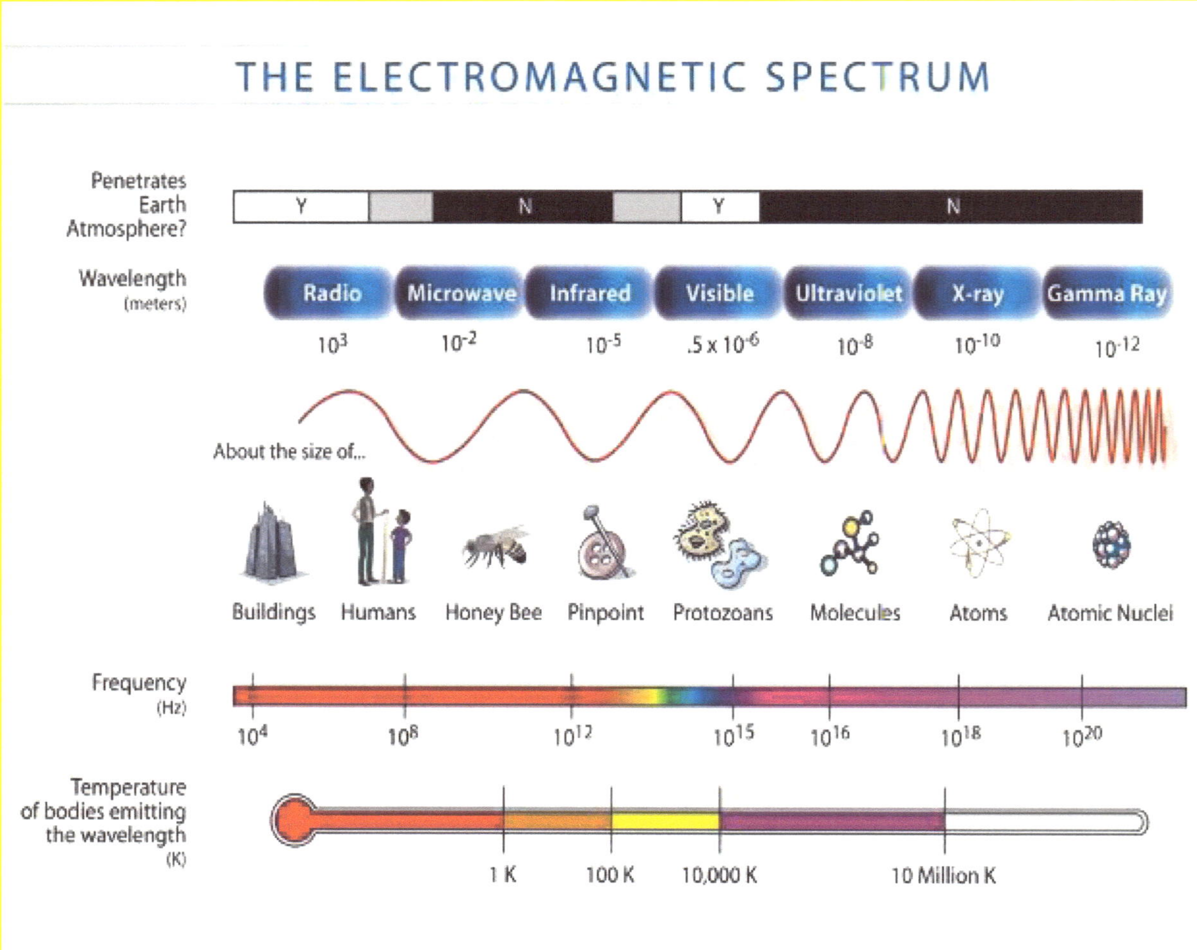

The Electro-Magnetic Spectrum (EMS) (2)

Importance of light:

- Photosynthesis
- Formation of chlorophyll (the green colouring matter found in plant

Problems of lack of enough light:

- Tall and weak stemmed plants
- Small leaved plants with pale colours
- Low crop yield resulting from i and ii above
- Improper feeding of poultry since they may not be able to see their
- feed properly
- Impaired ovulation, since some amount of light is thought to be

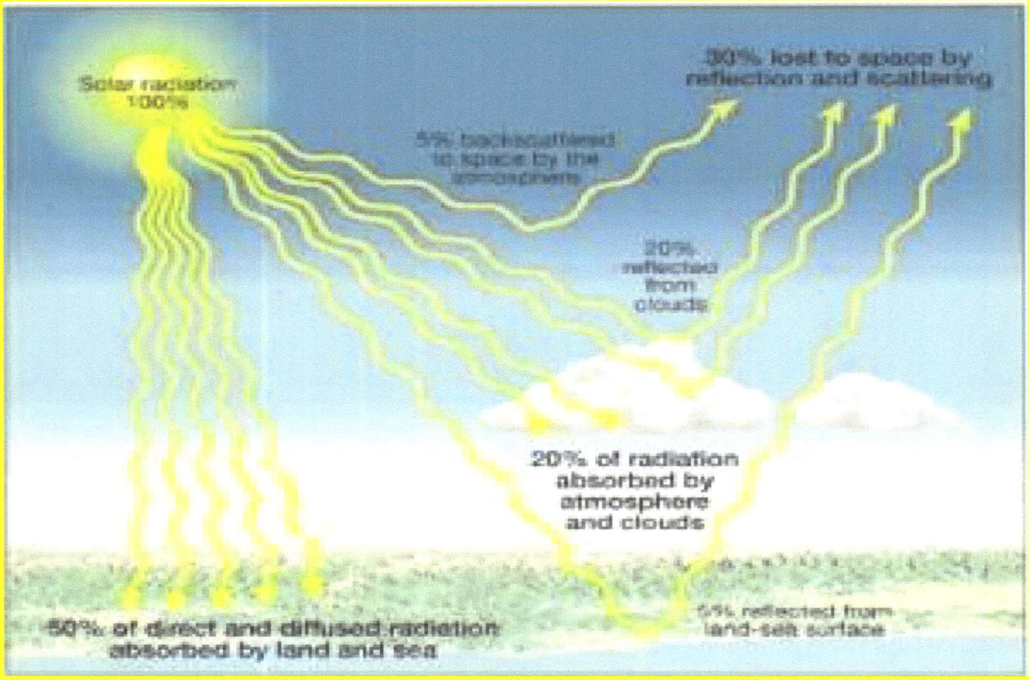

A: source: http://beyondweather.ehe.osu.edu/files/2011/02/web_heat-budget.jpg

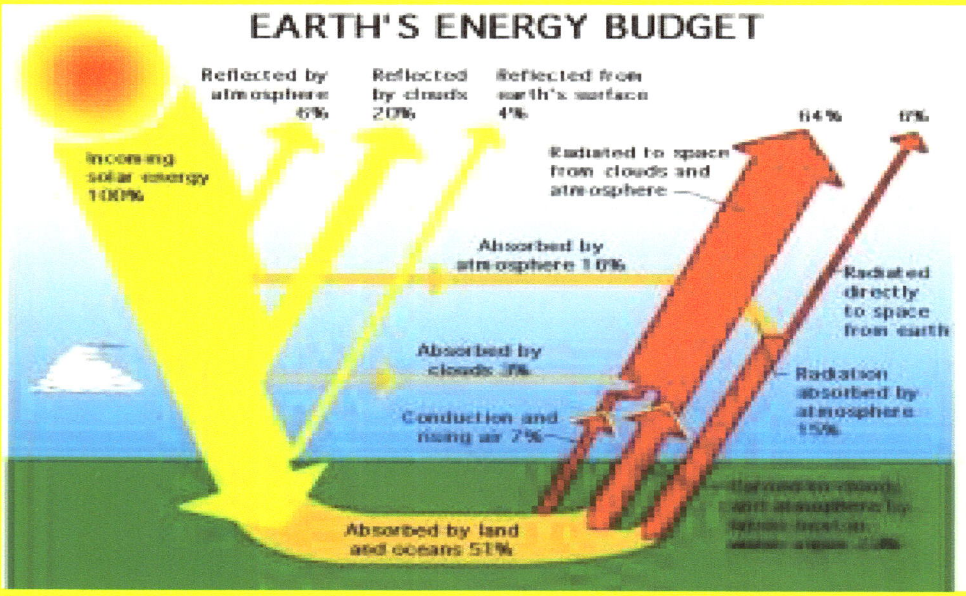

B: Source: http://beyondweather.ehe.osu.edu/files/2011/02/web_heat-budget.jpg

Figure 9a: The solar irradiation reaching the earth.

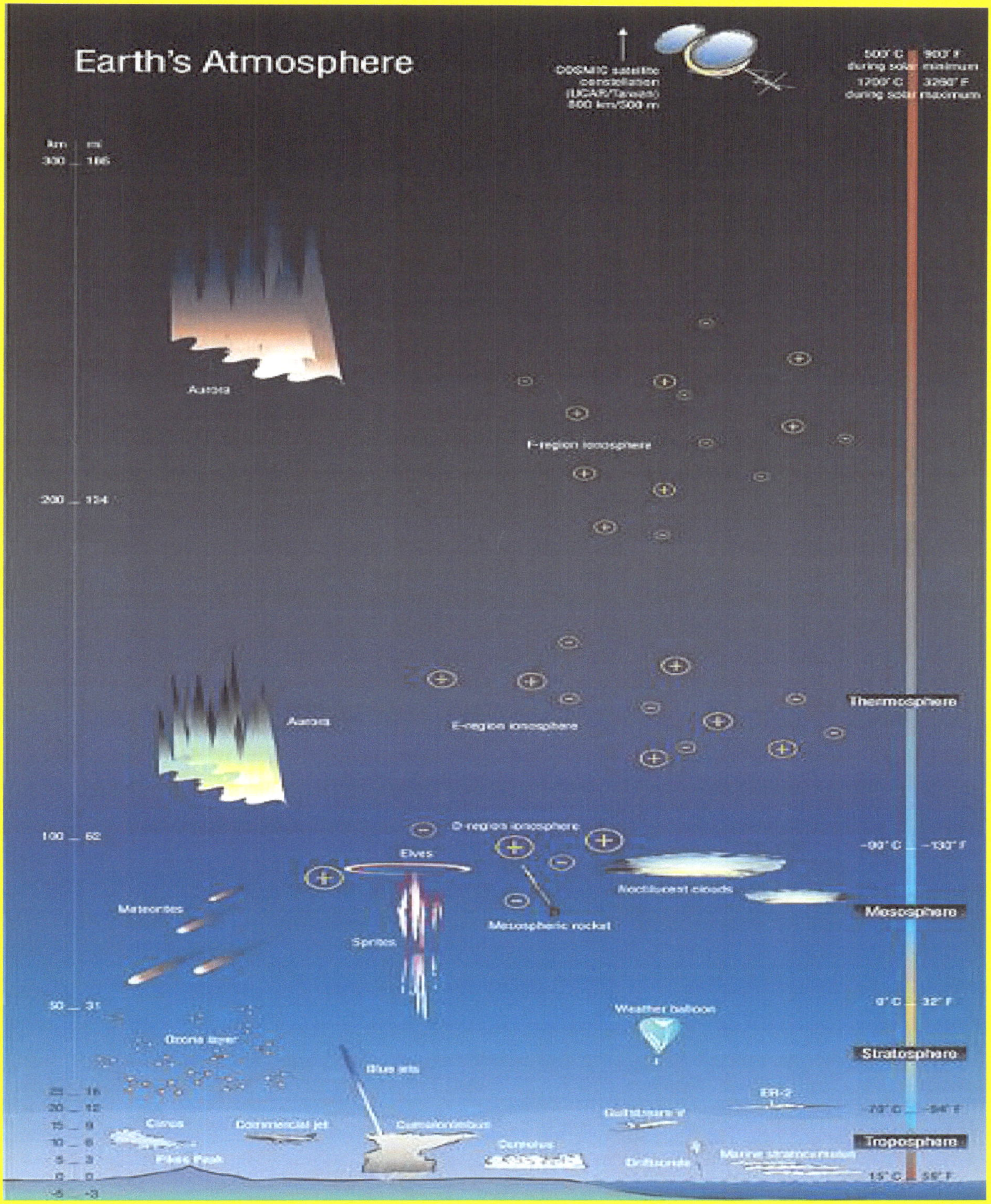

Figure 10: The solar irradiation reaching the earth.
Source: http://beyondweather.ehe.osu.edu/files/2011/02/web_heat-budget.jpg

Effects of Too much light

- Restlessness and discomfort among livestock;
- May lead to egg-pecking as a vice in poultry;
- May increase atmospheric temperatures as some is converted into
- heat, thus causing wilting in plants and desiccation in livestock;
- May distort the ovulation cycle in livestock as it causes discomfort /

How to cope with darkness / too low light

- Can provide supplementary lighting e.g. in poultry houses
- Can use greenhouses to manage some delicate light- loving plants

CHARACTERISTICS OF LIGHT

The characteristics of light include intensity, quality and duration.
Each of these is discussed below.

Intensity/strength/level of brightness/amount of light.

Thus a place may be dark (and thus has low light intensity, low level of brightness, or weak light), or bright (high light intensity, bright or high amount of light). Generally, the rate of photosynthesis increases with light intensity up to an optimal point beyond which there is no change. If it is too bright, some of the light tends to change into heat, thereby causing unbearable temperatures to plants and animals. This reduces rate of photosynthesis. If it is too much the plants wilt and may even die.

Quality/nature of light / radiations.

Light rays are made up of finer individual units / entities which can be scattered to make up the electromagnetic spectrum (EMS). Thus there is the visible part of the spectrum (Violet-indigo-blue-green-yellow-orange- red), as well as the invisible spectrum (e.g. ultra-violet and infra red radiation). Generally, green plants need red and blue lights for photosynthesis. Thus any green light is reflected, and this makes the plant look green. Any light meant for photosynthesis (even if it is artificial light) should thus have blue and red (high quality light / radiations, otherwise they are low quality light. Thus quality of light refers to the type of radiations (wavelengths) of light making up a ray. There are some types of light that facilitate ovulation, while others are either neutral or depressive. Some colours of light also make feed look appetizing even if they are actually not. They may thus increase the appetite of some livestock, thereby increasing yields.

SOURCE: http://encarta.msn.com/media_461532408/Electromagnetic_Spectrum.html

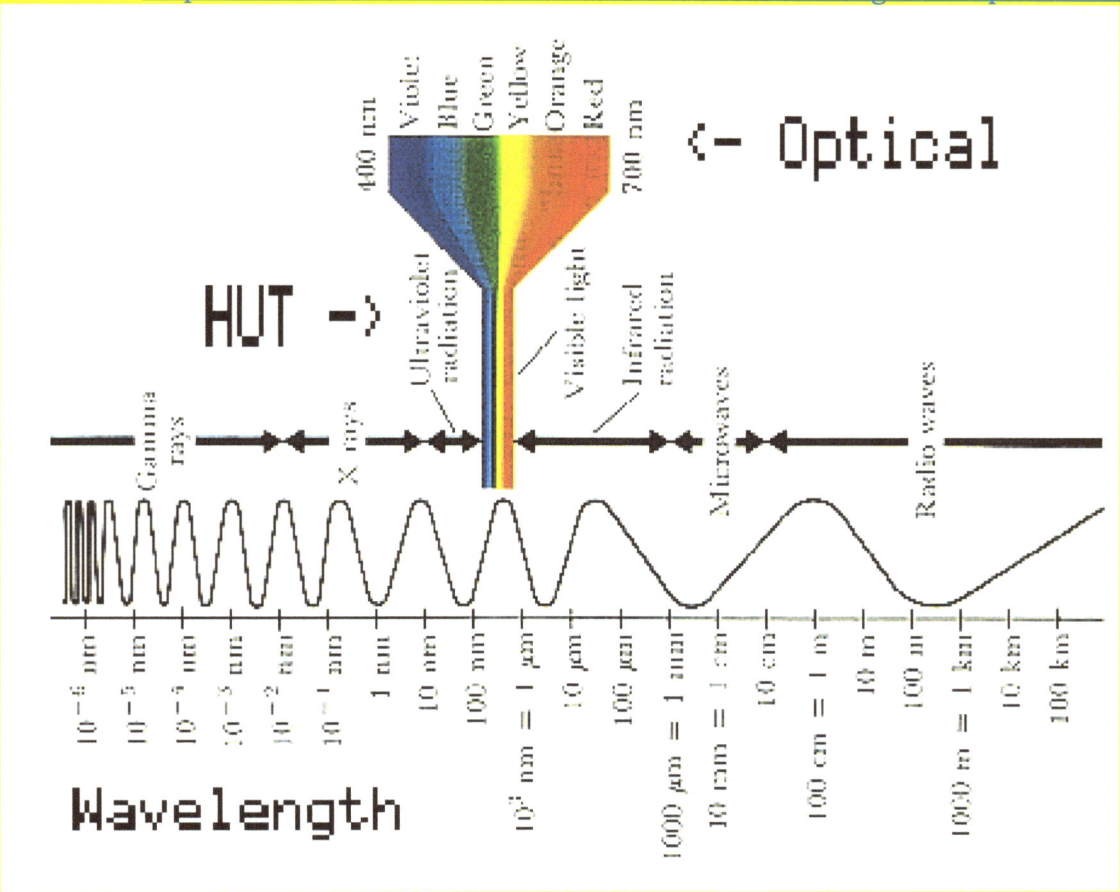

The visible part of the EMS
(source: http://beyondweather.ehe.osu.edu/files/2011/02/web_heat-budget.jpg

THE ELECTROMAGNETIC SPECTRUM (EMS)

The electromagnetic spectrum (EMS, in the figure above) includes radio waves, microwaves, infrared light, visible light, ultraviolet light, x rays, and gamma rays. Visible light, which makes up only a tiny fraction of the electromagnetic spectrum, is the only electromagnetic radiation that humans can perceive with their eyes.

Duration of light (photoperiodism)

This refers to the relative length of day in 24 hours. In some countries, day length changes with season so that in winter, days are short and nights long, while in summer, days are long and nights short. Different plants / crops germinate, grow, flower and produce viable seeds only under certain circumstances relating to the length of the day. The length of the day is called daylength, while this phenomenon (response of plants to length of day or night is called photoperiodism.

Effect of daylength on plants / crops:
Those which can only grow well to maturity and produce viable seeds only when the days are shorter than 12 hours are called short day plants (SDP). Those, which need longer hours more than 12 in a day, are called long day plants (LDP). Those which germinate, grow, flower and produce viable seeds regardless of day length are called day-neutral plants. Most crops in the tropics e.g. Africa are day neutral. Long day plants thus will remain vegetative (i.e. continue growing without flowering) when days are short- thus LDPs grown for seed should be cultivated only when days are long, whereas when days are short as they will not even flower. LDPs meant for leaves give best yields when grown under short days. Examples are:

- SDPs include pyrethrum, sorrel, and soybeans;
- LDPs include wheat
- Day neutral plants are tomato, okra, eggplant etc

3.16 PRECIPITATION:
This is water reaching the earth through various means and in different forms. It may fall in solid form called snow, as is common in many cold countries in Europe and America. Snowfall occurs if ground temperature is also very low, close to freezing point. Sometimes, it falls as hailstones, which are formed when water droplets in a cloud are rapidly cooled eg by being hit by strong air currents, as is often the case during thunderstorms. If it falls in liquid form, it is called rainfall. There are 2 aspects that can be measured in rainfall: (a) amount in millimeters per year (mm pa) (b) intensity i.e. how much rain falls within a given duration (millimeters per hour; 120 mm / hr would be high intensity rainfall, while 10 mm/hr would be low intensity rainfall. It is measured using a rain gauge.

Tipping bucket rain gauge

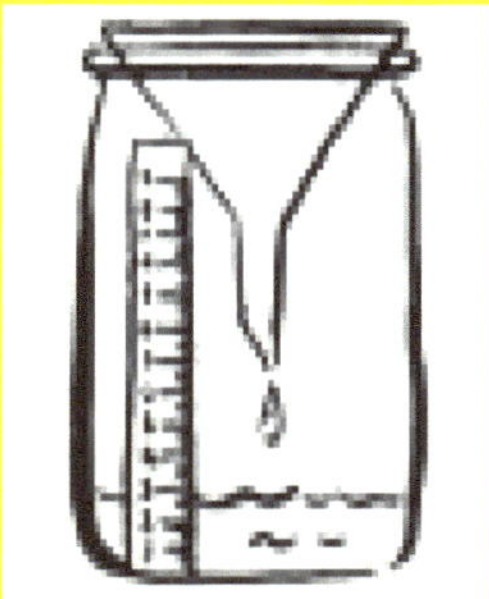

A

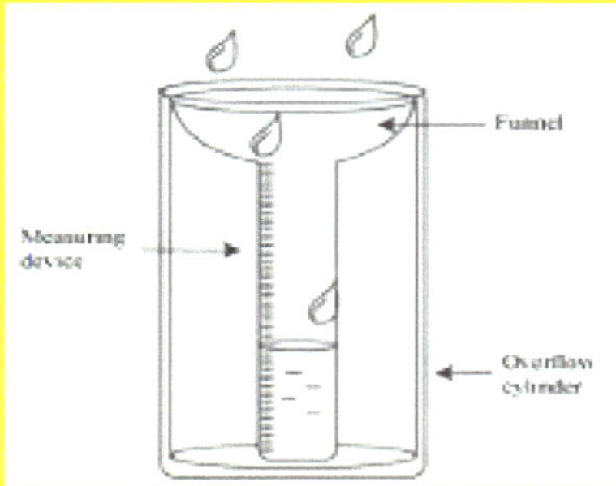

b

c

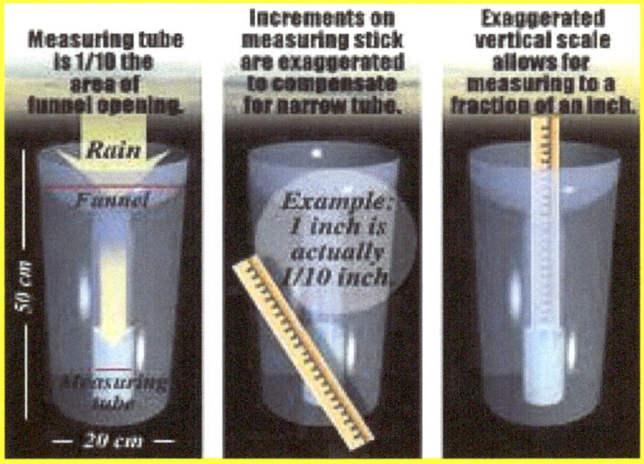

D, E Various forms of Rain gauge

Importance (useful effects) of rainfall

- Provides water for drinking, irrigation and for domestic use.
- Some get into lakes, rivers, ponds, oceans, underground etc;
- Brings a cooling effect on soil and the atmosphere, thereby
- increasing production in some plants and animals;
- Dissolves some minerals (fertilisers) so that they can be easily
- absorbed by crops;
- Ensures recycling of water in different places (i.e. accomplishes water cycle) ;
- Causes weathering thereby increase the depth of soils.

Negative effects of too much rainfall

- May cause flooding e.g. in black cotton soil areas (vertisols / seloko);
- May cause destruction of farm structures and crops;
- May cause restlessness in livestock;
- May cause death of livestock and crops;
- May cause soil erosion (sheet, splash and gully). Rainfall

of intensities exceeding 25 mm /hr are considered erosive;

- Can cause loss of plant nutrients from the soil by sending them to deeper unreachable layers of the soil, i.e., leaching;
- Increase humidity, thereby causing infection of fungal diseases to plants/ crops;
- Reduce soil and atmospheric temperature thus reducing rate of plant growth;
- Reduced temperatures may also cause restlessness in livestock.
- How to cope with the negative effects of very high rainfall
- Plant crops which like a lot of water / rain e.g. coffee, tea, coconut;
- Plant water-loving crops in flooded areas e.g. rice (Oryza sativa).
- Make shed for livestock, especially delicate and young ones;
- Plant delicate crops and seedlings in a greenhouse;
- Add fertilisers in soil to replenish leached soils;
- Plough across the slope, and construct water conservation structures such as bunds, terraces etc to reduce soil erosion.

Positive effects of High rainfall

- There is enough water and feed for livestock- thus there is more production;
- There is increased crop yield, especially from water-loving crops;
- Good soil cover, thus reducing soil vegetation;
- Can tap hydroelectric power from rivers, as they are permanent;
- Has a cooling effect thus depressing extreme effects of sunshine.

Very low rainfall: Effects

- High temperatures result, casing heat stress to livestock and crops;
- Crops face drought and moisture stress;
- Soil formation is low (weathering process, thus soils are shallow;
- Water becomes scarce, and inhibits crop and livestock production;
- Vegetation cover is sparse, leading to soil erosion e.g. by wind;

How to cope with effects of low rainfall

- Keep water in reservoirs e.g. tanks for general use in the farm;
- Irrigate crops;
- Mulch crops to minimize moisture loss;
- Plant drought resistant and tolerant crops;

- Shade crops from excessive heat;
- Water livestock more frequently;
- Time production period so that the little rain available can be useful to crops at their critical time of need;
- Practice water conserving techniques such as minimum tillage,
- zero tillage, dry planting etc.

3.2 THE WATER CYCLE

This is a process by which water changes location and form, i.e., moves from place to place in different forms. It involves a number of processes such as evaporation of water from the soil and surface water bodies, followed by rising, cooling and condensation of the water vapour to form clouds. This then falls as rain. The resulting water may run on the soil surface (surface runoff), seep deep into the ground (deep percolation or infiltration) or may go straight into a surface water body and the same process continues. Plants also take up some of the soil water or deep percolation and take it back to the atmosphere through its leaves in the process of transpiration. The source of energy in all this is the sun (irradiation or insolation). A diagram for these processes is given below.

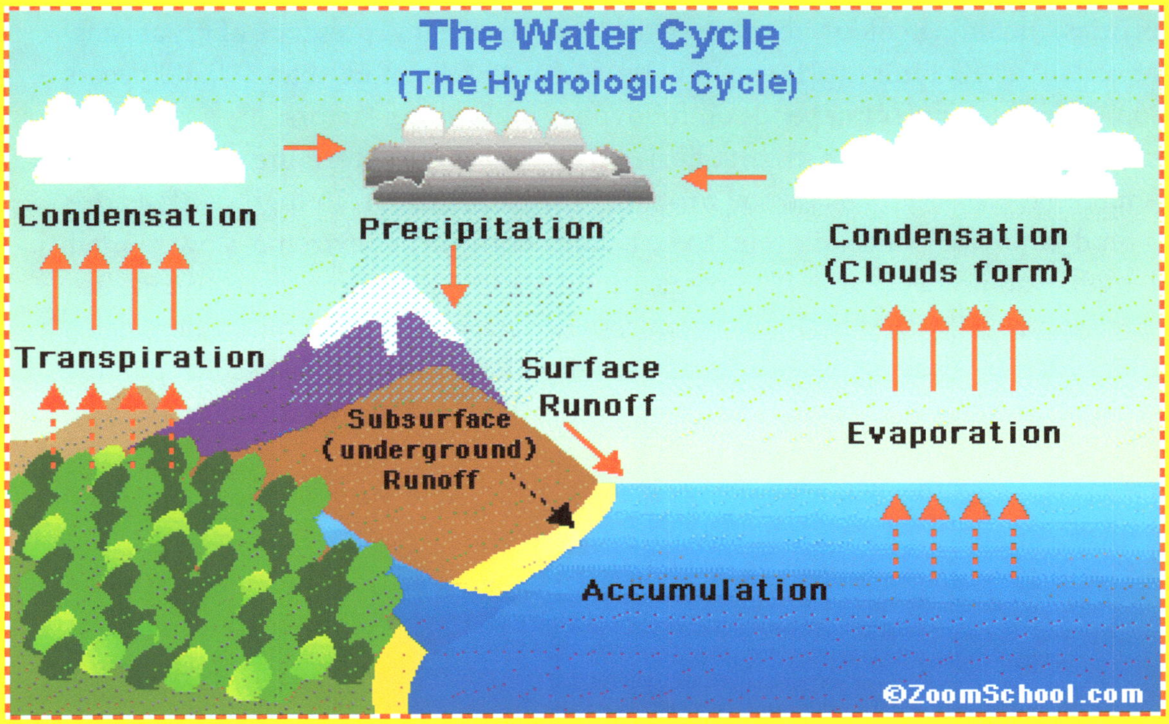

Soil is defined as the weathered top-most layer of the earths crust in which plants grow. It can also be viewed as the top-most thin, ever-changing and living layer of the earth's surface without which land life would be impossible. It develops from weathering of the parent rock, and by the accumulation of the top soil layer with humus. Soil provides the mechanical and nutrient support necessary for plant growth. It is that medium which has the potential to support plant life due to its complex composition of nutrients, air and living organisms, among others.

3.31 Soil Formation

The earth, like other planets, was initially a dry solid mass. Its topmost layer has been subjected to intense weathering process, over millions of years – breaking down to fine particles as a result of the forces of wind and water. This resulted to varying sizes of original rock – ranging from tiny mineral particles - clay particles to boulders. The fine earth fraction (sand, silt and clay) form what is refered to as the soil, the backbone of farmers activities. Soil formation starts at the surface and continues downwards, thereby forming layers (horizons). Different soils have different profiles with clear horizontal layers, also called horizons.

The horizons differ from each other in their physical, chemical and biological characteristics, including colour, texture, structure, consistency, presence of organisms, degree of acidity or alkalinity (pH), etc. Soil profiles may take more than 100, 000 years to develop. Soil development depends on following soil forming factors:
Climate. weathering forces such as rain, heat, ice and sunshine determine how fast or slow the material breaks down and soil formation processes occur. In high rainfall areas leaching tend to be high and soils tend to be acidic. At the other extreme, (low rainfall regions), high evapotranspiration (ET) results in salt deposits on the surface giving rise the so called saline soils. Also high temperatures encourage high physical & chemical weathering of rocks releasing minerals, high activity of microorganisms resulting in low humus content in tropical soils. Wind erosion occurs in arid lands without vegetation and sandy soils remain or are deposited.

3.32 Soil Organisms.

Plants and animals living in and on the soil influence soil formation. Include vegetation, which supplies the soil with organic matter & protects the soil from rainfall (raindrop impact), sunlight & wind; Soil organisms, the most important of which are the bacteria & earthworms. Important for soil OM decomposition, also the way soil particles get physical properties).   It gives a good idea of what to grow and where. The narrower the top soils, the younger the soil and the fewer the crops it can support, especially perennial crops. Deeper profiles can store sufficient nutrients, water for plant growth and important for firm plant support.

Plate 1: An excavated pit showing soil layers

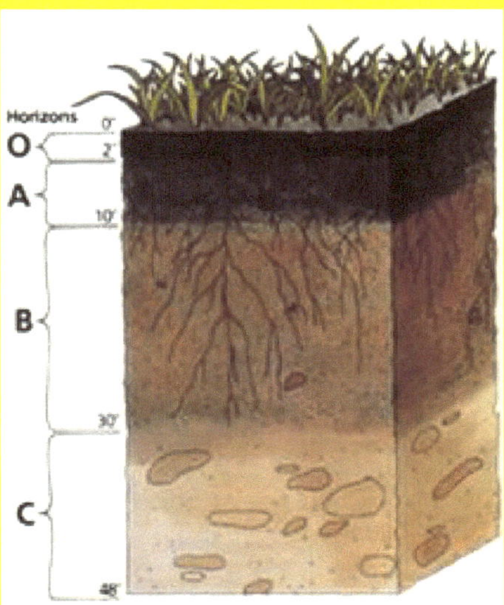

The Soil horizons can be summarized as follows from top to bottom

The "A" Horizons may be darker in color than deeper layers and contain more organic material, or they may be lighter but contain less clay or sesquioxides. It is a layer of eluviation – eluviates (= is depleted of) iron, clay, aluminum, organic compounds, and other soluble constituents. When eluviation is pronounced, a lighter colored "E" subsurface soil horizon is apparent at the base of the "A" horizon. It is well aerared and rich in plant nutrients – a repository of accumulative mineral nutrients.

Horizon B (Subsoil, or weathered layer)
This is the layer which lies underneath the A-Horizon and commonly referred to as "subsoil". This layer is mostly coloured brown by iron compositions and is still penetrable by roots. It accumulates clay,iron, aluminum oxides and organic compounds through leaching, a process referred to as illuviation. Accordingly, this layer is also known as the "illuviated" horizon or the "zone of accumulation". It varies in structure, consistency and color compared to to the A horizon above and the horizons below.

Plant roots penetrate through this layer, but it has very little humus. It is usually brownish or red because of the clay and iron oxides washed down from A horizon. it is usually compacted and less aerated. It may have hardpans – which impedes drainage and resist root penetration. The above 3 layers may be referred to collectively as the "solum". The layers below have no collective name but are distinct in that they are noticeably less affected by surface soil-forming processes

Horizon C
This is a partially weathered) parent material. This layer is little affected by soil forming processes (weathering), and the lack of pedological development is one of the defining attributes. The C Horizon may contain lumps or more likely large shelves of unweathered rock, rather than being comprised solely of small fragments as in the solum. The C horizon also contains parent material.

Horizon R
This is the Parent rock, or the unaltered bedrock. It comprises a layer of large unbroken rocks. This layer may accumulate the more soluble compounds. R horizons basically denote the layer of partially weathered bedrock at the base of the soil profile. Unlike the above layers, R horizons largely comprise continuous masses (as opposed to boulders) of hard rock that cannot be excavated by hand. Soils formed in situ will exhibit strong similarities to this bedrock layer.

water holding, providing nutrients, support, temperature and aeration. The soil provides the environment for plant root growth and development. A basic understanding of soils is therefore essential to crop production. Soil physical, chemical and biological properties are all important Soil Physical Properties. These include soil texture, soil structure, bulk density, and soil porosity. They all impact how soil, water, and air interact and influence the amount and availability of water for plants. Soil texture and structure are very important.

(a) Soil texture

Texture is defined as the relative proportion of sand, silt and clay fractions in a soil sample / soil mass. Also as the fineness or coarseness of soil – feel method. The texture is one of the most important soil properties affecting yield capability, soil formation and movement of inorganic and organic substances. It determines aeration, drainage, water retention (WHC), capillarity, workability and mineral nutrient content or retention.

A soil textural triangle is used to determine the overall textural class. It shows different textural classes based on the percentage by weight of each soil fraction i.e. based on particle size distribution, all mineral soils can be placed into one of the 12 major textural classes. 1. Sand 2. Loamy sand (more sand cf sandy loam), 3. Sandy loam, 4. Silt loam 5. Silt 6. Loam 7. Sandy clay loam, 8. Clay loam, 9. Silty clay loam (more sand cf silty clay), 10. Sandy clay, 11. Silty clay, 12. Clay .

These soils are classified as Coarse textured soils (or open or light textural soils) - are mainly coarse or sandy with low content of silt and clay eg sands and loamy sands, sandy loams. These soils leach quickly and do not hold water well. This can influence both fertility and the environment, as well as limit the plant's ability to obtain water for turgor, photosynthesis, and nutrient uptake.

Sandy soils have

- Good permeability, good aeration and easily workable.
- Possesses only a low content of inherent nutrients.
- Low adsorption of K and NH4 ions.
- High leaching of nutrients because of the low field capacity

(b) Soil structure

It refers to the physical arrangement of soil particles and aggregates with respect to each other. Aggregates are groups of individual soil particles adhering together. Type of structure refer to shape of aggregate. There are four types of primary structures – platy, prism-like (columnar & prismatic subtypes), block-like (blocky & subangular blocky subtypes) and spheroidal (granular & crumb subtypes). Describes the shapes of the clods into which the aggregates break.

The grade of aggregates describes how strongly the aggregates hold together (the degree of aggregation). Grades are termed as structure less, weak, moderate, strong and very strong depending on the stability of the aggregates when disturbed.

Soil structure is recognized as one of the most important properties of the soil mass. It has a primary impact on soil porosity and aeration. It determines

- Aeration – eg surface crusts
- Water entry/ infiltration, flow / permeability
- Water retention / WHC
- Temperature
- Impedance to roots = root penetration, & therefore important
- in utilization of nutrients by the crops.

Surface crusts impede water & air entry & germination. If water does not move through soils, air is excluded and plant roots die. If water does not enter the soil (infiltration) and instead runs off, the water is lost to the plant and erosion often occurs Aggregated soils types are generally the most desirable for plant growth. It contains many pores for water entry and aeration. The structure must be stable - resist breaking down by the slaking action of water and raindrop beating. Stable soil structure / aggregates result from cementation of particles by colloids – eg OM/ humus

Soil chemical properties

Important chemical soil properties include mineral nutrient content and availability, toxicity, Cation exchange in soils, soil pH / soil acidity.

Soil Sampling and Testing

Soil sampling and testing provides an estimate of the capacity of the soil to supply adequate nutrients to meet the needs of growing crops. The objective of soil testing is assess overall nutrient status (assess soil fertility status) and arrive at / provide proper fertilizer / compost /

characterise nutrient leaching process or pollution will require to be taken from several depths. In the sampling procedure, samples are taken evenly across the farm.

Soil samples submitted for analysis should be representative of the field or portion of a field. Avoid irregularities (eg old straw, hay or manure piles, waterways, saline spots, etc) and border effects. Soil variability is a major consideration in deciding how to collect a representative soil sample. Sample unique areas separately – old lime piles, wet aresa, flats / steep areas, eroded areas, areas managed differently, and consider textural differences. We can use - a grid pattern - usually the best way to be sure the entire field is represented. Use Random soil sampling (eg by zigzag method) for uniform fields with little variation .

Benchmark sampling is recommended for fields with more variability (hills, pot holes, etc.). it reduces the inherent variability of a field by reducing the area sampled. When picking a benchmark area, use observable features such as soil color and landscape (mark with a GPS) to identify where different soil types occur. Also the use of productivity, yield, plant growth vigor, aerial and/or topographic maps (slope, degree of erosion, drainage) and past treatment are essential to identify benchmark sites.

Select 15 to 20 sampling sites representative of the portion of the field to be tested. Laboratory testing of agricultural soils should include soil tests for soil pH, texture, soil structure, soil N- including nitrate-nitrogen, available phosphorus, available potassium and extractable sulfur, organic C / OM, salinity (electrical conductivity). If possible, the nitrate and sulfur analysis should be completed for subsurface soil samples (15-30 and 30-60 cm). Additional analyses for micronutrients (Boron, Chlorine, Copper, Iron, Manganese or Zinc).

The Living Soil

The soil (rich in humus) is home to and supports millions of living organisms. The soil life, the edaphon, comprises of soil flora and soil fauna. Being a living organism, the soil has a special place in organic farming. The soil organisms' activity is closely correlated to the decomposition of organic material. Microscopic fungi and bacteria are predominant, and play a role in soil formation and nutrient release (& maintenance of plant food stability) through decomposition processes. In the living soil, there is a healthy inter-play of living organisms, water, air, and mineral nutrients that results in healthy crops without recourse to synthetic fertilizers. In the living soil, disease causing organisms or pathogens are quickly overwhelmed into extinction as long as organic farmer maintains a stable environment.

Organisms in the soil differs and is dependent on climate, relief, vegetation, soil form, soil depth and season. These include Bacteria, actinomycetes, fungi, algae.

Bacteria.

Are Small, single celled organisms which are surrounded by slime. Because of their great number and their differences they are capable of converting enormous amounts of material. In anaerobic conditions, bacteria are solely responsible for changes Strongly adsorbed nutrients are dissolved and some protection from diseases is given.

Genera of Bacteria:

* N-fixing genera: Azotobacter, Rhizobium, Clostridium and Beijerinckia
* Nitrifying genera: Nitrosomas and Nitrobacter
* Others: Pseudomonas, Arthrobacter, Achromobacter etc.

Actinomycetes are also called the Ray fungi, and are a transitional form between bacteria and fungi. They take part in the decomposition process and in the humification process and can cause diseases. They form antibiotics and give the earth that fresh earth smell. Include genera: Streptomyces and Nocardia.

Fungi

Cover the soil with mycelium (hyphae) and thereby help with the rotting process. Strongly bound nutrients are dissolved, protection from diseases.Genera include: lignin attacking basidiomycetes, penicillium, aspergillus etc.

Algae

Capable of feeding on simple inorganic molecules, and play a significant role as pioneer settlers at extreme sites. Genera include: Green algae, diatoms, blue-green algae.

Soil Fauna

Examples – Protozoa, Threadworms, Earthworms

Unicellular Protozoa, like pseudopodia, ciliates and flagellates, which feed on dissolved organic material and bacteria.

Multicellular organisms:

Examples: - Threadworms (often plant parasites) and Earthworms (lumbriciden)

Earthworms

These are involed in the transformation of organic and mineral parts to most fertile earth. Earthworm faeces contains 5x more nitrogen, 7x more phosphate, 11 times more potash; and 2x more magnesia and calcium than normal earth.Earthworm routes improve the circulation of water and air. Formation of a stable structure with small and big cavities.

Wheel corollas (rotates) and tardigrada.

* Accelerate decomposition by removing sick plant parts from the soil surface.

Arthropods: include Wood lice (isopods), Mites (acarina), Insects

Insects: - Springtails (collembolen), Dipterans, Ants – termites.

Molluscs: Snails and slugs (gastropods).

Vertebrates: Moles, Rodents, Snakes, Reptiles.

The Dead soil

Soil life depends on the continuous replenishment of organic matter, soil air, healthily acquired mineral nutrients, and non-pathogenic living organisms. To maintain soil life organic farmers should adopt farming techniques that preserve and improve soil structure, fertility and organic matter content.

Dead soil simply refers to that soil which is totally deficient in humus, soil air and water, beneficial living organisms, and is poorly structured. It is too acidic or too alkaline ie the dead soil cannot support the growth and development of even those plants that are useless to farmers. Farmers us their indigineous knowledge to assess / detect whether the soil is poor or fertile, acidc, low orhigh in organic matter content. Examples include:

Soils high in OM content (fertile) – support growth of

- Pigweed – Amaranthus spp
- Wondering jew (Commelina bengalensis)
- Thorn Apple (Datura stramonium) – an annual poisonous weed
- Black jack (Bidens pilosa)
- Macdonald eye (Galinsoga parviflora)
- Poor soils – support growth of
- Double thorn (Oxygonium sinuatum)
- Striga weed (=Purple witch weed) (Striga spp) – parasitic on cereals
- Poverty grass – perennial arable grass weed

How we kill our soils

Naturally the soil will die through waterloging. But artificially soils are killed by Poisoning with synthetic fertilizers and pesticides. They lock up naturally available nutrients, increase or lower acidity – affect seedling development. High acidity reduce bacterial activity reducing the rate of OM decomposition. High or too low pH also affects N-fixing bacteria = low N. Lastly, burning of vegetation – kills billions of MOs both macro and micro, and OM / humus.

It is good to remember that earthworms whose casts are important sources of soil nutrients die, N-fixing rhizobia are eliminated, and soil air and water are eliminated too. Also the humus is destroyed – eliminating its beneficial effects as nutrients store / adsorption, water holding, soil structure destruction.

In addition burning encourages erosion whereby wind and water erosion blow away or carry soil downhill. Burning depletes soil of nutrients especially volatile ones – N, S etc. Burning is therefore avoided in organic farming. Ploughing or tillage disturbs the balance among soil insects, fungi, bacteria, viruses, among other soil life. It breaks up the root channels and soil aggregates. It also causes rapid oxidation / loss of OM – food for many soil organisms and crops. Tillage disturbance of soils is therefore minised in organic farming (OF).

2.6 MAJOR SOIL TYPES OF EAST AFRICA AND THEIR MANAGEMENT

Tropical Africa has a very wide range of soils which resulted from the variation in geology (parent material), in relief and climate. Soil resources vary from sandy to clayey, shallow to very deep and low to high fertility. However, most of them have serious limitations such as salinity/ sodicity, acidity, fertility and drainage problems. The major soils used in agriculture are Ferralsols, Nitisols, Vertisols, Andosols, Acrisols, Lixisols, Luvisols etc.

According the FAO/UNESCO (1976) legend, now the World Reference Base (WRB) for Soil Resources (1998; 2006). The following major soil types and their important characteristics are as described below.

1. Nitisols (Ultisols in USDA).

Are deep red and well-drained tropical soils with a clayey `nitic' subsurface horizon. Developed in relatively rich parent material (and marked by shiny, nutty structure elements. i.e. they are developed from volcanic rocks and have better chemical and physical properties than other tropical soils.

Types of soils

There are different types of soils, with specific names. These are outlined below, including Nitrosols, Feralsols, Andosols, Vertisols, planosols, etc.

Notrosols

Nitisols are strongly weathered soils, are 150 cm or more deep and show evidence of the movement of clay within the profile (i.e. clay illuviation). Have often high clay content (more than 35%). Although strongly weathered, they are more productive compared with other red tropical soils in the region. They are generally considered to be `fertile' soils in spite of their low level of `available' phosphorus and their normally low base status.

Have favourable physical properties (e.g. high aggregate stability of the nitic, subsurface horizon). Soils are friable or very friable & are porous through out. They have very favourable moisture storage capacity & aeration conditions. Most Nitisols are acidic (pH < 5.5) due to the leaching of soluble bases.

Management and use of Nitisols

Are used for plantation crops and food production (e.g. banana, tea and coffee). Nitisols have good workability due to their good internal drainage, stable soil structure and fair water holding properties. However, due to high degree of phosphorus sorption, optimum crop production require addition of P fertilizers & manure.

2. Ferralsols (= Oxisols, USDA)

Are deeply weathered, red or yellow soils of the humid tropics with a high content of sesquioxides (Fe/AL oxides) and very low cation exchange capacity. They are very old, highly weathered and leached soils, and therefore with a poor fertility, and fertility is restricted to the top soil, as the subsoil has a low cation exchange capacity. Phosphorous (P) and Nitrogen (N) are always deficient. Are dominated by low-activity clays, mainly kaolinite (1:1 silicate clays) (CEC < 16meq/100g clay; low BS) and are virtually devoid of weatherable minerals.

Management and use of Ferralsols

Ferralsols have good physical properties (depth, drainage) but are chemically poor. Their low natural fertility and tendency to `fix' phosphates are serious limitations. The natural fertility restricted to the A horizon & related to the OM content. In times may be droughty because of their low available water storage capacity. Liming and full fertilisation are required for sustainable sedentary agriculture.

3. Andosols (Young volcanic soils)

Also referred to as volcanic ash soils. These are soils developed in Volcanic material (volcanic ejecta - ash, pumice, tuff]. Typically black soils of volcanic landscapes; from Japanese an, black, and do, soil. Andosols occur in areas with steep slopes and high-rainfall. By rainfall over 1000 mm per year, Andosols are exposed to excessive leaching. They have low bulk density (<0.9g/cm3), are porous / high permeability and have a high water storage capacity. They are also acidic (low pH) due to the high leaching of soluble bases and to the high levels of Aluminium (Al). These conditions favour P-fixation, making it no longer available to the plants.

Use / management

Their major limitation = strong phosphate fixation. Otherwise, Andosols are naturally fertile soils with good physical characteristics i.e. easy to cultivate, good rootability and water storage properties. To improve agricultural production, liming and the use of P-fertilizers is necessary. Andosols are highly susceptible to erosion as they mostly occur on steep slopes. In these areas, they are mainly used for tea, pyrethrum, temperate crops and dairy farming.

4. Vertisols (Black cracking clays)

They are swelling and shrinking (churning) heavy clayey soils. Are soils developed in expanding 2:1 lattice clays (smectitic / montmorrillonitic clays as PM). It swells when wet, shrinks when dry. In dry periods these soils develop wide, deep cracks.

Other characteristics are:

- High CEC, high BS (Ca, Mg), low SOM, N content, the low permeability, drainage due to swelling and susceptible to water logging.- Tillage difficult when dry also hindered by stickiness when wet. Vertisols are productive soils if properly managed.- Are popularly known as black cotton soils. Because of the churning process, the A horizon is deep, homogenous & thoroughly mixed.

Management and use of Vertisols

Management practices for crop production should be directed primarily at water control in combination with conservation or improvement of soil fertility. Tillage is hindered by stickiness when the soil is wet and hardness when it is dry. The susceptibility of Vertisols to waterlogging may reduce the actual growing period. Excess water in the rainy season must be stored for post-rainy season use (water harvesting) on Vertisols with very slow infiltration rates.

5. Planosols

Planosols are soils with a bleached topsoil over dense, slowly permeable subsoil. The light-colored, surface horizon show signs of periodic water stagnation and abruptly overlies a dense, slowly permeable subsoil with significantly more clay than the surface horizon. Are imperfectly drained soils with a relatively coarse textured, light textured topsoil (part of which is whitish /albic or E horizon) and heavy textured, compact & hard B horizon. The transition between the 2 layers is abrupt. Are often waterlogged and have very low vertical & horizontal drainage. Artificial drainage is the main management problem of these soils. Soil fertility ranges from low to moderate therefore fertilization is needed for optimum crop production.

Management and use

Root development hindered severely by oxygen deficiency in wet periods. The low hydraulic conductivity of the dense subsurface soil makes narrow drain spacing necessary. Surface modification such as ridge and furrow can lessen crop yield losses from waterlogging. Planosols may be planted to paddy rice. Fertilizers are needed for good yields.

6. Acrisols, Alisols, Lixisols and Luvisols

These kinds of soils occur in the coffee zones in the sub-humid areas, on undulating to hilly topography. They show an increase of clay content in the sub-soil (B-Horizon). The sub-soil is often not very porous, impeding root spreading. They have a relatively low water-storage capacity, compared with Nitisols. Acrisols and Alisols in wet areas have a low pH (acid), Al and Mn toxicities and low levels of nutrients and nutrient reserves. These soils have poor structure and need erosion-control measures. The soils respond well to fertilizers (especially N, P and K) and to the use of soil organic matter.• Alisols - soils with high cation exchange capacity and much exchangeable aluminium. Acrisols – are strongly leached, red and yellow soils on acid parent rock, with a clay accumulation horizon, low cation exchange capacity and low base saturation

• Lixisols with a low cation exchange capacity but high base saturation percentage.

• Luvisols – are base-rich soils (USDA= Alfisols) with a distinct clay accumulation horizon, i.e. are soils in which clay is washed down from the surface soil to an accumulation horizon at some depth. Argic horizon / illuviation

Other soils, amongst others of minor occurrence and importance, in East Africa, include Arenosols, Fluvisols, cambisols etc discussed below.

7. Arenosols

Are sandy soils or soils developed in residual and shifting sands, found in desert areas, beach ridges, inland dunes, and areas with highly weathered sandstone. have low fertility, very low WHC, very low SOM, but well drained, good aeration, high bulk density.

8. Fluvisols – are young alluvial soils which show stratification or other evidence of recent sedimentation. Are soils developed in alluvial deposits; from L. fluvius, river. Has an AC profile. Found in periodically flooded areas of alluvial plains, river fans, valleys and (tidal) marshes. pH neutral, and rich in bases. Flood control, drainage and/or irrigation are normally required.

9. Cambisols (USDA = Inceptisols, young soils).

Soils with beginning horizon differentiation evident from changes in colour, structure or carbonate content; from L. cambiare, to change. Are in a transitional stage of development, from a young soil to a mature soil.' Phaeozems (Prairie soils; USDA = Mollisols). These are the dusky red soils of prairie regions with high base saturation but no visible signs of secondary carbonate accumulation. They are dark soils rich in organic matter; found in wet prairie (grassland) vegetation.

Histosols –. Are organic soils, properties due primarily to organic matter, not mineral matter. May be Peat or Muck soils. Have very high OM content in top soils. Develop under wet, boggy or swampy conditions.

Anthrosols – are man-made soils. These soils vary widely in properties and appearance and can occur in any environment. Their properties are strongly affected by human intervention e.g. cultivation, addition of other materials such as manure, intensive fertilization.

Solonchaks – are soils with a high content of soluble salts. Also called saline soils or salt-affected soils. They are largely confined to the arid and semiarid climatic zones and to coastal regions in all climates. 'ECe>15 dS/m; Excess salts affect plant growth due to high osmotic potential, reduce water uptake by plants. 14. Solonetz – Are soils of ASALS with a high percentage of adsorbed sodium ions.

Solonetz that contain free soda (Na_2CO_3) are strongly alkaline (field pH > 8.5). Also referred to as `alkali soils' and `sodic soils' . High levels of exchangeable sodium ions affect arable cropping, either directly (Na-toxicity) or indirectly, e.g. dispersion of soil aggregates and structure deterioration.

3.6 PRINCIPLES OF LAND USE

3.61 LAND USE PLANNING:

Farm planning and the introduction of better farming systems are made difficult as every member of the community has a right of the use of the community land. It may therefore allow some lazy individuals or inexperienced farmers the use of the land that may lead to misuse of lands. The system also prevents a good and enterprising farmer from expanding and developing his land, as transactions that may enhance rapid development of the land cannot be undertaken. A farmer may not have the same interest in the proper care and development of the land; this can lead to serious erosion hazards as a result of improper land use overgrazing and absence of erosion control measures.

3.62. CONCEPT OF AGROECOLOGICAL ZONES

MEANING OF AGRO-ECOLOGICAL ZONES

Agro-ecological Zoning (AEZ) refers to the division of an area of land into smaller units, which have similar characteristics related to land suitability, potential production and environmental impact. An Agro-ecological Zone is a land resource mapping unit, defined in terms of climate, landform and soils, and/or land cover, and having a specific range of potentials and constraints for land use (FAO, 1996). The essential elements in defining an agro-ecological zone are:

- The growing period or moisture supply – i.e. quantity and distribution of rainfall.
- Temperature regime (altitude is also important as it affects temperature and relief characteristics)
- Soil mapping unit - mainly soil fertility, erosion hazard etc

These parameters used relate to the climatic and edaphic requirements of crops, and to the management systems under which the crops are grown.The purpose of zoning - as carried out for rural land-use planning - is to separate areas with similar sets of potentials and constraints for land use or development. Specific programmes can then be formulated to provide the most effective support to each zone. There are several systems for describing Agro-ecological zones in the Tropics. In Kenya two are used:

The FAO classification for tropics generally, and

An older Kenya version which is only applicable in Kenya.

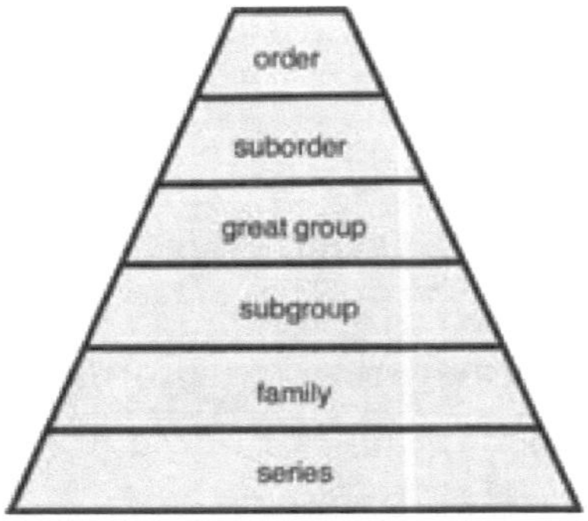

FIGURE 13. The Hierarchical Soil Classification System

3.63. THE KENYAN SYSTEM

The agro-ecological zones, also known as agro-ecological climate of Kenya are shown in the figure and Table below. Kenya has a total area of about 582, 646 square kilometers of which 11,230 or about 1.9% is covered by water. The dry land mass is commonly divided into six agro-ecological zones as indicated in Table 2.

Table 1: Kenya agro - ecological zone classification (Source IGAAD 1990)

ZONE	Mean annual rainfall (mm)	Land area ha (000)	% of total land area	Cumulative %	Classification potential	Climatic class
7	<150	-	0	0	None	Desert
6	< 300	10,573	18.4	18.4	Very Low	Arid
5	300 – 500	27,041	47.4	65.8	Low	Semi Arid
4	500 –700	7,314	12.8	78.6	Low to Medium	Marginal
3	700 –800	2,426	4.3	82.9	Medium	Semi Humid
2	800 -1000	3,296	5.8	88.7	High to Medium	Sub Humid
1	> 1000	6,407	11.2	99.9	High	Humid
0	Other	1733	0.1	100		
	Total	58790	100	100		

Figure 11: The % distribution of the agro-ecological zones in Kenya

LAND CLASSIFICATION

There are eight (I-VIII) land classes according to the United States Department of Agriculture (USDA) system. These are named: I, II, III, IV, V, VI, VII, and VIII, with characteristics given below.

Figure 12: The USDA Land classes: Source: Conte et al, 1997 pp 370.

3.7 LAND CAPABILITY CLASSES AND SUBCLASSES:

Capability class is the broadest category in the land capability classification system. Class codes 1, 2, 3, 4, 5, 6, 7, and 8 are used to represent both irrigated and nonirrigated land capability classes. Each of these has peculiar features that are based on suitability of the land for arable farming. The main aspects include: slope of land, soil depth and rockiness. 1.Class 1 soils have slight limitations that restrict their use.2.Class 2 soils have moderate limitations that reduce the choice of plants or require moderate conservation practices.

3.Class 3 soils have severe limitations that reduce the choice of plants or require special conservation practices, or both.

4.Class 4 soils have very severe limitations that restrict the choice of plants or require very careful management, or both.

5.Class 5 soils have little or no hazard of erosion but have other limitations,impractical to remove,that-limit-their-use-mainly-to-pasture,range,forestland,-or-grazing-land.

6.Class 6 soils have severe limitations that make them generally unsuited to cultivation and that limit their use mainly to pasture, range, forestland, or wildlife food and cover.

7. Class 7 soils have very severe limitations that make them unsuited to cultivation and that restrict their use mainly to grazing, forestland, or wildlife.

8. Class 8 soils and miscellaneous areas have limitations that preclude their use for

The subclass represents the dominant limitation that determines the capability class. Within a capability class, where the kinds of limitations are essentially equal, the subclasses have the following priority: e, w, s, and c. Subclasses are not assigned to soils or miscellaneous areas in capability classes 1 and 8.

Further details about characteristics and use of each land class are given in the table below

Table 2: United States department of Agriculture (USDA) Land classes.

Class of land	%Slope	Other characteristics	Use
I	< 2	All good features e.g. good texture, good drainage, good aeration, no swamp, no soil erosion and no soil capping	Can be used for a wide range of crops every year
II	2-5	Lacks one or more of the good qualities of class I. There is a possible danger of soil erosion	Still capable of growing crops but not quite so well
III	5-8	Lacks more of the best features of arable land. Has danger of soil erosion	May not be capable of growing crops every year
IV	8-12	The poorest arable land;	Cultivation not possible without complex soil conservation measures e.g. terracing
V	-	Swampy, wetlands close to river beds	Vegetable growing in raised beds in some places; Rice production in some places
VI	12-20	Shallow soil, rocks present, High danger of Soil erosion	Grazing or tree planting
VII	More than 20	Great danger of Soil erosion, very rocky, very shallow soils	May be used only for grazing and sometimes tree planting
VIII	Very steep	Shallow, infertile soils, rock outcrops present,	No agricultural value. Any available natural vegetation should be left to reduce runoff or storm water; Tourist attraction e.g. hill / mountain climbing

3.8 THE LAND TENURE SYSTEMS

Land tenure is the system of rights and obligations governing the acquisition, holding and disposal of lands by law. In other words, it is the ownership of land and the conditions governing that ownership. In Botswana and Tanzania, Land tenure system is largely state. In Kenya, freehold land tenure holds. In All cases, however, 9, 99or 999-year land lease is common for investors. The 999-year lease was coined by the colonialists during their reign in most African countries to guarantee them use of land for a long, long time.

Types of land tenure:

These include: Freehold, leasehold, communal and state ownership.

Free hold land tenure:

This is the private ownership of land (Dipolase in Setswana). The owner has the title deed to the land and has the right to use or occupy the land in any way he thinks will benefit him. A title deed is a document issued by a government to confirm ownership of a piece of land by an individual. This gives details about location, size, and dimensions of land. It also has the owner details and signature of the government agent. The owner of the land is secure and can make any number of development projects such as fencing, sinking boreholes, etc on the land. The owner can also use the title deed to get loans for developing the land. However, the land can be misused, as the power has too much freedom to use it in any way. It can be overgrazed, over cropped (continuous cropping), or may be exposed to erosion without good care. The owner has security of tenure and feels secure on land use. If most of or all land is under freehold, the government may not have land for emergency cases and national programs.

Also, some investments may be too large and involving to be effectively undertaken by an individual. If there is unfairness on land ownership, so that there are marked differences in land size owned by individuals, it may lead to acrimony and some community members can invade land. This has happened in Kenya on many occasions e.g. with the Criticos family which owned large tracts of land in Coast Province, Kenya; and the Kenyatta family which owns large tracts of land across the country, estimated to be about 5% of the whole country, the Moi and Kibaki families owning a large chunks of land in Kenya and all in high potential areas. This unfairness in land distribution can lead to vices such as arson, land clashes etc. These are living examples in Zimbabwe as well, and will continue to spread to other parts of Africa. It has been observed that major wars in Africa will be fought over land ownership and land use, while the next world war will be over energy.

Leasehold land tenure:

This is the renting of land by the owner (or landlord) to a tenant farmer for a specified period. The tenant pays rent for using the piece of land to the landlord who may be a private individual or the state. Any form of development by the tenant has to be approved by the landlord; and most of the time, the developments cannot be long term. The tenant can easily be exploited by some landlords who constantly increase their rents. Some worthless tenants may also claim the land before the end of lease term. This may inconvenience the tenant. There is therefore a need to make lease agreements in writing. This lease agreement should include: Date of lease; duration of lease e.g. exact date of expiry of lease; details of tenant; details of landlord; signature of both parties; size of land; and exact locations of key landmarks in the land. The short-term use rights limit any long-term development projects, takes into account such issues of land ownership, compensation, tenure interference with lifestyles of the host community etc.

Communal land ownership:

This is the ownership of land by the community as a whole, rather than by any one individual. The land is usually allocated, free of any charges, by the head of the community as long as the individual receiving the plot satisfies any rules that may have been worked out by the clan or village. Since the land does not belong to the individual, it cannot be used as a security for obtaining loans. Communal ownership and use of property commonly lead to tragedy of the commons- a saying meaning the tendency for all users want to derive maximum benefit and utility from the property without any care given in terms of management etc. Such property therefore always tends to be overused e.g. land is overgrazed. In such cases the whole community suffers, but they do not care as long as they suffer together. Usually long-term development and investment is not possible in the land. The advantage is that everyone has a right to use land as long as there is proof of being a member of the community.

State ownership / State land (Lefatese la ga Goromente):

This is the ownership of land where no one has a right to the land; all the land belongs to the state. Individuals may be allowed to settle on the land for specified period of time but may be removed when the need for the land arises. Compensations may be paid for any investments made on the land.

Advantages:

- Land is made available to more people;
- Land is available and can be used for emergency purposes e.g. refugee camps, hospitals, etc;
- Land is available and can be used for national interests e.g. schools, hospitals, resettlements research etc;

Disadvantages:

- Land may be abused by the government e.g. using as a tool to reward political cronies and loyalty at the expense of more needy cases;
- Land grabbing may take place;
- Land may be misused by communities, e.g. by overgrazing the open fields, as the government may not have the machinery to restrict use of its land all over the country;

MODULE 4: CROP HUSBANDRY

4.0PLANT NUTRITION AND NUTRIENTS
4.1THE PLANT
4.2 REPRODUCTION IN PLANTS
4.3 GERMINATION
4.4 TROPISMS
4.5 LAND PREPARATION

4.0 PLANT NUTRIENTS

Concept of soil fertility

The fertility of a soil can be defined as the inherent ability of the soil to supply nutrients to plants in adequate amounts and in suitable proportions.

Sources of plant Nutrient Elements

- air & water,
- Soil solids – through weathering
- Fertilizers
- Organic manures, / organic residues, OWs
- Liming
- Fixation of N_2 – lightning, bacteria (Azotobacter, Rhizobia0, BGA

Most of the elements are derived initially from the weathering of minerals and taken up by the plant roots. They are then returned to the soil in plant litter which is decomposed to release elements which are again taken up by the plants.

4.01 ESSENTIAL ELEMENTS / ESSENTIAL PLANT NUTRIENTS

The soil holds a wide range of mineral particles / elements which are used as plant food – nutrients. Though crucial, they vary in their importance as plant food – with some playing minor role as plant food (e.g. not essential but beneficial elements such as Co, Na etc. An essential plant nutrient is a chemical nutrient / compound necessary for the normal growth and development of the plant. There are 16 elements essential for plant growth which include 3 structural elements (non- mineral) & 13 mineral elements.Structural elements, such as C, H, O. play a major role in plant cell Structure, while mineral elements include N, P, K, Ca, Mg, S, Fe, Zn, Cu, Mn, B, Mo, Cl.

Criteria for Essentiality

Arnon & Stout (1939) proposed the term 'essential mineral nutrient/ element' and concluded that for an element to be essential for plant growth and development, three criteria must be met. The plant must be unable to complete its life cycle in the absence of the mineral element. Life cycle includes both vegetative & reproductive stages. The nutrient must be directly involved in plant metabolism / nutrition of the plant – either structural or metabolic Reactions i.e. directly involved in basic physiological processes e.g. photosynthesis, respiration, protein synthesis, transpiration, etc. The role / function of the nutrient must be specific / not replaceable by another mineral element.

Beneficial elements

They stimulate growth and are not directly involved in plant metabolism. Examples – Na, Co, Si = major beneficial elements; others are V, Al, Ni, Se.They compensate for toxic effects of other elements, maintain O.P etc.

Classification of essential nutrient elements

Commonly based on plant need i.e. amount required by plant, and not related to amount of nutrient present in the soil. Macronutrients – required in large amounts -(> 1000 ppm) = 9 Micronutrients – required in small amounts -(< 1000 ppm) = 7. The macro elements are required in relatively large amounts whereas micro elements are required in small amounts. All the essential elements are equally important for proper plant growth. The elements must be present in the correct proportions. When there is a deficiency or excess of any one element plant growth is seriously affected and the plants develop symptoms of nutrient deficiency or toxicity.

Categories

Required in relatively large amounts (Macro nutrients)		Required in small amounts (micronutrients / Trace elements)
Supplied by Air & Water	Supplied by Soil solids	All supplied by soil solids
C, H, O – category 1	N, P, K – category 2	Fe, Zn, Cu, Mn, B, Mo, Cl.
	Ca, Mg, S – category 3	Co, Na, Si, V – beneficial elements

Categories of elements-

Non-mineral nutrients – C,H,O – are essential nutrients supplied by air and water. Are the major constituents of plant tissues constituting over 90%. Main components of carbohydrates, proteins and fats.Major nutrients (primary macro nutrients) – NPK –essential nutrients required by plants in slightly larger quantities. Are called fertilizer elements

Secondary macro nutrients – Ca, Mg, S – essential nutrients required by plants in fairly large quantities (less cf NPK). Ca, Mg are added to acid soils in limestone / dolomite and are called lime elements. Except for Fe, and Mn (some cases eg acidic soils), trace elements are found sparingly in most soils & their availability to plants is often very low. The 3 general situations where micronutrients are most apt / likely to be a problem are: Sandy soils, Organic soils and Alkaline soils.

4.02 IMPORTANCE OF MACRONUTRIENTS IN PLANTS

Nitrogen

Nitrogen is an important component of amino acids. Amino acids are the building blocks for all proteins. Proteins are useful as structural components (e.g. in cell wall, cell membrane and nuclear membrane; phycobillin and carotene) formation; metabolic protein component (such as enzymes, pigments, hormones, antibodies etc.) hereditary chemical constituent (DNA, RNA). It is needed for making plant proteins and for normal growth and repair of worn-out tissues.

Deficiency Symptoms

- Chlorosis – leaves turn pale yellow to yellow in color – lower tips first.
- Since nitrogen is mobile within plants, it tends to move from older to younger tissue when in short supply.
- Stunted growth – slow, poor plant growth
- Early maturing
- Premature leaves fall – defoliation

Phosphorus

- Promotes root development
- Necessary for seedling development, flowering, seed and fruit formation
- Crop maturation - hastens ripening of fruits, maturity and promotes seed production
- Nitrogen fixation - Assist nodulation in legumes
- Improvement of crop Quality e.g. vegetation, forages. Increases palatability of forage and vegetables.
- Strengthens plants – help prevent lodging in cereals
- Increases resistance to certain diseases and pest attack.

Symptoms of P Deficiency

- Purplish coloration on foliage of some plants (older leaves first) = anthocyanin pigments
- Slow growth; stunted plants.
- Inhibited root development
- Delayed maturity.
- Poor fruit or seed development.
- Slender stalks
- Poor yield in potatoes

Potassium

- Increases plant resistance to disease.
- Increases / improves quality of fruits products - sugar content of fruits, size of nuts
- Prevents lodging by strengthening stalks.
- Important in plant water uptake and balance through effect on Osmotic potential for maintenance of plant turgor.
- Potassium enhances yield of tubers and root crops (promotes root growth)
- Essential for translocation of sugars / carbohydrates transport
- Starch synthesis – conversion of soluble sugars into starch especially for storage / grain filling process
- Enzyme activation. K required for activation of starch synthetase
- N uptake (especially nitrate uptake) and protein synthesis in plants

Deficiency symptoms

- Tip and marginal chlorosis. Margin and veins are scorched brown.
- Marginal necrosis = death of leaf tips and margins
- Chlorosis along leaf margins
- Leaf curl
- Slow growth
- Bluish, dark green leaves
- Weak stems and stalks – crop lodging, premature leaf fall
- Small fruit or shriveled seeds = Reduced yield

CALCIUM

Form taken up by plants as the calcium ion (Ca++) or Ca2+

Functions

- Constituent of cell walls – in form of ca- acetate = a structural nutrient
- Formation of middle lamellae
- Affect / maintains membrane permeability
- Regulate cell acidity
- Influences uptake of plant nutrients, water
- Cation balance for anion transport
- Formation and functioning of root nodule bacteria
- Protein synthesis
- Plant vigour & stiffness
- Growth, proper functioning of meristems, esp root tips.

- Leaves and root development
- Regulates translocation of CH_2O through phloem vessels
- Immobile in the plant

Deficiency symptoms

- Show first in young leaves
- Dying back of tips = deformed terminal leaves. = Tip burn" of young leaves – celery, lettuce, cabbage.
- Death of terminal buds / growing points = apical meristem = poor;
- Root tips also affected.
- Distorted leaves with irregular margins = curl upwards
- Abnormal dark green appearance of foliage
- Premature shedding of blossoms and buds
- Weakened stems.

N.B. Once deposited in plant tissues, Calcium is not remobilized. Therefore, young tissue is affected first under conditions of deficiency. Since Calcium is not mobile, the requirements of a crop for Ca^{++} after early fruit set are commonly supplied in the form of nutritional sprays. Water-soaked, discolored areas on fruits – blossom-end rot of tomatoes, peppers, and melons; bitter pit or cork spot of apples and pears

MAGNESIUM

- Form = Plant uptake in the form of magnesium ion Mg^{++}

Functions

- Central constituent of chlorophyll molecule = essential for photosynthesis eg

Mg^{2+}

- (NB: 15-20% of leaf Mg content is present as a central atom of Chloroplast).
- Enzyme activator of many plant enzymes required in growth processes
- eg in CH_2O metabolism
- Regulate uptake of P & translocation of P in plants, carrier of P Especially into seeds

- Synthesis of oil in oil crops eg groundnuts, soybeans
- Component of ribosomes thus important in protein synthesis
- Mobile in the plant

Deficiency Symptoms

- Interveinal chlorosis in older leaves
- Leaf margins curl with dead spots (Curling of leaves upward along margins).
- Reddish purple color
- Marginal yellowing, with green "Christmas tree" area along mid-rib of leaf

- N.B. Magnesium is mobile within plants and can be readily translocated from older to younger tissue when it is deficient. Plants need magnesium to balance the generally high use of potassium from fertilizers and manure.

SULPHUR

- Forms: Sulphur is absorbed as SO42-; SO2 through leaves in areas where the atmosphere has been enriched with sulphur compounds

Functions

- Synthesis of amino acids - Sulphur is a constituent of 3 amino acids (cystine, methionine and cysteine) and is therefore necessary for
- protein synthesis.
- Essential for nodule formation on legume roots hence assist in
- N-fixation
- Formation and activation of enzymes eg coenzyme A
- Important in synthesis of plant hormones - vitamin hormones eg biotin (Vitamin H); others thiamine (Vitamin B1) & other metabolites
- Present in oil compounds (glycocides) responsible for the characteristic odors & flavor of plants such as mustard garlic, onions.
- Immobile in plant

Symptoms of S Deficiency:

- Retarded growth rate and delayed maturity.
- Young leaves light green to yellowish color. – chlorosis in young leaves
- Small and spindly plants.
- (- low protein content – a symptom?; not a visible sign)

Note: the symptoms of specific mineral deficiency are related to the role of that mineral in the plant & its relative mobility within the plant.

QUESTION: State the soil and environmental conditions under which sulfur deficiencies are most likely to occur in the field (5 marks).
Guideline response: Waterloging – under anaerobic conditions, SO_4 is reduced to H_2S
Fe^{2+} in soil – H_2S may escape to atm or reacts with Fe^{2+} to form amorphous FeS which undergoes conversion to pyrite (FeS_2)Large C/S or N/S ratio in the soil – immobilization of S
Presence of $CaCO_3$ – enhances pptationm of SO_4

MICRONUTRIENTS

IRON
Form – Fe taken up by plants as ferrous ions (Fe^{++}), also Fe^{3+}; organically complexed Fe (chelates)

Functions
- Formation of chlorophyll in plant cells.
- Component of various flavoproteins (metalloflavoproteins).
- It serves as an activator for biochemical processes, such as respiration,
- Photosynthesis, and symbiotic nitrogen fixation. Occurs as ferredoxin, which plays a role in electron transfer in photosynthesis and in the reduction of N_2 (gas) in N-fixation.
- Redox reactions in cells

Symptoms of Fe Deficiency:
Interveinal chlorosis of young leaves. Veins remain green except in severe cases.
Twig dieback.
In severe cases, death of entire limbs or plants.

Iron deficiency can be induced by high levels of manganese or high lime content in soils (= lime induced chlorosis - bec bicarbonate interferes with Fe metabolism). Deficiencies are common in high pH soils, in situations with poor aeration, or when high levels of zinc or manganese are present.

ZINC

Form - Zinc (Zn) is taken up by plants as the zinc ion (Zn^{++})

Functions

An essential constituent of several important enzyme systems in plants.

It controls / Involved in biosynthesis of IAA - indoleacetic acid, an important plant growth regulator.

Role in protein sysnthesis

Immobile

Terminal growth areas are affected first when zinc is deficient. Deficiency is most common on soils with neutral or alkaline pH that are sandy or have low organic matter content (including cut areas) and are very high in available phosphorus.

Symptoms of Zn Deficiency

- Decrease in stem length (shortening of internodes) and rosetteing of terminal leaves (called little leaf disease)
- Reduced fruit bud formation.
- Mottled young leaves (interveinal chlorosis).
- Dieback of twigs after first year.

COPPER

Form = ionic form (Cu^{++}).

Functions

- Serves as an activator of numerous plant enzymes
- Plays a role in the development of plant pigments that influence color.
- Chloroplast contain a cu-containing protein called plastocyanin without which photosynthesis cannot take place.
- A copper deficiency interferes with protein synthesis.
- Copper deficiency has been observed on trees and vines growing on organic soils and sands.

Symptoms of Cu Deficiency:

- Stunted growth.
- Dieback of terminal shoots in trees.
- Poor pigmentation.

- Wilting and eventual death of leaf tips.
- Formation of gum pockets around central pith in oranges.

MANGANESE

The major form is = Mn^{++}.

Functions

- Serves as an activator for enzymes in plant growth processes – respiration, N metabolism.
- Assists iron in chlorophyll formation.
- Plays a role in Nitrate reduction from NO_3^- to NH_2 for preparation of essential amini acids.. NO_2^- is gotten rid of, converted very fast to NH_2 because its toxic i.e acts as activator of nitrate reductase & hydroxylamine reductase.
- Involved in photosynthesis – formation of O_2. Essential for the Hill reaction of photosynthesis, in which water is split into hydrogen ions & oxygen gas

High concentrations of manganese in plants may induce iron deficiency. Manganese is generally required with zinc in foliar spraying of commercial citrus. Other tree crops may show deficiencies.

Symptoms of Mn Deficiency:

Interveinal chlorosis of young leaves.

Gradation of pale-green leaf coloration, with darker color next to veins.

No sharp color distinction between veins and interveinal areas as with iron deficiency.

BORON

Boron (B) is predominately in solution as boric acid, H_3BO_3, and is taken up by plants in this form.

Functions

Differentiation of meristematic cells.

Regulating metabolism of carbohydrates in plants. Also translocation of sugars i.e. may form complexes with sugars & facilitate their transport within the plant as borate complexes Nearly all fruiting crops have a high demand for boron in the early fruiting stage. Foliar applications are an effective way to meet this demand at this critical growth state. Recent research has shown that boron is effective in increasing pollen viability, thus contributing to increased fruit set. When boron is deficient, cells may continue to divide, but structural components are not differentiated.

Symptoms of B Deficiency:

- Death of terminal growth (shoot tip), causing lateral buds to develop, producing a "witch's broom" effect. (tips of laterals also die)
- Thickened, curled, wilted and chlorotic leaves (thick coppery texture of leaves).
- Soft or necrotic spots in fruit or tubers.
- Reduced flowering or improper pollination.
- General disintegration of internal tissues resulting in abnormalities such as 'heart rot' in sugar beets.

MOLYBDENUM

Form = taken up as the molybdate ion (MoO4=).

Functions

- Required for the utilization of nitrogen. Nitrate nitrogen (NO3-) is converted to amino acids (NH2) by the nitrate reductase enzyme; this enzyme requires molybdenum. Fe absorption & transport
- Required for the symbiotic fixation of nitrogen by legumes.

Symptoms of Mo Deficiency:

- Stunting and lack of vigor. This is similar to nitrogen deficiency, because of the key role of molybdenum in nitrogen utilization.
- Marginal scorching and cupping or rolling of leaves.
- "Whiptail" of cauliflower.
- Yellow spotting of citrus

CHLORINE

Electrolyte necessary for osmotic pressure and acid- base balance Necessary for the Hill reaction of Phs in which water is split into hydrogen ions & oxygen gas Activates the oxygen producing enzymes of Phs Disease control. Chloride containing fertilizers greatly depresses take-all root rot infections in wheat. Also stripe rust = ammonium chloride cf ammonium sulfate.

4.03 FERTILZERS

Fertilizers are soil amendments applied to promote plant growth. They are usually applied directly onto the soil, but can also be applied onto leaves (foliar feeding). Fertilizers are broadly divided into organic fertilizers (composed of enriched organic matter—plant or animal), or inorganic fertilizers (composed of synthetic chemicals and/or minerals).

Inorganic or Chemical Fertilizers

These are any chemical substance added to the soil to supply nutrient elements required in the nutrition of the plant. The main nutrients added in fertilizer are nitrogen, phosphorus, and potassium (major plant nutrients) but other nutrients are added as soil amendments eg Ca, Mg, S and others in smaller amounts (trace elements). The use of inorganic fertilizers is based on fertilizer recommendations determined by soil & plant analysis.

Advantages of using Inorganic chemical fertilizer (compared to manures)

- High nutrient content in a given quantity/ weight.
- Easy to handle because relatively small quantities are needed.
- No risk of spreading insect pests, diseases and weeds.
- Small amount used/ needed during application.
- Are cheap to apply in terms of labour expense.

Disadvantages

- Are expensive.
- Technical skills is required in their application.
- Certain fertilizers may cause/ induce soil acidity.
- They do not improve soil (physical) properties e.g. structure, water holding capacity e.t.c.
- Supply nutrients within a restricted period of time.
- It is used up quickly by plants.
- It is easily lost by leaching.
- The high costs of fertilizers, lack of knowledge on proper use, poor transport and marketing infrastructure in rural areas serve as disincentives for fertilizer use by poor smallholder farmers in Africa (Buresh and Giller, 1998).

Classification of chemical fertilizers based on Nutrient carried / composition

Straight fertilizers – carry / supply only one major element e.g. AS,TSP,

General Characteristics

Most nitrogen fertilizers are very soluble in water and are therefore easily washed down the soil beyond plants' rooting zone, i.e., leached. This tendency for them to be easy prey to leaching requires that they be applied cautiously. They are often applied in split, i.e. a little at a time, at some reasonable interval in the form of top-dressing. For example, 3 – 4 week interval is common for most vegetable crops. When exposed to the environment many nitrogen fertilizers are hygroscopic. Hygroscopic substances absorb water, but do not become liquid. Deliquescent solids absorb water up to a point when they turn into liquids. When in relatively hot environments, nitrogenous fertilizers are suspected to loose some of their nutrients by volatilization. They therefore need dry and cool stores. When in contact with plants, they cause desiccation / scotching, wilting and subsequently, death because of the high osmotic pressure they exert. They behave like the common table salt placed on a wound on human skin. We obviously feel pain- and this is what the plant also feels -silently- leading to wilting. The plant suffers without bitterness, but nevertheless, dies if the osmotic force is extreme.

Ammonium fertilizers are volatile and tend to lose their active ingredient when left in an open environment. They therefore need to be slightly covered by soil after application. Leaching and volatilization loses normally reach as much as 50-70%. Thus fertilizer use efficiency of most nitrogenous fertilizers range from 30% to 50%, with the lower side mostly experienced with the use of nitrate fertilizers while the higher side experienced with ammonium fertilizers. This figure increases further if application is wrongly timed, and wrongly done. Split application normally improves the use efficiency.

The positively charged ammonium ions (NH4+) are attracted by the negatively charged soil colloids, thus reducing the leaching losses. This is because of the basic principle that like charges repel, while opposite charges attract.

On a similar vein, the negatively charged nitrate ions (NO3-) are repelled by the similarly charges soil colloids, greatly increasing leaching. Since soil colloids are generally negatively charged, nitrates tend to experience more leaching losses than the ammonium fertilizers. This is because the soil colloids trap and hold the latter, and repel the former. Nitrogen fertilizers therefore need cautious application since it can cause more havoc to the crops and the environment. Leached nitrogen nutrients pollute underground water resources, thereby reducing its value.

GENERAL CHARACTERISTICS

General characteristics of phosphatic fertilizers

- Supply only P as major element. Available P expressed as P_2O_5
- Absorbed as phosphate – $HP0_4^{2-}$, H_2PO_4-
- Soluble in water & are readily absorbed by crop roots.
- Less soluble Compared with N fertilizers
- Promote root development- applied next to seed at sowing
- Less leached – can be applied early at planting
- Not mobile in soil, remains in the root zone where applied.
- No / less scorching effect on crop foliage
- Fixed in acidic soils – place in root range for max utilization.
- Persist in the soil – got considerable residual / carry-over effect for a number of years…2-4.
- May benefit crops grown in subsequent years.

Examples of straight phosphatic fertilizers are:

Single superphosphate (SSP) - It is obtained when rock phosphate is reacted with sulphuric acid. It is 85% - 90% water soluble, white granules, and contains ~ 21% P2O5, range 16-22% P2O5.

Double superphosphate (DSP). This contain 42% P2O5. TSP – 46% P2O5; (concentrated superphosphate, CSP). Synthesized by treating (high grade) phosphate (rock phosphate) with phosphoric acid.

Ground Rock phosphate - contains 27-41% P2O5; 11.5-17.5% total P.

Raw rock phosphate comprising complex apatites. Its mined then ground finely for direct application to the soil. When finely ground, its attacked by acid soil & P slowly become available to crops mineralized slowly, e.g minjingu rock phosphate.

Basic slag. It is a by product of open-hearth steel production and effective on acid soils due to its alkalinity.

Potassic fertilizers

They are obtained by mining underground salt (KCl) beds, also brine from salt lakes. Kainit & manure salts are most common crude K sources.

General characteristics

- Soluble in water – more soluble than P but less soluble than N fertilizers

- Easily leached
- Have scorching effect
- Most soils in EAfrica abundant K, so not so much is in use
- Absorbed as K+, and available K is expressed as K2O equivalents

- Examples
- Muriate of Potash (KCl) – Most widely used potassic fertilizer.
- are creamish, lightish granules, which contains 60-62% K2O.
- Sulphate of Potash (K2SO4) – Manufactured from KCl, hence more expensive than
- MP. Are fine yellowish salt and contains 50% K2O (53% K2O) are used for crops
- sensitive to chloride.

Nitrate of Potash, KNO3
contains 44% K2O and used for crops ; sensitive to chloride, eg tobacco. High chlorine content in tobacco ; results low quality of leaf.

4.04 OTHER MACRONUTRIENTS

- Calcium: Is needed for formation of cell walls (as Calcium pectate) and plant leaves.
- Magnesium: Needed for formation of chlorophyll, the green coloring matter in plants. Deficiency results in severe yellowing of leaves, which reduces photosynthetic surface area of the plant, resulting in stunted growth.
- Sulphur: Needed for protein formation in a plant. Is an essential constituent of some essential amino acids (Cysteine, Cystine, Methionine etc), some of which help form the first class proteins. Deficiency results in retarded growth and low yield.

MICRONUTRIENTS
These include Iron (Fe), Manganese (Mn), Copper (Cu), Zinc (Zn), Boron (B), Chlorine (Cl), Molybdenum (Mo).etc.
Molybdenum: Controls the use of nitrates by plants; it is like a catalyst in nitrogen metabolism.
Copper:Helps a plant to change nitrogen into proteins (through the anabolic process of Protein synthesis).
Boron:Helps a plant to use other nutrients properly (i.e., a catalyst in the metabolic processes).

Compound fertilizers (or Mixed fertilizers)
Have more than one fertilizer element (i.e carry two or all the 3 of NPK). Are also known as mixed / complex fertilizers. . Can be made chemically or mechanically.

Fertilizer grade or analysis
Refers to the minimum guarantee of the plant nutrient content in the fertilizer in terms of % total N, % available P as P2O5 (phosphorus pentoxide), and % soluble K as K2O (potassium oxide). The grade / analysis thus gives % by weight of NPK.

The balance consists of those substances that contribute little to plant growth or may be soil amendments / conditioners. Are referred to as filler material (= a make-up weight used to balance nutrients in 1 tonne)

Compound fertilizers are divided into 2
Incomplete compound fertilizers – where 2 major / fertilizer element – mainly ammonium phosphates, considered P sources but also contain significant N.

Monoammonium phosphate, (MAP) $NH_4H_2PO_4$, given configuratively as 11:52:0.
Diammonium phosphate, $(NH4)_2HPO_4$, DAP – 18:46:0.
Others are: 20:20:0; 17:17:0 and 23:23:0

These are largely used in Kenya due to adequate K in most soils. Use of K is largely based on soil tests Complete cpd fertilizers / complete fertilizers. They contain all the 3 primary macronutrients. Examples are: 20:20:20; 20:10:10; 17:17:17; 19:19:19; 10:20:10; 15:15:6+4 – Mg (4% MgO); and 25:5:5:5 – Sulfur esp for tea

Bulk Blending
Mechanically mixing fertilizers to make a compound fertilizer, such as done by MEA Ltd, Kenya

Advantages of bulk blending
Economical – finished raw materials of NPK fertilizers can be produced in different large plants & combined in the market. Versatility – practically any fertilizer grade or ration can be produced thus offering convenience

Disadvantages
Different materials - must be of the same size, density, shape – to ensure homogenous mixing. Materials must be chemically compatible should not chemically combine to form harmful cpds Segregation (after mixing) – may affect uniform applicn in field

Rules followed in bulk blending

NH4-N Fertilizers should not be mixed with lime containing materials / alkaline materials. This is because N loss as NH3 ie NH4 cpd + lime release NH3 Exception – lime ammonium nitrate Do not mix fertilizers that form insoluble product – result in low availability. Water soluble phosphates should not be mixed with lime containing fertilizers bec form secondary & tertiary phosphates which are insoluble.

Hygroscopic fertilizers Vs non hygroscopic – if mixed will absorb water more. Can be mixed in dry weather just b4 distribution. Nitrate containing fertilizers – can cause fire when mixed with substances that can burn A mixture of NH4-NO3 fert with Chloride results in an increase in temperature.

Formulation of bulk blending

Determine the fertilizer grade of the compound fertilizer when the following materials are Mixed: 400 kg AS -20%N; 400 kg SSP – 20%p2O5; 200kg MP – 60%K2O

N supplied = 20% x400 = 80 kg; P2O5 supplied = 20% x 400 = 80 kg
K2o supplied = 60% x 200 = 120 kg; In 1 tonne = 80 -80-120 = 8:8:12 expressed as %.

Qn2 calculate the amount of each of the following materials required to prepare a cpd fertilizer of the grade 5:10:10. SA = 20% N; SSP= 20%P2O5; KCl = 60% K2O. what is the weight of the filler material?.

Foliar feeds

- Fertilizers formulated to be used to correct deficiencies noticed during plant growth
- Liquid fertilizers – usually in form of sprays - applied to foliage by direct spraying. Examples – Wuxal, Murphy foliar feed, Rapid – Gro.
- Used commonly to correct micro-nutrient deficiencies during plant growth.
- Has been used for high value crops citrus, pineapples, bananas, sugarcane.
- Avoids nutrient fixation problems in soils.

Fertilizer placement

The aim is to put nutrient within feeding range of roots for maximum Benefit. Fertilizers are placed in or on the soil, preplant, at planting & after planting. Some (minor) applied to the leaves as foliar spray.

Methods of surface placement

- Broadcasting dry materials
- Top dressing
- Spraying liquids
- Applying fertilizer in irrigation water = fertigation

Methods of placing fertilizer in the soil include

- Band placement
- Plow down
- Side dressing
- Injection
- Application in irrigation water

Foliar spray – minor amt of fertilizers applied to leaves

Choice of placement technique / method depend on

- Type of crop / crop rotation
- Soil test level / degree of nutrient deficiency
- Mobility of nutrient in soil
- Equipment available

4.05 ORGANIC MANURES

The term "Organic" means derived from living matter. organic manures are also referred to as organic fertilizers. They are derived from decayed plant/animal matter, and once applied to the soil they become part of the soil organic matter(OM). The nutrients are released into the soil through the process of microbial decomposition / decay. Organic manures perform two important functions:

(1)Provides the plant with all essential nutrients including trace elements which are not contained in commonly used chemical fertilizers.

(2) They also act as a soil conditioner i.e. by improving physical properties of soils e. g. drainage and aeration of heavy soils , the structure of soil and also moderate soil temperature, improve H2O holding of sandy soils.

Naturally occuring organic fertilizers include manure, slurry, worm castings, peat, seaweed, guano and sewage sludge (toxic metal accumulation is a disadvantage and limits its use.

Types of manures.
Classification of manures is done according to source, hence we have got the following types of manures.
Green manure; Farmyard manure; Compost manure; crop residues and other organic wastes; and Sewage sludges.

Advantages of using manures in comparison to fertilizers.
Less expensive.- Cheap to make and affordable to peasant farmers.
Provides trace/ micro- nutrient elements- besides macro- nutrients
Increases/ improves humus(O.M) content of the soil.
Addition of manure to the soil increases microbial activity.
Acts as a soil conditioner and
Improves drainage and aeration of heavy soils (clay).
Improves water holding capacity of sands.
Improves structure(binds) of both sandy soils and clay soils.
Moderates soil temperature.
Chemical properties improved
Improved CEC
Improvement of chemical compounds / nutrients
Mobilization of nutrients through complexation or chelation
Reduces soil erosion.

Long lasting effect- benefits the crop for a longer period of time (- i.e. nutrients are not all readily available for plant uptake.

Disadvantages

- Bulky.
- May harbor insect pests and diseases.
- Have less nutrients in a given quantity (nutrient element content is low).
- Low concentration of nutrients, particularly macro- elements required in large quantities cp. Rate of application =20- 40 mg/ha or 9-18 tons /acre.
- More labour is needed in preparation.

Green manure
Crops established in the field and are ploughed back into the soil when theyreach flowering stage and still green (at blooming). Crops used include Groundnuts, Cowpeas, Sunhemp (crotalia juncea), soyabeans, clovers, lucaena, sunflower, desmodium, vigna, centrosema,

stylosanthes, tropical kudzu, flemingia shrub. They possess the following characteristics:

- High N-content/ N- fixing crops – mainly legumes.
- Rapid growth = Because they are mainly grown in a short rainy season with limited moisture supply.
- Highly vegetative or leafy, hence have low C: N ratio for ease of decomposition.
- They provide adequate soil cover which help smother weeds, prevent evaporation and for erosion control.
- Well developed, extensive roots.
- Improve soil structure= high infiltration

Disadvantages of legumes

- Adoption rates are low because crops used are mostly food crops hence it becomes
- difficult to convince peasant farmers to incorporate them back into the soil at flowering.
- It is difficult in turning the manure crop back into the soil(Unless done mechanically.)
- Green manure crop may use a lot of water and leave very little for next crop. Manure crop may become hard+ fibrous before being incoperated into the soil such that becomes difficult for bacteria to decompose thereby causing N-immobilization.

Farmyard manure (F. Y. M.) or Muck

Refers to animal wastes (dung, urine) mixed with crop residues as animal beddings and left to rot into brownish- black mass. The following is the process of making FYM

Farmyard manure makes the bulk of organic fertilizers. It consist of dung / droppings / Urine – of farm animals, Litter/ straw as animal beddings. However, if cow dung is used alone without crop residues it is referred to as cattle manure. Guano is an example of poultry manure.

Quality of F. Y. M.

Chemical analysis of FYM shows that its quality (=nutrient content) varies considerably according to: Food given to animal ; Type of animal (+age); Age of manure; How the manure is stored; and Nature of the litter.

The process of making compost manure using the heap method

- Select a well drained and sheltered site.
- Clear the site from bush
- Mark out an area measuring 1.2m- 2m square.

- Fix posts at the corners of measured area. The posts should be 1.2m- 2m square. Provide side protection of the materials by building the sides with planks of wood Dig a trench around the composting area to facilitate drainage.
- Collect the materials to be used for composting.
- Chop the vegetable matter into small pieces.
- Lay down the fibrous/ hard materials first
- Place a layer of chopped vegetable material on top of the foundation layer.
- Place a layer of well rotten compost manure or farmyard manure or top soil to introduce micro- organisms.
- Add a layer of wood ash/ lime to reduce acidity. To hasten decay, small quantities of fertilizers e.g. S.S.P. may be added (Fortifcation)
- Repeat the same layers until the desired height is reached.
- Apply water to the pile when it is dry to keep it moist (50- 70%H2O)
- After desired/ final ht. is obtained, cover the heap with dry leaves/ or soil.
- Pass thro' the heap a no. of sticks right across as the pile is being made to facilitate aeration. Remove the sticks after pilling
- Put a shed/ shade to cover the heap
- Turn over the compost after 3- 4 weeks

The materials should be turned over 3 times before it is ready.

ECOLOGICAL EFFECTS OF INORGANIC FERTILIZERS:
A CASE OF SULPHATE OF AMMONIA (SOA)

This is a very popular top dressing fertilizer in paddy fields. It supplies crops with both Sulphur (as Sulphate) and Nitrogen (as ammonium). Both these are known to be amino acid components, and therefore, prerequisites to enhanced growth and vigor of crops. No doubt, therefore, the SOA has a positive role in boosting rice yield. However, unfortunately this is only a short- term gain, but not sustainable.

The negative side of SOA

It is on record that for every 100kg of SOA applied to a soil, 45kg of Calcium is lost from the soil as part of the drainage water (Afullo, 1995, as cited from Russel, 1983). As a divalent cation, Calcium carries with it both nitrates and sulphates in the ratio of 2:1. This means that 67% of the lost Calcium goes as Ca(NO3)2 and 33% goes as CaSO4. The Calcium originates from the soil cultivated, while both the anions originate from the applied fertilizer (SOA).

Effects on the soil

Major soil stabilizers, i.e. those components that help hold together and sustain its structure, are divalent cations such as calcium and magnesium, as well as organic matter. Of these three stabilizers, Calcium is the stronger cations. It is thus required in quantities larger than magnesium, as it has stronger cementing qualities. Therefore, a Calcium-deficient soil is very unstable. By inference, a paddy field, which generously utilizes SOA, is bound to lose its stability and therefore structure in the long run. It exhibits relatively high Sodium adsorption ratio (SAR) and exchangeable Sodium Percentage (ESP), rendering it sick. It will have loose soil particles which are readily blown by wind, and when it rains, by the surface runoff. A sick soil cannot produce a healthy crop. It has magnesium hazards, low water infiltration rate, reduced aeration and porosity, and causes high surface run–off. Such are the qualities of a structure less, loose and deflocculated soil.

With low infiltration capacity and loose (deflocculated) nature, the resulting scenario becomes conducive for surface erosion. Such soil can be very expensive to use for economic agricultural purposes. Its rehabilitation requirements render it a bad soil- as it becomes many times more expensive to repair (rejuvenate or restore). It cannot therefore sustain the very high, maximum yields that it previously could. To reclaim such a soil, thousands of tones of lime and organic matter are required. Both are very labor intensive to avail in the right form and appropriate quantities. Before starting to use Sulphate of Ammonia, therefore, one needs to think twice as the short – term benefits are very fast overcome by the long-term reclamation costs.

Effects on microorganisms

Independent, free atmosphere nitrogen fixers, Azotobacter; symbionts (Rhizobia), Cyanobacteria, some fungi (yeast) and actinomycetes are nitrogen fixers. They only work effectively when there is less soil nitrogen and get suppressed when soil has excess nitrogen. Therefore natural soil organisms need a natural environment to operate effectively in. This conducive environment should be provided for sustainable soil utilization. This can be enhanced by adding some organic matter which some of the microbes require as their source of carbon (i.e., as food).

Effects on the receiving water body

A sink is a site where matter is accumulated or deposited. Just like our kitchen or bathroom sinks. In this case the disposal point of the irrigation wastewater / effluent becomes the sink. These may be water bodies such as lakes, rivers, oceans, ponds etc. The

$Ca(NO_3)_2$ and $CaSO_4$ rich waste – water cause ecological disasters in the sinks. The sulphate makes the water have a poor taste. The nitrates, on its past, is known to be deficient in most water bodies. Thus, as a limiting factor in most aquatic systems it is the major controller of floral growth and vigor once it is available. Once it reaches an aquatic body, it triggers off a series of reactions. To say the least, this is described as eutrophication- described in another section of this chapter. Nitrates levels as high as 40 mg /L have been recorded in the wastewaters of West Kano irrigation scheme in Kabonyo, Kisumu district, Kenya. This waste from the Kano plains is discharged back to Lake Victoria. This is the second largest fresh water lake in the world, and the third largest lake. It is therefore not a surprise to have the current ecological disaster in Lake Victoria – Water Hyacinth. To date, it is still a problem.

Effects on water users

Infants of less than six months of age are very susceptible to nitrate – rich waters. They suffer from blue baby syndrome- sometimes called methaemoglobinaemia. Report have also indicated that drinking water with high Nitrogen levels expose both adult and infants humans beings to early onset of hypertension. Hypertension means high blood pressure. This stresses the heart, as it is overworked, leading to possibilities of heart attack or its twin brother, heart failure. In oxygen – deficient, eutrophic aquatic systems, reducing conditions favor the existence nitrates (NO_3^-) as opposed to nitrite (NO_2^-). The nitrites (NO_2^-) are capable of forming N-nitroso compounds that are known to be serious carcinogens.

Besides, there are risks of human beings catching botulism arising from clostridium botulinum. C. botulinum is a thermostable bacterium that tends to inhabit eutrophic environment. Therefore, poor disposal of nitrate – rich waters poses a serious threat to the lives of human beings. Medical expenses become very high.

ORGANIC FERTILIZERS (MANURE)

These are also called manure, and play crucial roles in both soil health and plant growth. In the soil they play such roles as:

Source of food for soil fauna, both micro and macro – organisms. They are the source of carbon for many soil saprophytes, pathogens and parasites, all of which together make up a healthy lithospheric Biosphere. They are the environmental indicators of a healthy soil. The end products of decomposition, otherwise called humus, helps the soil in various ways. When directly used by the plants humus provides essential elements/minerals, thereby reducing incidences of nutritional diseases and conditions associated with plants.

Decomposed organic matter, being largely in colloidal state, has very high surface area. In this respect it helps in soil ionic and PH (acidity and alkalinity) balance and stabilization. Since many crops grow best in neutral soil (pH between 4.5 and 7.5), extremes of acidity or alkalinity are avoided, thus improving crop performance in growth, vigor and yield. Humus, when serving as soil – ionic stabilizer, selectively absorbs/adsorbs and releases specific ions - especially cations. In this regard therefore, any specific elements existing in the soil in toxic concentration are removed from the soil solution. On the other hand, other minerals that could be rendered deficient by virtue of the high concentration of their counterparts in soil solution then become available since less of them are adsorbed by the organic colloids. On a similar vein, ions that actively contribute to soil pH are selectively absorbed or released, depending on the existing soil state. Thus an appropriate soil pH balance is maintained almost throughout as long as the organic matter exists in the soil. Either way, from ionic or soil pH point of view, the humus serves as a cushion, absorbing soil shocks, thereby making it stable or at equilibrium. For stable food supply, stable soils are a prerequisite.

Organic fertilizers/matter are poor absorbers and poor releasers of heat. The albedo (percentage of incident solar radiation reflected) is increased by the presence of organic matter because The cuticle's shiny nature renders it a good reflector.
The active palisade layer that is highly rich in chloroplasts only exists in healthy plants. When the leaves have died, they have less ability to absorb radiation. The chloroplasts lose their shape and functionality integrity.

Even when a lot of heat manages to penetrate a soil, the discontinuity that characterizes an organic – matter rich soil offers no conduit for the heat to penetrate deeply. The discontinuity of the organic soil is attributable to the large voids that exist, especially macro pores. Since these contain air (a poor conductor of heat), the soil temperature is prevented from going to extremes. Organic matter, in this regard helps stabilize soil temperature. This is an important characteristic because in marginal to semi arid agriculture zones, net moisture availability in the soil can be favorable where the organic matter level is high. This is because the heat does not penetrate deep in to root – zones, thereby reducing evaporation and moisture stress.

Organic matter stabilizes soils by reducing their erodibility. This is because it contains Calcium and related cations that have cementing properties. Calcium exists in large amounts in the leaves of plants as Calcium pectate- both in leaf petioles and in cell walls.

Magnesium, the next major soil-cementing cation after calcium, is found in the plant leaves as a component of conjugated proteins, the chlorophyll. This is the chemical that renders chloroplasts able to absorb and trap sunlight, thereby enabling photosynthesis to occur in green plants. Lastly, organic matter-rich soils are better conservers of both soil and water. With lots of micropores, the soils readily take in transported silt from runoffs. This improves water seepage into the soil, as well as conservation of soil moisture. It can therefore guarantee long term sustained impressive crop yields.

Organic manure applied to the soil

The food wastes are increasingly being separated and used for making organic manure, as shown here.

Organic manure in a supermarket for urban farmers

SOME BASIC Food for thought CALCULATIONS

(1)What is the % of Nitrogen, Phosphorous and Potassium in each of the following fertilizers?

(i) 2:3:2 (30)(ii) 5:6:1(24)(iii) 3:3:0(27)

Hint: For 3:5:2(40), % N=3/10X40=12%; %P = 5/10X40 = 20%; % K=2/10X40 = 8%.

(2) How much fertilizer (SOA) would you need for 1 ha of Kale (Brassica spp) spaced at 45cm X 60 cm, if the recommended rate of application is 20 g/plant? Also calculate SOA requirements for

(i) 1.5 ha of maize at 60 cm X 60 cm spacing, application rate being 30 g/plant

(ii) 2 acres of money-maker variety of tomato spaced at 90cm X 60cm, at a fertilizer rate of 1 ton per ha.

4.1TYPES OF ROOT SYSTEMS:

(1) Tap root e.g. carrot; (2) Fibrous e.g. millet and other grasses; (3) Tuberous e.g. sweet potatoes; and (4) Prop roots e.g. maize plant after flowering

FUNCTIONS OF DIFFERENT PARTS OF PLANT

Pith: Gaseous exchange (aeration) of the root and stem;

Phloem: Transports manufactured foods from the source (leaves and storage organs) to where they are wanted for either storage (sinks) or use; Put simply, it is for translocation.

Xylem: Transports water and dissolved minerals from the root to other parts of the plant.

Cambium: Separated the phloem and xylem, and also causes secondary thickening (increase in girth or stem diameter) of dicot stems.

Endodermis: Protects the vascular bundles.

Cortex: Storage point for manufactured foods, especially starch. It also offers support to the stem and root.

Epidermis: Protects the inner part of plant.

Root Hair: Absorbs water and mineral salts from the soil into the plant.

Figure 17: The sections of plant roots

Fig : Internal structure of a leaf, stem and root for a monocot and a dicot plant

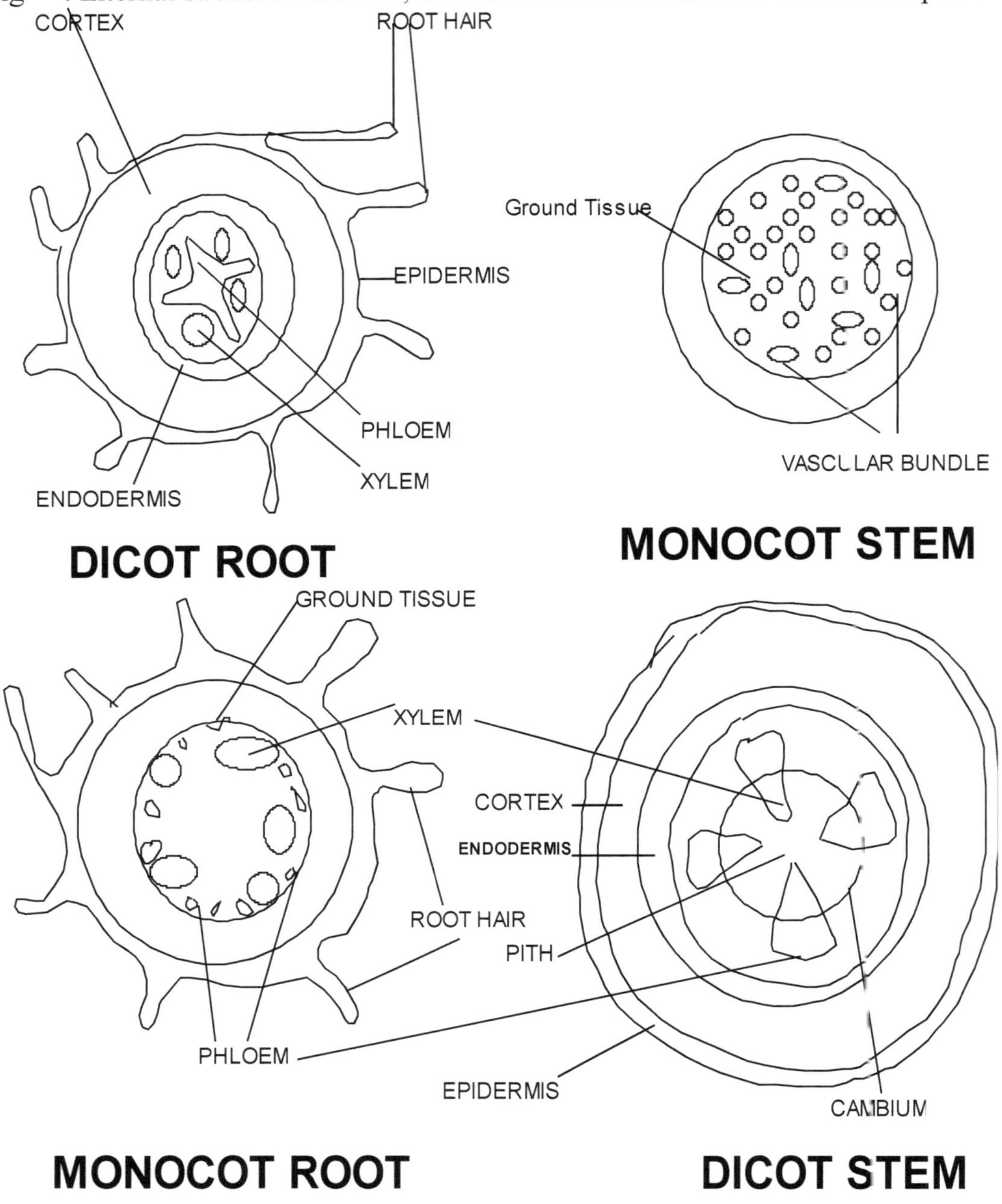

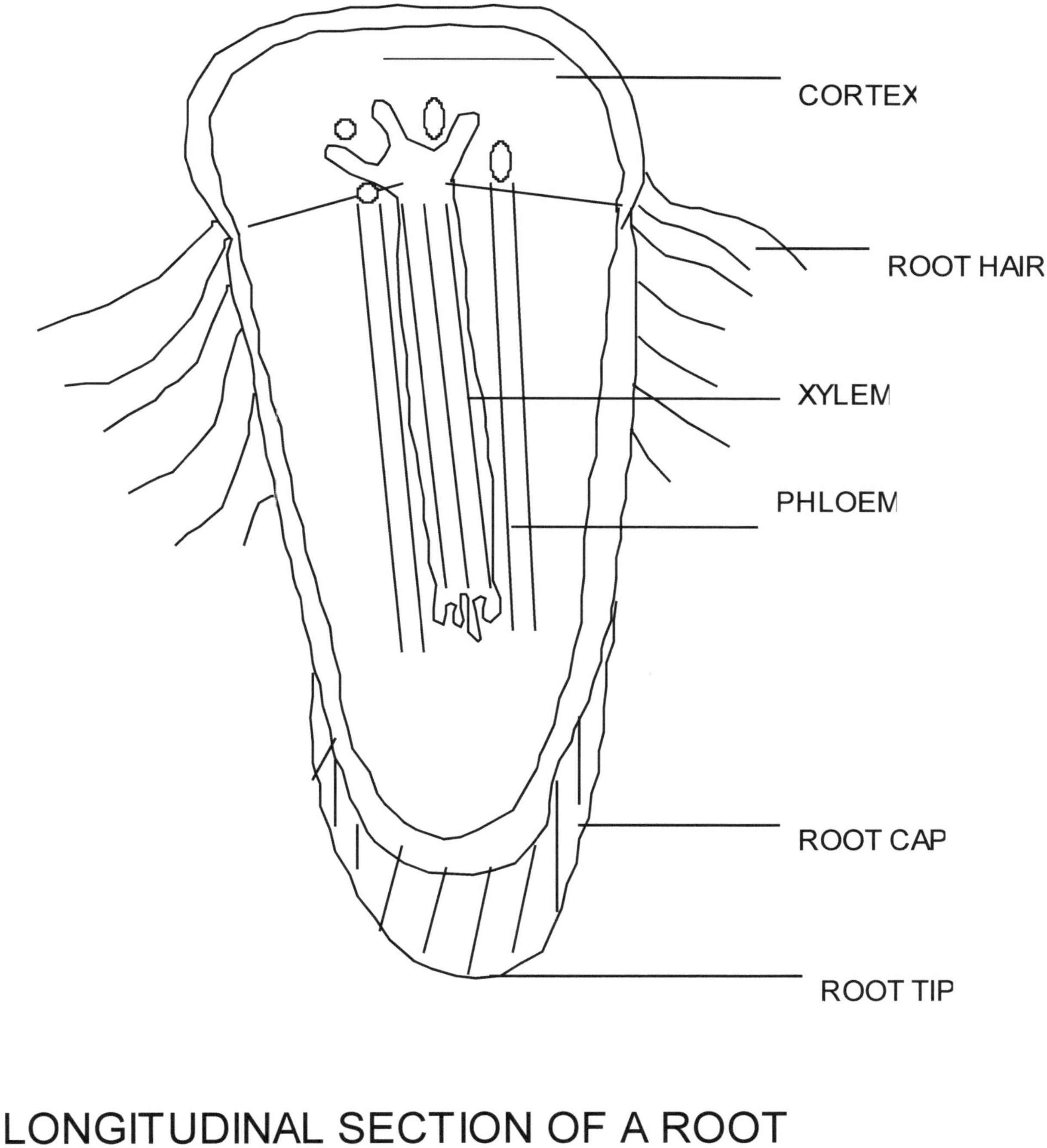

Figure 18: The longitudinal section of a root

Section of a stem

Table 3: Major elements required by plants and their uses.

Mineral and source	Function	Deficiency	Comment
Macronutrients			
Nitrogen. LAN / CAN, SOA, urea	Protein formation; growth	Yellowing, slow or no growth of plant (stunted plant)	Too much makes plant soft, weak and easily attacked by pests & diseases. Highly soluble (high leaching losses), & should be applied in splits by topdressing. An overdose scotch plants
Phosphorous TSP, DSP, SSP, Rock phosphate	Root establishment and expansion / development; Causes early flowering and ripening of fruits.	Purple patches on leaves, slow growth, shallow roots susceptible to falling over (lodging), and general poor health due to mineral deficiency	Should be applied as basal dressing either at or before planting. Has high residual effects, thus can skip a year before applying again. Too much can cause premature ripening
Potassium	Used to produce starchy parts of plant; development of strong stems	Poor growth, weak stems and scorched leaf edges	Too much causes slow germination and slow ripening
Calcium	Cell wall development (Form calcium pectate)	Weak leaves and stems susceptible to disease and pest attack.	Helps stabilize soil structure and pH.
Magnesium	Formation of chlorophyl	Yellow leaves with green veins; stunted growth	Help stabilize soil structure and pH
Sulphur	Formation of proteins	Yellow leaves and slow growth	Needed for protein formation in a plant. Is an essential constituent of some essential amino acids (Cysteine, Cystine, Methionine etc), some of which help form the first class proteins. Deficiency results in retarded growth and low yield.
Micronutrients			
Manganese	Proper growth. Activates enzymes in cells	Yellowing of leaves	
Molybdenum	Utilisation of Nitrogen and Nitrogen fixation by bacteria in the root nodules, intake of nitrates.	Nitrogen deficiency and stunting	Controls the use of nitrates by plants; it is like a catalyst in nitrogen metabolism.
Iron	Chlorophyll formation	Yellow leaves with green veins	Required to activate enzymes which produce chlorophyl

4.11 THE PLANT LEAF

FUNCTIONS OF DIFFERENT PARTS OF THE PLANT LEAF.

Epidermis:

This forms a layer around the leaf one cell thick (epithelium). It helps protect the leaf from damage and infection. It also reduces water loss. The cells are transparent.

Palisade layer (Palisade mesophyll):

This is a layer of sausage-shaped, cylindrical cells beneath the epidermis. They have a lot of chloroplasts (green colouring matter) that contain chlorophyll. The chlorophyll is a pigment that absorbs light and makes it possible for the plant to make its food.

Spongy layer (spongy mesophyll): These are loose-fitting cells beneath the palisade layer. They have vacuoles and large air spaces between them to enable air to circulate freely between them. They connect to the stomata through which air comes in and out.

Vascular bundles: These comprise the xylem and phloem vessels. The xylem transports water and dissolved minerals, while the phloem transports manufactured foods. The xylem vessels and sieve tubes are right inside the midrib of the veins of the leaf.

Stomata: These are small openings or pores on the leaf surface through which water comes out. Each stoma is enclosed by two guard cells that shrink to allow water to pass through and swell to close the pore and reduce water loss.

Guard cells: Are "watchmen" situated at both sides of the stomata to open and close it as need arises. The guard cells therefore control rate of transpiration.

4.12 PHOTOSYNTHESIS

Food manufacture in a plant takes place in the leaves, and is called Photosynthesis. Photo means light, synthesis means to make / produce or manufacture. The requirements for photosynthesis are CO2, water and light. CO2 diffuses via the stomata into the plant leaf from the atmosphere. Water is absorbed by the plant roots from the soil, while light is absorbed by the chlorophyll from the sun. Chlorophyll is the green colouring matter found in the leaves of green plants. They trap the solar energy and have it changed into a form that can be stored for later use even when it is dark. Photosynthesis results in production of starch (initially in the form of glucose), and a byproduct such as oxygen. The reaction is as shown in the equations below, with further explanation in the boxes which follow.

TRANSVERSE SECTION OF A LEAF

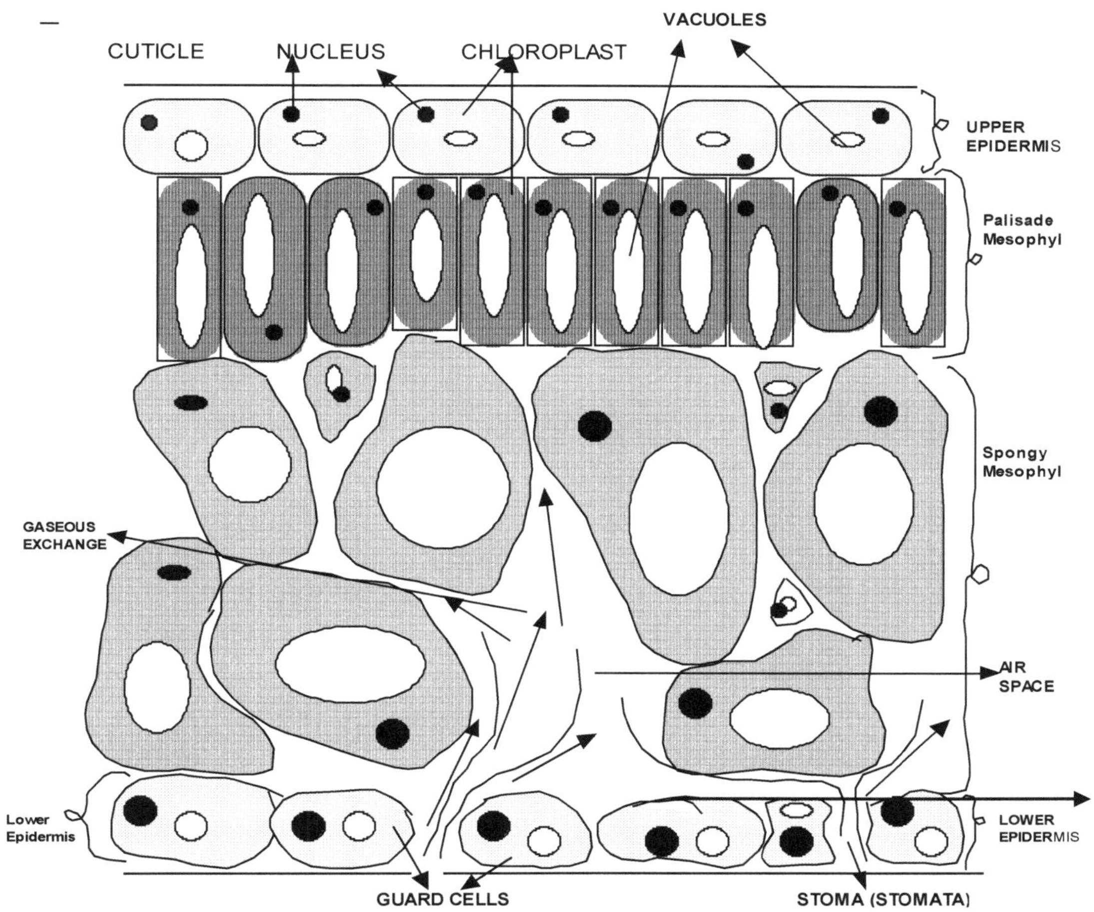

Figure 19: The plant leaf

$CO_2 + H_2O \xrightarrow{\text{Sunlight, Chlorophyl}} C_6H_{12}O_6 + O_2$

$6CO_2 + 6H_2O \xrightarrow[\text{Chlorophyl, sunlight}]{} C_6H_{12}O_6 + 6O_2$

Word equation:

Carbon dioxide + Water $\xrightarrow{\text{Sunlight, Chlorophyl}}$ Carbohydrates + Oxygen

The products are used in a number of ways:

Starch: Is the main product of photosynthesis. Is transported (translocated) to the storage organs such as cortex, roots, seeds, fruits etc, or to actively growing plant parts for use in respiration and for absorption of minerals in the roots. Excess id converted into cellulose which is stored as wood. This gives the plant the support it needs.

Oxygen: diffuses out through the stomata into the atmosphere. Some may be retained for respiration, as the bulk of the O2 plant needs for this process often comes with CO2 from the atmosphere by diffusion.

4.13 RESPIRATION

Is the oxidation of carbohydrates (glucose) to give energy. Through it, energy is released by the cells as the oxidized carbohydrate give out energy. The equation is as follows:

$$C_6H_{12}O_6 + O_2 \longrightarrow CO_2 + H_2O + \text{ENERGY}$$

$$C_6H_{12}O_6 + 6O_2 \longrightarrow 6CO_2 + 6H_2O + \text{ENERGY}$$

Word equation:

Carbohydrates (Glucose) + Oxygen ⟶ Energy + Carbon dioxide + Water

The reactant, carbohydrate comes from the photosynthetic process, while oxygen diffuses into the plant through the leaf (stomata). Energy is the main product. It is used by the plant in translocation, active transport, response to stimulus etc. The Carbon dioxide is re-used in photosynthesis, while excess is released through the stomata into the atmosphere. Water, in the form of vapour, is also released through the stomata into the atmosphere. Some water may, however, be used again by the plant when there is drought. Such water is called metabolic water.

4.14 TRANSLOCATION

Means transport of manufactured foods within the plant. This occurs in the Phloem. The food is transported in dissolved form (mostly glucose), and taken for storage, or use where it is needed. This creates more space at the points of food manufacture so that the process can continue. Suppressed translocation can stop photosynthesis.

The movement of water and dissolved mineral substances into and out of the plant

Water is absorbed by the root hair by osmosis into the root, and it is then transported along the xylem upwards to leaves of the plant. Since minerals are dissolved in soil water, the two are always absorbed together as a solution. Water thus enters by the process of osmosis, while the minerals enter by the process of active transport. Water thus absorbed reaches the leaves as a continuum or transpiration stream (continuous flow) by the process of transpiration pull. Once it reaches the leaf, it moves towards the stomata where it is expelled out in the form of vapour, thereby cooling the plant. The transpiration pull keeps this process continuous.

4.15 OSMOSIS

Osmosis is the movement of substances from a region where they are more (more concentrated) to a region where they are fewer (less concentrated), through a semi-permeable membrane. The semi-permeable membrane allows smaller particles to pass but disallows larger ones. Thus it is a form of diffusion, except that in this case, a partially / selectively permeable membrane is required.

4.16 ACTIVE TRANSPORT

Active transport means transporting substances from where they are fewer to where they are more, i.e., across a concentration gradient. It is like climbing a hill, and therefore needs energy. Mostly, the root already has higher concentration of minerals, and thus to absorb more from the dilute soil solution means it has to use energy.

4.17 TRANSPIRATION:

Is the process by which water is lost by the plant through the leaves. It is important in two ways:

It cools the plant: As the latent heat of vaporization makes the water to change into vapour and escape into the atmosphere, it carries excess heat within the plant with it.

It enables the plant to absorb more water (which is always a solution), and therefore with it, minerals - which means that through it, the plant gets extra minerals on a continuous basis. Excess minerals are added onto the barks of trees etc, and kept there until they peel off or the plant dies.

4.2 REPRODUCTION IN PLANTS

Reproduction in the production of young ones (offsprings) so that a species continues to exist. It is meant to ensure continuity and guard against extinction. Plants reproduce by 2 methods: sexual and asexual. Sexual methods involve fusion of two sex cells (gametes). These are the ovule and the pollen grain to produce a seed. Non- seed propagation of plants is called asexual. Thus types of reproduction in plants are: (i) Sexual or (ii) asexual.

ASEXUAL:

Involves the use of vegetative parts of a plant (e.g. leaves, roots, stems) to produce another independent plant. This may happen artificially or naturally.

(a) Artificial methods: budding, grafting, cutting and layering;

(b) Natural methods: stem tuber (potato, yams), Bulb (onions), splits (pyrethrum), suckers (banana, pineapple), cutting (sweet potato, cassava), layering (sugarcane), corms (the ginger), rhizomes (couch and star grasses), stolons (star grass), root tubers (cassava, sweet potatoes) and bulbils (sisal).
(NB cassava is a root tuber, but is often propagated using cuttings)
Asexual methods of plant reproduction involve the use of vegetative parts of a plant (roots, stem, leaves etc) to produce a new independent plant. The methods can be natural or artificial. In artificial methods, man has to come in- otherwise such a method may not occur. These include budding and grafting. However, layering and cutting can be both natural and artificial. Natural methods take place without the intervention of man e.g. suckers, bulbs, corms, root tubers, stem tubers, runners rhizomes etc. Some of these methods are described below.

4.21 SEXUAL REPRODUCTION IN PLANTS:
The perpetuation /multiplication of a plant by use of seeds (i.e., the fertilized egg in the fruit). The flower is the reproductive organ in this case, and produces male and female gametes (sex cells) which unite/fuse at fertilization time to give rise to a seed contained in a fruit.

FUNCTIONS OF THE DIFFERENT PARTS OF A FLOWER
Male (anther and filaments); Female (ovary, ovules and style).
Male flower (androecium/stamen
It consists of the filament that acts as a support to the anther. The anther produces pollen grains that are the male sex cells. The pollen grains are stored inside pollen sac.

FEMALE FLOWER (GYNOECIUM)/PISTIL
It consists of the stigma that is often sticky and supported by the style. The pollen grain lands on the stigma during pollination. The ovary is the part where fertilization takes place. It contains the ovules, which have female sex cells. The female sex cells fuse with male gametes during fertilization to form a viable seed. The ovary will mature into a fruit.

FUNCTIONS OF PARTS OF A FLOWER
Pistil:These are the female parts of a flower, and compise the style, stigma and ovary.
Stamen: These are the male parts of a flower, and comprise the anther and filament.
Sepal (calyx) : The outermost part which protects the flower i.e. when still in the bud

Petal (corolla) : Bright coloured and this bright colour attract insects. It also encloses the make and female parts.

Anther: Have 2 lobes each with 5 pollen sacs where pollen grains are produced. Pollen grains are the male sex cells.

Filament:Acts as a support to the anther

Stigma: Stigma is the sticky and hairy ection of the female part, with a purpose of trapping pollen grains. Pollen grains land on stigma during pollination. Therefore it receives pollen grains and transfers them down the style.

Style : Tube - like structure, which transport pollen grains to the ovary.

Ovary : Has little roundish bodies called ovules. Ovules when fertilized by the male gametes will develop into seeds. Ovules are attached to the ovary. Maize flower

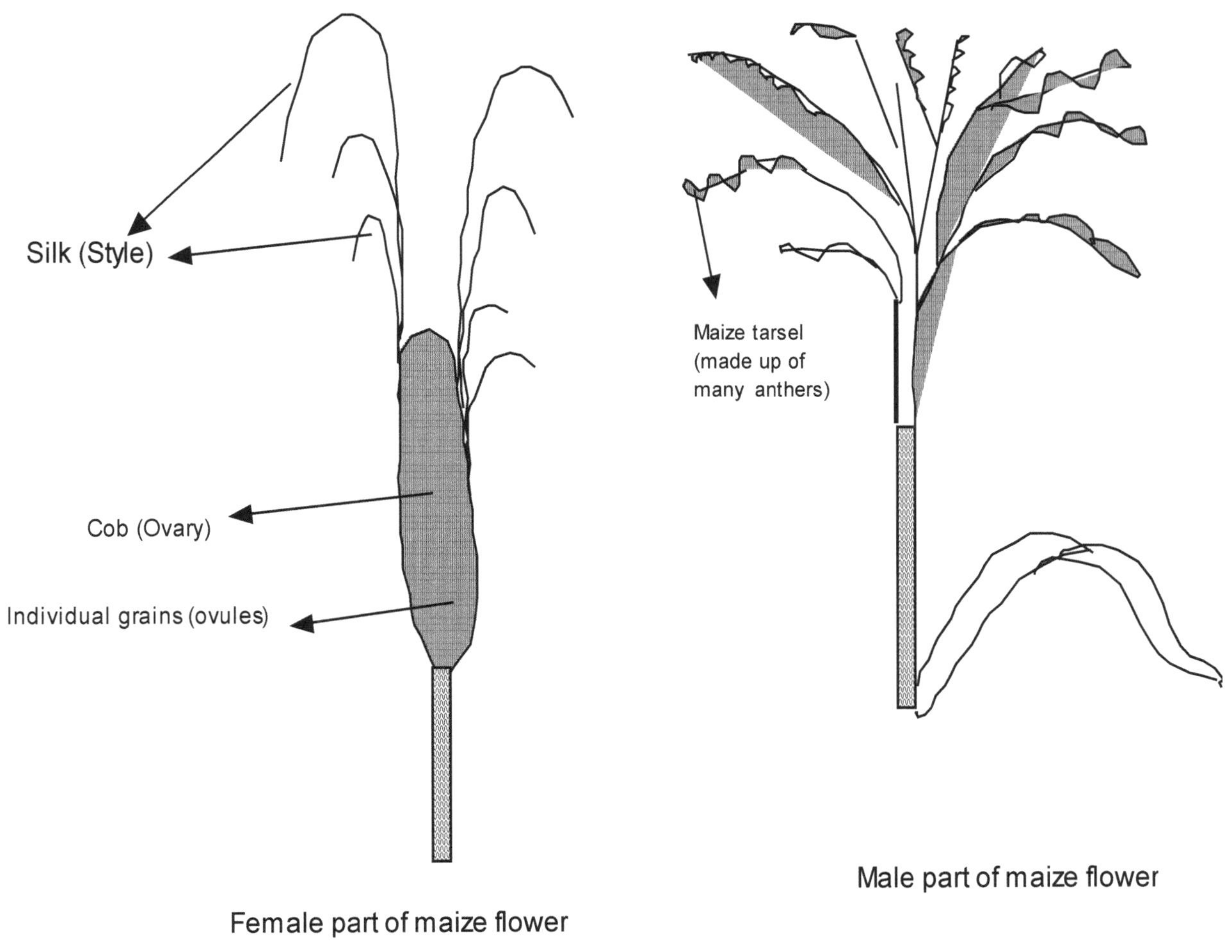

Figure 20: The reproductive organs of maize plant

THE MAIZE FLOWERS:

The male and female flowers of maize are borne on separate inflorescences but on the same plant. The male inflorescence, the tassel, is borne at the terminal end of the stem. The female inflorescence, the ear on the other hand, is borne in the axil of one of the foliage leaves about mid-way along the stem. Maize is cross-pollinated by wind. The male part is known as the tassel. It contains pollen grains. The pollen grains are contained in some hard coated bracts. The female part is known as the ear/cob. It consists of feathery styles known as silks. The silks trap pollen grains during pollination. The figure below shows insect pollination of a flower.

Figure 21: Insect pollination

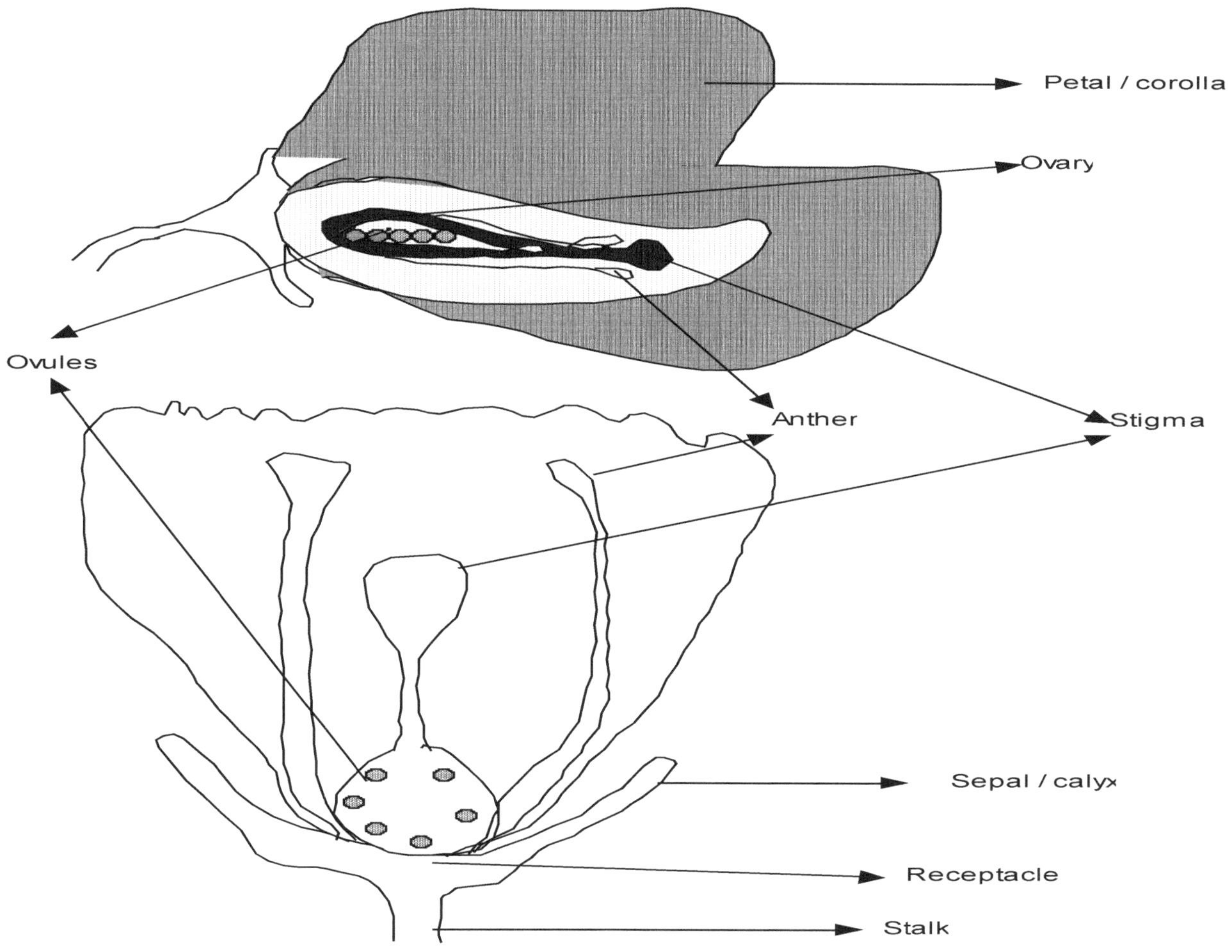

Flower of insect pollinatad plants (Bean, cowpeas)
Figure 22: The flower

4.22 ASEXUAL REPRODUCTION IN PLANTS

INTRODUCTION:

Plants can be propagated using either artificial or natural methods. The artificial methods are: layering, budding, grafting and cutting. However, cutting and layering can also be natural methods. The artificial methods are hereby described.

LAYERING:

Involves taking a branch of a mature plant and bending it so that a section of or the whole of the stem is buried into the soil. This is held in place using pegs. After some time, the part covered with the soil starts producing roots and new shoots. Then the pegs can be removed, and the part of the stem still holding this new young plant onto the mother plant is cut. Thereafter, a new independent plant results. This is described in the diagrams below.

GRAFTING

This is the artificial securing of a shoot (scion) from one plant onto the stem or stock of another of the same genus. Examples- citrus (lemon, lime, orange etc are compatible with one another). It may be done in a plant with good rooting qualities providing the rootstock, and another with good shooting qualities providing the scion. The two have to be of the same stem size, and are tightly joined together and tied with a waterproof material. Some candle wax may be placed at the joint to ensure it is watertight. As the new shoot develops, all others arising from the rootstock have to be persistently pruned; they should not be allowed to grow otherwise they will produce the same old poor quality fruits which compete with the new shoot for nutrients- resulting in smaller and lower quality products. The resulting plant will have combined qualities of both parents- good root and shooting qualities, but the fruits are seedless. Whereas this is good for the consumer, the future of a whole society/ country can be at stake as there will be no planting materials (seeds) for producing new plants. This may seriously interfere with biodiversity, as well as food security in a country. The principle of grafting is the same as in budding, except that the budding process uses the bud instead of the stem (scion).

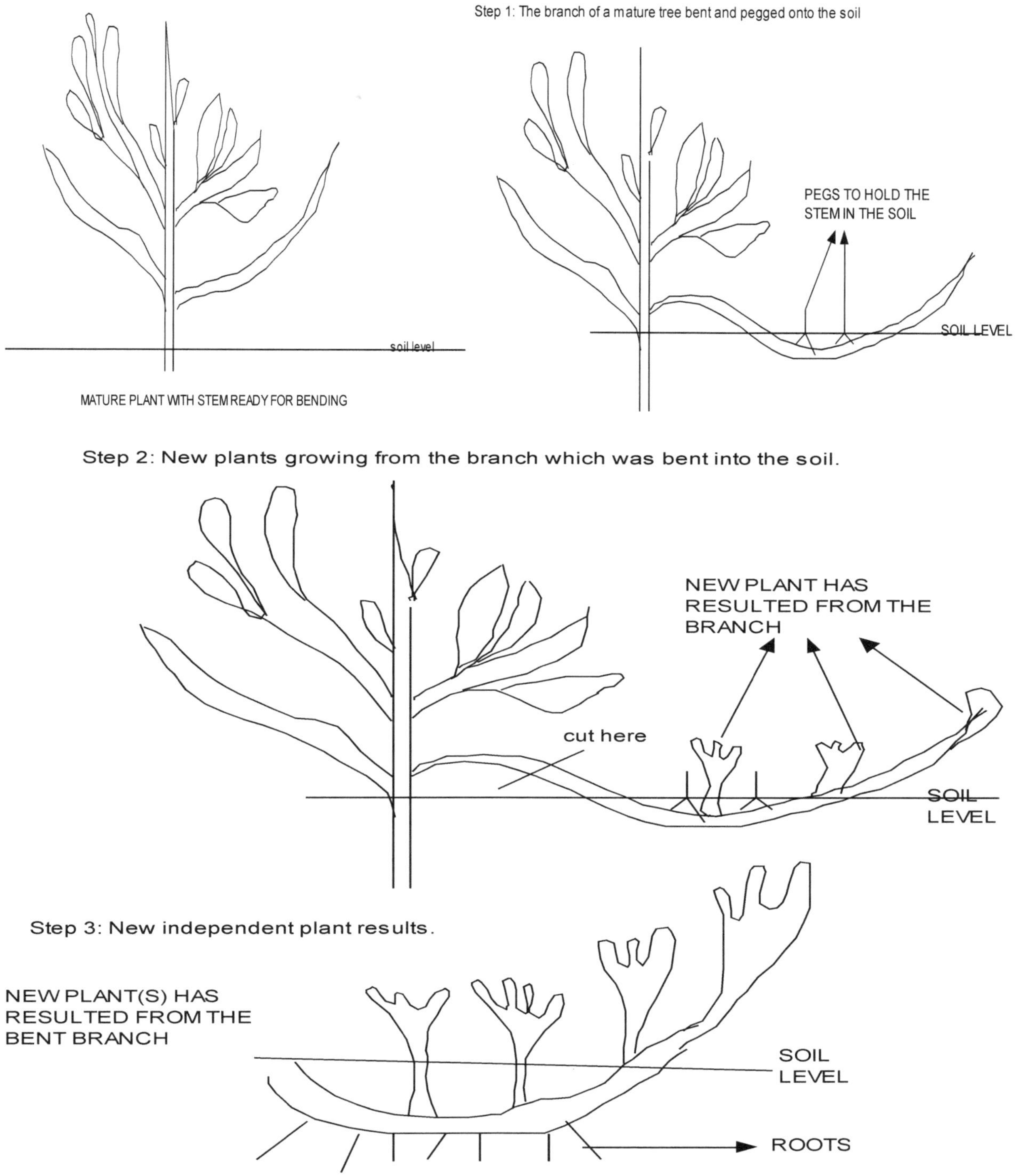

Figure 23: An illustration of layering

BUDDING

This is the artificial securing of a bud from one plant onto the stem or stock of another of the same genus. Examples- citrus (lemon, lime, orange etc are compatible with one another). It may be done in a plant with good rooting qualities providing the rootstock, and another with good shooting qualities providing the bud. The resulting plant will have combined qualities of both parents- good root and shooting qualities, but the fruits are seedless. Whereas this is good for the consumer, the future of a whole society/ country can be at stake as there will be no planting materials (seeds) for producing new plants. This may seriously interfere with biodiversity, as well as food security in a country.

CUTTING:

Stems, leaves or roots cut into sections (e.g. sugarcane setts of 45 cm length) and placed in small trenches on the ground in vertical, slanted or horizontal position. New roots and shoots arise from the cutting. Cassava, sugarcane, sweet potato etc are propagated this way.

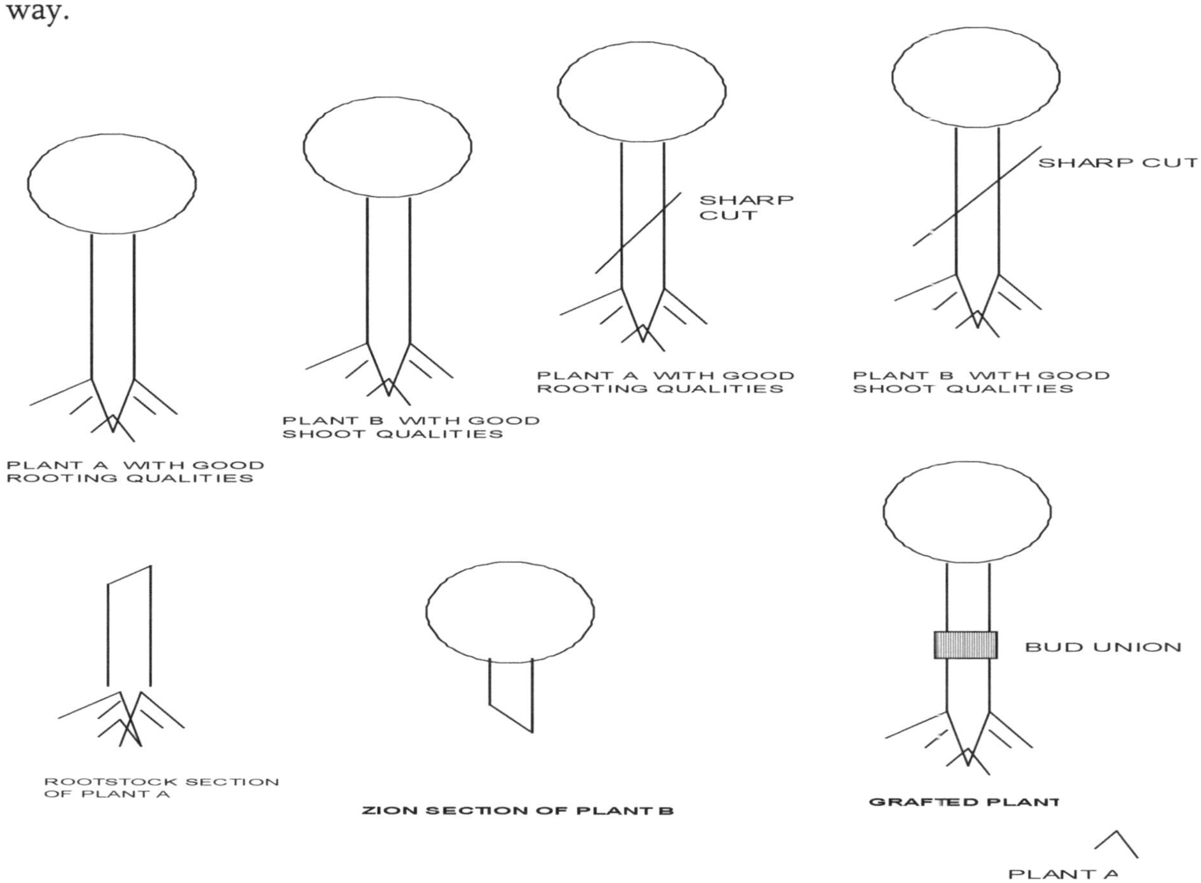

Figure 24: Grafting of Plants

4.221 NATURAL METHODS OF PLANT PROPAGATION

SUCKERS:

Examples: Bananas, sisal etc. The main stem is a false stem (pseudostem) develops a bud at the base. These grow until they form new sets of roots around the base. Thus a number of shoots form as a cluster of plants. The shoots, then called suckers, can be uprooted and planted. But they normally grow on their own.

A banana plantation/ A banana head

BULBS:

Example: onions. Buds develop in axils of leaves that make up the bulbs. The new plants feed on the fleshy leaves of the bulbs. As the bulbs enlarge, they become separated from the parent plant and grow new roots from the base.

A bulb crop

CORMS:

Are thickened underground stems that serve as storage organs. New growth occurs when a bud develops on the side of the parent corm and gradually a new individual plant is formed e.g. coco yam

The yam plant

ROOT TUBERS

Example: Cassava. These though are not propagated by tubers, but by stem cuttings. The swollen edible tubers are the root tubers.

STEM TUBERS

Examples include Irish potato and yam. The stem tuber is the swollen part of the plant which acts as a storage organ. It is a growth from the plant stem, and they produce buds. They can be propagated by cutting the tubers into sections, as long as each section has an eye (better 2 eyes), which is actually the bud from which new shoots and toots emerge.

ADVANTAGES AND DISADVANTAGES OF SEXUAL AND ASEXUAL REPRODUCTION

SEXUAL:

ADVANTAGES:

(1) There is variation between the parent and offspring, thus giving the offsprings hybrid vigour, i.e., traits which may make it grow stronger and resist diseases, pests, and drought more effectively (2) Seeds can be stored for a long time without risks of getting spoilt (especially since they are dry and smaller), and only planted at the right time (3) It is

possible to transfer good traits from one plant or variety and introduce it into another one (through controlled crossing by plant breeding and bio-technology), thus increasing crop yield and growth vigor.

DISADVANTAGES:
(1): there are variations in the young one so that it does not exactly resemble the parent. This may be due to fertilization, and subsequent acquisition of new genes from another plant (2): It at times leads to a very slow crop establishment arising from seed dormancy; (3): The plants grow slower since they have very little food reserves for their use before they start photosynthesizing; (4): seed-borne diseases are easily transmitted to the offspring .

ASEXUAL:
ADVANTAGES:
(1): Very fast establishment of offspring due to the presence of enough food reserves (2) very fast growth since the offspring may already be having well established root system for use in absorbing water and mineral salts from very deep parts of the soil (i.e., the plants are more hardy) (3) the offspring is exactly similar to the parent, i.e., they breed true to type (4) planting material is easily available (5) when carefully used, the root and shoot apices can give rise to disease-free clones that are useful in bio-technology Example is the use of tissue culture used in some parts of Kenya to produce disease free bananas which also mature at least 4 months ahead of the other types of bananas. (6) useful in propagating seed-less plants such as bananas.

DISADVANTAGES:
(1) May transmit diseases and pests to the offspring (2) May lead to over-crowding and low yields if not well managed (3) Sometimes the planting material is not easily available when required, e.g. sweet potatoes cuttings after a long spell of drought. Normally, farmers use a combination of the above methods. A good crop can be produced by: (i) choose right cultivar (ii) right planting depth which eases both germination & emergence (iii) correct spacing (iv) correct planting time, e.g. at onset of rains (v) protecting crop against enemies e.g. pests, weeds and pathogens (vi) good cultivation practice & management

4.223 POLLINATION
Is the transfer of pollen grains from the anther to the stigma. Forces called agents can do it. These agents include wind, insect or man. Pollination can also be self or cross.

MAN: Can do pollination through crop breeding processes- ie taking some pollen from a plant of desired quality and smearing them on the stigma of another plant (same species) of desirable qualities. Through this process, man has come up with many varieties of crops. This is the subject discussed under crop and livestock improvement and genetic engineering.

INSECTS : Transfer pollen from the anther and / or to the stigma of a plant flower they have visited to collect nectar from the receptacle. The pollen may stick on their legs, feelers, mouthparts or wings. By continuously luring insects with food and good smell, plants have for centuries continued controlling the insects like their puppets, while they continue to procreate.

Figure 25: Insect pollinated flower.

Figure 26: Blue Damselfly pollinating a flower

WIND

Transfer pollen from the anther of a flower which is very exposed, and whose pollen are loose and light. The pollens then fall on stigmas which are also relatively exposed with very high surface area. Normally thousands of pollen are transferred this way to increase their pollination chances. There are clear differences between insect and wind pollinated flowers. These are given below:

Table 4: Differences between insect and wind pollinated flowers

Nature of flower	Wind pollinated	Insect pollinated
Smell	None	Strong and sweet
Size	Small	Large
Nectar	Absent	Present
Colour of petals	Dull	Bright
Pollen grains	Light and many, with no spikes	Heavier and fewer, with spikes
Stigma	Exposed and feathery	Not exposed and sticky
Stamens	Hang outside the petals	Inside the petals
Anthers	Loosely attached to filaments	Very firmly attached to filaments

4.23 ERTILIZATION, SEED AND FRUIT FORMATION

During pollination the pollen grain lands on the stigma and pollen tubes develop pollen tubes are tubes which develop through the style to the ovary. It is through the pollen tube that the pollen nucleus will travel down until it reaches the egg nucleus. Fertilization will then take place as a result of the fusion of the male and female nucleus. A viable seed will be formed and the ovary will develop into a fruit. The figure below shows brightly coloured insect pollinated flower. After the pollen fall onto the stigma, it germinates just like a seed may germinate in a soil. Then its root, called the pollen tube, is sent through a cylindrical tube called pollen tube. This tube goes through the style into the ovary. When the tube reaches the ovary, its tip opens up to release the nucleus of the pollen grain. This nucleus fuses with the ovule. Similar process occurs with other pollen grains so that if it is a fruit with many ovules, each is fertilized by a different pollen nucleus. The fertilized egg then multiplies and grows to form the seed. The ovary then forms the fruit. The rest of the materials making up the flower such as the petals then wither and fall off, while the sepals continue holding the fruit at its base. Once the fruit grows to maturity, it ripens and is ready for dispersal.

SEED AND FRUIT DISPERSAL

Seeds and fruits may be dispersed by man, other mammals, birds, wind / air or water. Dispersal is meant to reduce the competition between the plants of the same family as it ensures they are widely spread. Those dispersed by man are largely fruits that man eats and throws its seeds far from the mother plant. Other animals such as birds do the same. But some fruits are spiky so that they stick onto the hair and fur of passing animals. Examples here are the black jack, thorn apple etc. Then they later fall somewhere, or are removed further away from the original parent. Wind / air dispersed seeds and fruits are hairy and light. Water dispersed fruits are spongy and can float for long before they land on rich ground for germination. Weed seeds are often scattered from the point of production/ mother to new grounds either by self or via an agent; this helps reduce intra-specific competition (i.e. competition among members of the same family- mother and children).

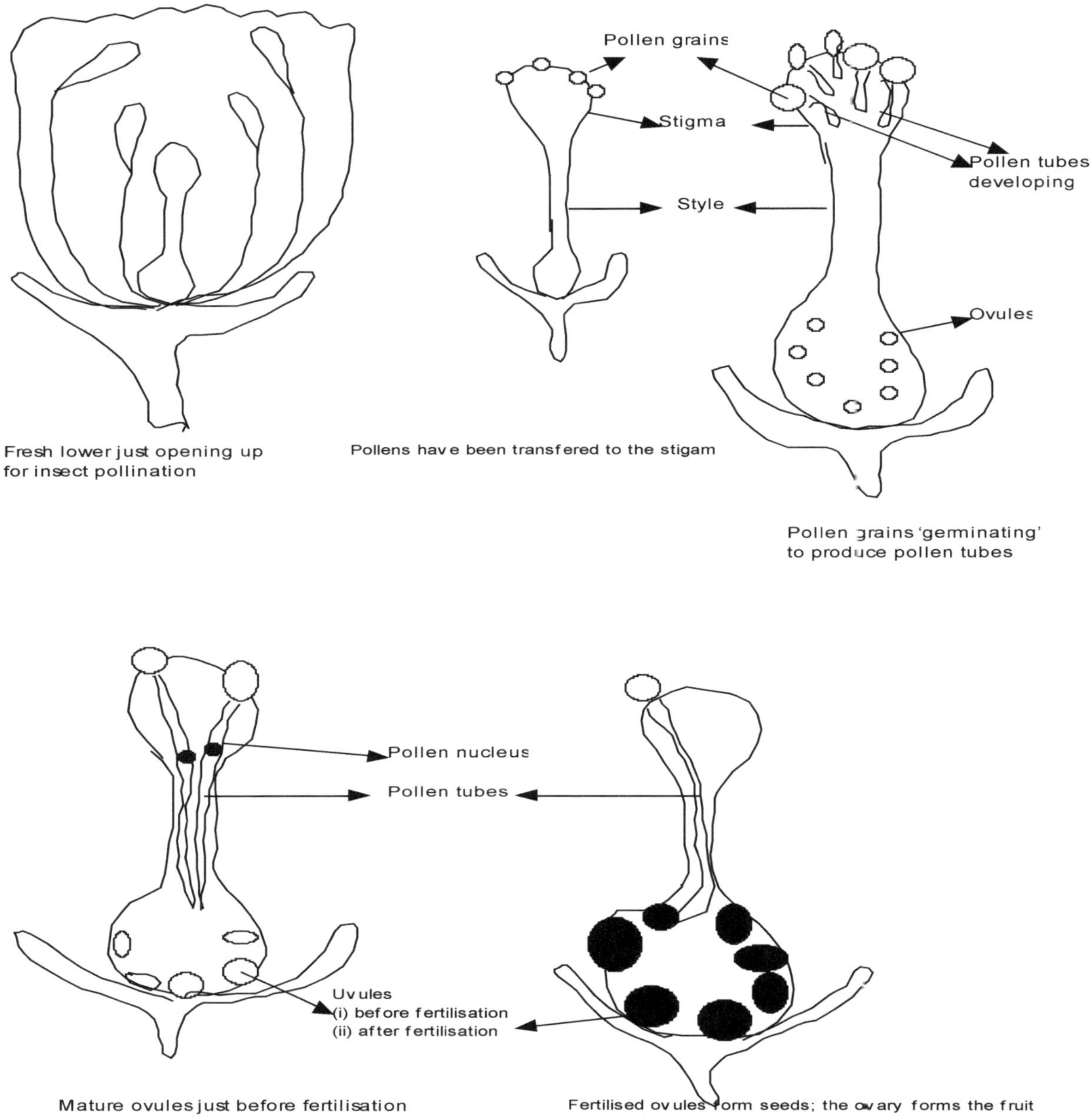

Figure 27: The process of fertilization, including seeds and fruit formation

Figure 28: Rice seeds broadcasted on a nursery

AGENTS OF DISPERSAL

These are: wind, water, animals and self. Wind-dispersed seeds are hairy, light, and stream-lined (i.e. shaped in a way that the wind offers minimum resistance to movement). Water-dispersed seeds are light and have a spongy (air-filled) cover that keeps them light and floating. They also have a hard testa to help them avoid premature germination and rotting while still floating. Thus can be transported long distances by moving water. Animal-dispersed seeds are often edible fruit with hard seed coat, e.g. guava, to avoid digestion while in the animal digestive system. Some have hooks, spines, and thorns e.g. double thorn or mosetho by which they get attached to passing animals. These are then dropped elsewhere far from the mother. Self-dispersed seeds often are enclosed in pods that have a line of weakness along its edges. When temperatures are high, the pods explode spontaneously, or at times when in contact with other animals (e.g. when man is harvesting)

SEED STRUCTURES

A seed is made up of an embryo together with a store of food which will provide energy for germination at early stages of growth. Food is stored in the endosperm in some seeds e.g. maize others have their food stored in cotyledons e.g. beans, groundnuts.

Radicle: Embryonic part that grows and develops into the root system of the plant.

Micropyle: Is an opening in the integuments through which the pollen enters at fertilization. It also allows water to enter the seed to the embryo before germination.

Testa/Seed coat: Protects the seed from microbial infection.

Embryo: The earliest stage of development of an organism. Can also be defined as the stage during which an organism develops from a fertilized egg to an independent functioning individual.

Plumule: The leafy part of the embryonic shoot. These leaves are attached to the embryonic stem, of which the part above the attachment of the cotyledons is called epicotyl and the part below, the hypocotyl.

Helium: Is a scar left by the stalk which attached the ovule to the ovary.

Cotyledons: Store food reserves which are stored in the early stages of germination. Some plant seeds such as those of grasses, cereals have only one cotyledon (monocotyledons) while other plant seeds e.g. groundnuts have two cotyledons (dicotyledonous). In some seeds food is stored in the endosperm surrounding the embryo. Such seeds are called endospermic seeds e.g. maize seeds. Others have their food supplied in the form of cotyledons especially in beans, which form the main part of the seed but later become the first leaves. Such are called non-endospermic seeds e.g. groundnuts.

4.3 GERMINATION

This is the process by which a seed develops into a new plant.

CONDITIONS NECESSARY FOR GERMINATION.

(i) Viable seeds: Seeds must be living and of good quality for germination to occur - certified seeds are normally used to ensure that the seeds are of good quality, certified seeds are those that have been tested and proved to be fit enough for growing. The seeds are often coated or dressed with a pesticide formulation to avoid being attacked by pests and other harmful organisms.

(ii) Oxygen: After being planted the seed is not able to photosynthesize. The seed contains carbohydrates, proteins or fats which will be broken down by respiration in the presence of oxygen to provide energy. The energy will be used in the growth process.

(iii) Suitable temperature: Most plant seeds will germinate well when the temperature is warm. The optimum temperatures between 20 –300C are the most suitable for germination. When temperatures fall below 100C, the germination process may be slowed down.

(iv) Water: Water is essential for germination. It has to enter the seed before germination can take place. The young roots also should be kept most to prevent them from drying and dying.

TYPES OF GERMINATION

(i) Hypogeal: This is whereby cotyledons remain below the ground during germination and act as the newly emerged plant e.g. maize.

(ii) Epigeal: It is where the cotyledons are lifted above the ground during germination and act as the newly emerged plant e.g. bean.

THE GERMINATION PROCESS:

The seed coat split/rots; The new root or radicle starts to grow downwards; and The new shoot starts to grow ; and Upwards. The embryo uses the food store in the seed in order to grow.

4.4 TROPISMS

Tropism is the response of plant(s) to stimulus. Stimulus (plural stimuli) is an external condition such as light, water etc to which organisms respond. There are different types of tropism depending on the stimulus. These are: Phototropism, Geotropism, Hydrotropism and Haptotropism. Phototropism is the response to light; Geotropism is the response to gravity; Hydrotropism is the response to water and Haptotropism is the tendency to attach to solid stuff such as walls and firm plant shoots by weak stemmed plants. This is to enable them get support, and thus grow to the full view of light. This is a most useful characteristic of smaller plants growing in thick forests where shading is a problem. They twine around the stems of hard wood plants and grow until they reach light. Different parts of plants respond differently to different stimuli. The root is therefore said to be positively geotropic, positively hydrotropic and negatively phototropic. The shoot, on the other hand, is negatively geotropic, positively phototropic, positively haptotropic and negatively hydrotropic. Another type of tropism is thigmotropism- response to touch. The Mimosa plant commonly exhibits this. A table showing the differences among the different tropisms is given below.

Table 5: Types of tropisms and stimuli

Name of tropism	Phototropism	Geotropism	Hydrotropism	Haptotropism
Stimulus	Light	Gravity	Water	-
Response of root	Negative	Positive	Positive	Negative
Response of shoot	Positive	Negative	Negative	Positive

4.41 SOILS AND LAND ISSUES IN GRICULTURE:

Many factors lead to degradation of lands and subsequent impoverishment of soils. Gully (or donga) erosion is a common cause of this. Very commonly, poor or improper crop spacing e.g. with too wide spaces between crops gives a chance for e falling raindrops to cause havoc.

4.5 LAND PREPARATION

Land preparation is essential to make the field ready to sustain a crop for the entire lifetime. Clearing and destumping are the first processes that most fields undergo in the entire preparation chain. However, either of the two can be skipped if the original condition of the land is capable of directly being ploughed. The and mustbe cleared, stumped, ploughed, and harrowed before being planted on.

4.51 CLEARING AND STUMPING.

Clearing: Means the removal of the above ground material to pave ways for other activities in a field. This may be done by use of fire (burning), slashing, chemicals (herbicides), use of a bull-dozer, axe, panga etc. As can be seen in the figure below, it is not possible to cultivate a field unless it is cleared. Figure 30 below further shows a partially cleared forest in Naro Moru in the Mount Kenya forest region.

Figure 29 : The deep valleys created by erosion Figure 30: The degraded lands.

Figure 31: Forests in the Mount Kenya region, Kenya.(Left- natural forest; Right- Planted forest (Plantation).

REASONS FOR CLEARING:

a. Create space for planting; b. Facilitate the cultivation process; c. Remove weeds and parasites; d. To mark boundary, as there would be less conflict with the neighbours; e. Control notorious pests such as squirrels, monkeys, some birds, etc; f. To create space for constructing a farm structure (see below).

Clearing sites hitherto occupied by trees and forests often gives rise to stumps (see below). These can seriously hinder farm operations, and thus have to e removed through a process called destumping (or simply stumping).

4.52 DESTUMPING

This means removal of stumps by uprooting. Stumps are the belowground remnants of plant parts, especially after clearing, though it can also be viewed as a clearing process. It can be done using hoe, jembe, axe, panga, pickaxe, systemic herbicides (chemicals), bulldozer etc.

Figure 30: Left: Cuttings from a forest in Naro Moru, Mount Kenya region;
Right: Stumps readily destroy farm equiment

Figure 31: Cleared ground with a tree stump
Figure 32: The uprooted tree stump near a construction site

REASONS FOR DESTUMPING

Avoid spoilage of farm equipment during the cultivation process;
Make it easy to do other subsequent farm operations e.g. cultivation;
Control of some root-borne pathogens such as the Armilea melea (root rot disease often associated with coffee, tea etc)
Create more space for the crop;
Reduce competition for water and minerals (as the live stumps would still continue absorbing these as long as they are intact);

METHODS OF SOIL PREPARATION BY PLOUGHING, DISCING AND HARROWING

4.53 PLOUGHING: Meaning: This is a primary tillage operation often done after clearing.

The importance of / purposes are:
Improve water penetration into the soil, i.e. Improve drainage; Improve aeration;Burry crop remains into the soil; Uproot weeds; Expose pests to extreme heat, light and predators; Prepares land for subsequent operations; Sometimes may also accomplish destumping role;

Tools and equipment used in primary cultivation:
Disc plough; Mould board plough (Ox or donkey drawn as shown above); Hoe / Mogoma / Jembe (for manual operation); Can use a shovel or spade if soil is very light and sandy.

Figure 33: Primary cultivation using donkey plough. Figure 34: A Disc plough.

Primary cultivation should be done at least a month before the rains begin, to give the weeds and pests enough time to dry and die, as well as to allow the soil to weather. Late cultivation especially after rains begin is an effort in futility, as the effectiveness of the process is minimized- and thus the inputs wasted.

Figure 35: A tractor – commonly used for ploughing, harrowing, rotovating, among others

4.54 HARROWING: Meaning: Is a secondary tillage operation often done after ploughing.

functions / importance:
Buries new weeds which may have started coming up after ploughing; Exposes the pests even further; Improves aeration and drainage of the soil; Creates the appropriate tilth for planting (i.e., breaks large soil clods); Final stage of preparing the ground for planting (though not always);

TOOLS AND EQUIPMENT USED
Any type of harrow e.g. disc harrow, tandem harrow etc. Hoe or jembe / mogoma are commonly used in the work is manually done. Manual land tillage has proved very effective in highland parts of Kenya e.g. Marakwet where the hilly terrain does not allow alternative methods of land preparation. The Wagisu (Gisu community) from Uganda and parts of Kenya commonly helps in this errand, as they are known to be hard workers. Where ox ploughs are used, there is at times less effectiveness and work has to be redone by hand, or else the crops come out very poorly. The Luo community of Kenya is serious victims of the ox-plough complex and inefficiency, a scenario that has contributed significantly to permanent hunger in the region. It is a challenge to them to seek alternative ways of doing it lest the whole community is swept away. Harrowing should be done just before or at planting time, about a month after primary cultivation. Sometimes it is also the stage of basal dressing, i.e. addition and incorporation of phosphorous-based fertilizer or manure into the soil in preparation for planting.

Single tandem disc harrow

4.55 . DISCING: Is a secondary tillage operation often done in very heavy soils. A discer is used to break the very large lumps of soil, before subsequent operations are done in preparation for planting. It is commonly done in sugarcane farms where the soils are very heavy clays.

4.56 ROTOVATION: This involves the cultivation of soils under water. It is a common practice in rice fields, as shown in the figure below. Generally, due to the heavy nature of the soils, and the wet manipulation of the soils puts the soil structure at risk, as is common with soils cultivated under inappropriate soil moisture content.

Figure 36: The rotovated rice field

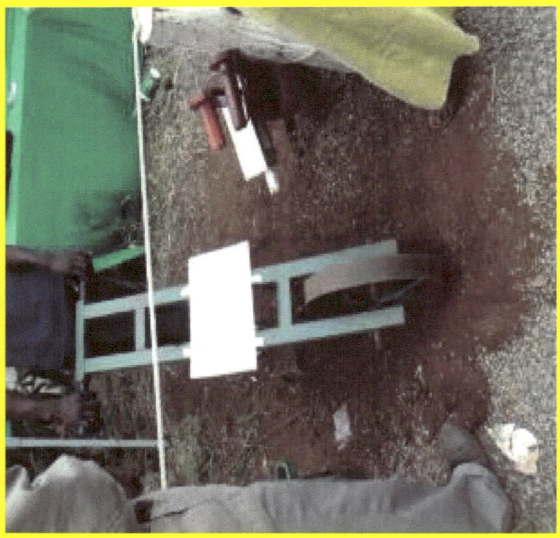

Figure 37: Ox plough Figure 38: Planter

4.57 DEFORESTATION

Meaning: This means large-scale cutting down of trees without replacement.

The undesirable effects of deforestation: Deforestation may lead to:

1: Loss of rare plants (reduced biodiversity);
2; Loss of rare animal species (reduced biodiversity)
3: Reduced soil fertility resulting from soil erosion;
4: Bare, less attractive sites
5: Changes in the water cycle and rainfall patterns as forests normally get back deep water to the atmosphere;
6: Wind destruction of farm structures becomes worse, as windbreaks are removed;
7: Reduced CO_2 sinks, leading to global warming. This is because the CO_2 which could have been used for photosynthesis are left in the atmosphere, contributing more to increased atmospheric global temperatures. CO_2 is a greenhouse gas.

HOW TO ACCOMMODATE UNAVOIDABLE DEFORESTATION

Sometimes, deforestation is inevitable due to:
To create land for settling people e.g. squatters; To establish a major infrastructure e.g. road, water works, effluent treatment works, dam, or hydroelectric power station; For large scale agriculture; Accidental fire.

In these circumstances, it is useful to get the best ways of accommodating the significant effects. In Kenya, any large-scale project which may warrant deforestation, among other key significant environmental impacts, are expected to undertake a mandatory environmental impact assessment (EIA). The impacts are predicted, prioritized, and mitigation measures proposed in the EIA report. All this is done before licensing. However, a number of government agencies have not been faithful to this law, assuming that they cannot be reprimanded by their same master of which they are part. This is a challenge which must be addressed for sustainability of such projects. Some ways of mitigating effects of deforestation include:
Establishing alternative forest / woodlot elsewhere in the neighborhood;
Selecting endangered species of flora and fauna and establishing them elsewhere;
Avoiding catchment areas for such activities by carefully conducting an alternative site analysis (part of the EIA process); Only do it if you must.

4.6 FACTORS LIMITING CROP AND ANIMAL PRODUCTION

The factors limiting crop and animal production in Africa are numerous, and are more or less related to traditional farming approaches. Poverty and the general economic challenges of the country. Are the main ones. The strategies for transforming agriculture have to address such challenges as low investment and productivity, under-capitalization of agriculture and research, low land and labour productivity, inadequate investments in irrigation, inadequate use of mechanization and agro-chemicals, poor infrastructure, weak institutions and corruption, lack of funding for agricultural research, inadequate use of yield-enhancing technologies, weak linkages between agriculture and other sectors, unfavourable policy and regulatory environments, climate change, among others. This chapter discusses these and other challenges.

(a) Aridity and inadequate investment in irrigation.
About 80% of the country's land area is arid and semi arid with low potential for agriculture. Irrigated agriculture has not kept pace with food security needs and not been taken seriously in the past by the government and remains underdeveloped. Some arid soils such as Solonchaks are saline and affect crop growth, and require reclamation even under irrigated agriculture. Pastoral communities lose many of their cattle to prolonged drought in arid and semi arid areas due to lack of pasture and water supplies.

(b) Widespread subsistence, traditional farming and poverty.

African agriculture remains largely traditional and concentrated in the hands of smallholders and pastoralists. The farmers are poor and risk averse (do not invest in yield enhancing technologies). Given the dominance of rain-fed agriculture, yields are low and farmers can be trapped in a cycle of poverty and food insecurity for decades. This is at the expense of embracing commercial agriculture development, marked by specialized & high productivity, important for both domestic food security & export market. Use of Poor Tools is a common feature for smallholder subsistence farming, examples being the hoe, jembe, cutlass, rake etc.

(c) Land tenure and land reform.
Communal land tenure discourages investment on land. Land subdivision for settlement and urbanization in fertile high potential areas has negated any gains in intensive smallholder farming in high potential areas (SHFs till smaller units / land resource poverty increases). This sis because land owned by inheritance encourages land fragmentations over generations. Increase in population puts pressure in land use, increases the various alternatives to which land can be put. Addressing land policy issues, especially security of tenure and equitable land distribution is a priority, will make land available to prospective and genuine farmers. Improving land use by restricting the encroachment of cultivation into fragile ecosystems, rampant urbanization, and sub division for settlements are key to agriculture development.

d) Unreliable rainfall and climate change

Over-reliance on rain-fed agriculture is a challenge for development. Unreliable rainfall often leads to crop failure and decreased yields. It affect planning for field operations such as when to plant by farmers. Of late this has been attributed to such factors as climate change resulting from environmental/ forest degradation (increased global warming due to increased emission of green house gases) as well as destruction of water towers etc. the government should invest in irrigation, and research in climate change adaptation. Both drought and flooding destroy crops and livestock and reduces yields.

e) Declining soil fertility and land degradation.

Continuous farming without nutrient replenishment and rampant erosion, common in African smallholder farms, depletes soil nutrients leading to decreased land productivity. In smallholder subsistence farmers, low inorganic fertilizers use are related to high cost / expensive and low availability in rural areas. This leads to poor crop yields. But many soils in high potential areas are acidic and may limit crop yields, aggravated by continuous monocropping and N-fertilizer use.

Soil degradation or erosion is mainly caused by loss of vegetation, land exploitation, especially overgrazing and shifting cultivation. insecurity of land tenure and lack of aggressive promotion of soil and water conservation increases land degradation. Increased encroachment into marginal fragile ecological areas, converting to intensive agriculture use encourages land degradation.

f) Lack of capital and under-capitalization of agriculture and research.

Low investment in agriculture by government and by the individual poor farmers leads to low crop yields and productivity. This affects uptake of yield-enhancing practices and techniques. Yield-enhancing practices include mechanization, use of agrochemicals, (fertilizers and pesticides), and increased use of irrigated land.

Many farmers are resource poor especially SHFs. They lack capital, and credit to buy inputs – fertilizers, pesticides, high yielding seeds, and for real agricultural investment / development. Moreover, commercial agriculture requires substantial capital to acquire land, improved breeds, equipment etc and the lack of such capital due to poverty ensures farmers continue to engage in subsistence farming and its perpetuation. The government should invest at least 10% of their public expenditure to agriculture in order to enhance food production and accelerate economic transformation. The government should create an enabling environment for the provision of adequate financing and credit facilities by both the public and the private sectors for the funding of agricultural production, processing and marketing. This should include the promotion of microfinance institutions and insurance schemes to serve farmers.

g) Inadequate market access and poor infrastructure

Promoting agricultural investment and productivity requires improved market access and adequate service infrastructure, including better road networks, communication, rural electrification and water supply. Access to inputs, and their pricing (seeds, fertilizers, drugs etc) and access to markets for farmers produce in rural areas all depend on good infrastructure, eliminates exploitative middlemen. Besides, they are required to start agro-processing, value addition for increased incomes by SHFs.

Improved rural infrastructure and marketing are essential to enhance the competitiveness of our agricultural products, for improved agro-industrialization and domestic and regional trade by strengthening the linkage between the agricultural and non-agricultural sectors, especially in rural areas. This would help the landless and the unemployed obtain jobs in agro-industries and associated small- and medium-sized enterprises (SMEs) in rural and urban areas. Also tariffs - competitive power and road/rail freight tariffs are important.

h) Poor Storage and Processing Facilities

Farmers experience high post harvest losses due to poor storage or inadequate storage facilities. Exploitation by middlemen due to lack of proper storage facilities reduces profits to farmers. The prices of agricultural products fall too low immediately after harvest, because the farmer cannot store or process them properly till they can have better prices. Consequently, farmer may not be adequately rewarded for his labour. Most agricultural produce are easily perishable and require preservation / processing / proper storage or else entire products are offered for marketing immediately. The government should improve or provide storage and processing facilities like crib, barns, shelters millers, grater, etc. excess farm products e.g. recent milk glut should also be purchased by the government so as to prevent glut and bring them to the market during scarcity.

i) Government policy and lack of incentives

Unfavorable policy and regulatory environments will not spur agricultural growth. Poor incentives related to prices, markets, and property rights such as high taxation, liberalized fertilizer markets, lack of subsidy for farm inputs, security of land tenure discourages production activity. The government must provide incentives and support to farmers and agro-businesses, including improved access to credit, subsidized fertilizers (currently doing this), low taxation regimes, promotion of entrepreneurship and development of technical skills in general.

j) Poor crop husbandry practices and low agriculture education.
Lack of adequate knowledge and practice of good farming methods is a hindrance to agriculture. Due to low education of many farmers, they plant late, don't control weeds, use poor spacing, don't control diseases and pests, cannot recognize diseases etc. Education, especially agricultural education and technical skills development programmes, including knowledge on value addition, especially targeting women farmers will improve labour and agricultural productivity.

k) High cost of inputs and Lower prices of agricultural products.
Cost and access to affordable agricultural inputs e.g. improved seeds / livestock, fertilizers, pesticides, drugs, farm machinery is a problem to many SSFs. In contrast, agricultural products are often lowly priced and this leads to low income and profits to farmers, and results in low motivation for further investment in agriculture.

l) Inadequate extension staff and weak research dissemination and adoption.
The government has not provided adequate skilled personnel to advice farmers, i.e. to improve research technology dissemination and adoption. Inadequate use by famers of yield-enhancing technologies such ISFM (integrated soil fertility management practices) are linked to inadequate extension (lack of staff or effective approaches / extension programs). This leads to delay in technology transfer, and farmers do not apply recent technology in farming.

m) Low land and labour productivity.
Low land productivity is due to low use of irrigated land, low use of fertilizers and machinery, improved seeds, livestock etc. Labour productivity is lower in Kenya mainly because most of our agriculture is manual or semi-mechanized. Low labor productivity is due to lack of knowledge / education and technical skills (esp. for female farmers in rural areas who make majority), other factors are current the production system (subsistence / small scale), weather, lack of complementary inputs such as improved seeds and fertilizers, and the use of modern faming technology.

n) Fluctuation in market prices: World market prices for agricultural products often fluctuate and farmers occasionally incur huge losses when the prices of major cash crops such as coffee, tea, horticultural crops fall in international markets.

o) Cultural, traditional beliefs / religious prejudices.
Muslims for example do not encourage pig production for pork, whilst their Christian counterparts do not engage in tobacco growing, regardless of ecological potential and profitability.

p) Psychological factors/ attitudinal setbacks and rural urban migration.
Many able bodied men, well educated do not want to engage in direct agricultural production as farmers; they prefer white collar jobs in cities even after acquiring agricultural production skills. They regard agriculture as a dirty profession.

q) Pest and diseases.
Prevalent pests and diseases cause serious yield losses to farmers and are expensive to control. Expenses incurred increase cost of production. Examples include Coffee berry disease, East coast fever, Newcastle, wheat rust etc. storage pests e.g. weevils etc.

r) Natural calamities
Include hailstorms, floods, droughts. East Africa is prone to periodic Elnino floods; Lanina long droughts. These are destructive to crops, and other agricultural enterprises. It also disrupts settlement and economic / farming activities. Farmers should be trained in early pests/ disease identification and control, including which chemicals to use. Credit to purchase pesticides for SHFs is important.

Reference
APN, 2008. Africa's Development: Promises and Prospects. Geneva, Switzerland.
UNECA, 2008. Trends in land and labour productivity by regional economic community, Addis Ababa, Ethiopia.

4.7 AGRICULTURAL PRODUCTION SYSTEMS
Two common agricultural systems used for producing crops and livestock by farmers, are subsistence and commercial farming

DIFFERENCES BETWEEN SUBSISTENCE AND COMMERCIAL FARMING
Subsistence agriculture is the production of food primarily for consumption by the farmer's family. The farmer gets his food requirements and cash for basic needs. Commercial agriculture is the production of food primarily for sale off the farm. It is

geared towards profit maximization. Five principal features distinguish commercial from subsistence agriculture:

purpose of farming; percentage of farmers in the labor force; use of machinery; farm size; (and) relationship of farming to other businesses

Table 3: Comparison between commercial and subsistence farming

	Commercial Agriculture	Subsistence agriculture
1	Large scale farming is adopted	Small scale farming is adopted
2	Mechanization is practiced	Traditional farming is being practiced. Use simple tools.
3	It is ideal for monocropping, intensive animal production and fish farming	Ideal for mixed farming, bush fallowing and shifting cultivation
4	Highly capital intensive which can be gotten through the bank	Low capital requirements which can be provided by friend and family members
5	Labour is highly specialized and Expensive	Labour is from friends, family members and cheap.
6	Marketing of product is specialized	No special marketing skills are required.
7	Research into aspects of production required	No research activity is required
8	Highly scientific with no superstitious belief.	Highly traditional with superstitions in nature
9	Yield and often returns are high meant for every body	Yields are low and meant for family members
10	Processing of final product is common	No processing of any form is practiced
11	It requires good record keeping so as to make good decisions	No record keeping is done and so no improvement is expected

SUBSISTENCE FARMING

This is a system of farming whereby a farmer cultivates crops and rears animals in order to produce food for use by himself and his family only. In crop production, the subsistence farmer concentrates on arable food crops such as maize, sorghum, millet, cassava, beans, groundnuts, cowpeas etc.

In livestock production the subsistence farmer keeps a few cattle, goats, sheep, rabbits or pigs. In the arid regions, the subsistence livestock farmer is mainly nomadic and moves with his animals from place to place in search of water and pasture. When involved the establishment of plantation agriculture, the subsistence farmer uses traditional methods. The farmer may plant such crops as sugarcane, banana, tea, coffee and citrus trees.

Subsistence agriculture include the following:- Shifting cultivation, Intensive subsistence farming and Pastoralism. Subsistence mixed farming is also practiced by smallholder farmers especially in developing countries.

A. Characteristics of Subsistence Farming (see table above)
Labour is provided by the farmer and his family.
Only small area of land is used e.g. 2 acres.
Crude implements like hoe, jembe, panga /cutlass, digger, baskets are used.
The produce from the farm are not necessarily for sale.
The yield is also very low compared to mechanized farming.

B. Problems of Subsistence Farming
Control of pests and diseases is difficult.
Illiteracy of the farmers.
The crude tools used by the farmers.
Labour supply is erratic and unreliable.
The practice of subsistence farming wastes land and there is no means of replenishing soil e.g. shifting cultivation which is migration to another fresh land.
The process bores down farmers because the work is unchallenging and there is little or no profit since the produce are not for sale

6.4.1 Shifting Cultivation

Shifting cultivation is an agricultural system in which plots of land are cultivated temporarily, and then abandoned. This system often involves clearing of a piece of land followed by several years of farming until the soil loses fertility. Once the land becomes inadequate for crop production, it is abandoned - left to regain fertility /reclaimed by natural vegetation.

It involves slashing & burning vegetation cover, cultivating for 3-5 yrs (average 3 yrs cultivation) and abandoned after it loses fertility, and allowed to rest. The most common form is slash-and-burn, but other cultivators (few) employ land clearing without any burning.

Common Features: Hand tools ; No draft animals; Long rotations; Low population density; and Land owned by community

The length of cropping & rest period depends on land pressure / population density & tenure arrangements.

Involve small plots, cultivated plots are scattered, cultivation done with simple tools.

Practiced by primitive people - Shifting agriculture (slash & burn) is now mostly practiced in the tropical world by primitive tribes in the forested highlands of:

South America—Amazon basin

Congo basin of Africa -, linked to animal husbandry wherever possible. The tsetse fly limits cattle production

SE Asia—sharp line between shifting cultivation and wet rice.

It is also still found in sparsely populated areas of Tanzania, Zambia and northern Mozambique and in the empty area of the Zaire basin" (FAO, 1984).

The relationship between the time the land is cultivated and the time it is fallowed are critical to the stability of shifting cultivation systems. This determines whether or not the shifting cultivation system as a whole suffers a net loss of nutrients over time.

The main purpose of the fallow period in shifting cultivation is simultaneously to improve soil fertility and the soil's capacity to resist erosion.

Advantages

Shifting Cultivation is sustainable if the fallow is long enough.

Land is left to regain fertility – soil organic matter (SOM) accumulation, nutrient recycling, nutrients are extracted from the subsoil, soil fauna increases, acidity is reduced, soil structure and moisture characteristics improve.

large amounts of P & K are released into the soil in available forms on burning the vegetation in ashes.

It may reduce the incidence of soil-borne diseases, play a role in pest and weeds control during fallow periods, wind and water erosion is much reduced.

Disadvantages

Crop yields are always low, therefore not enough food.

Not cost effective.

Burning – leads to degradation of vegetation, loss of volatile nutrients e.g. C, N, and S.

When plots are left fallow for long, this is economically wasteful (of land resource)

Reduced biological diversity & Loss of valuable species of economic/ medicinal value. i.e. destruction of forests constitutes a tremendous loss of valuable resources.

Deforestation - More forest area brought under cultivation. Deforestation increases global greenhouse effect, global warming and climate change.

Rapid depletion of soil nutrients, increases in acidity, increased soil erosion, increased local aridity and desertification.

Low remuneration of shifting cultivation relative to its labour requirements and to shifting cultivators' labour supply. It is also low because he cannot get a good price for his produce because there are no markets for it.

Intensive Subsistence Farming.

In this system, land is intensively cultivated & the crops are rotated. The population is high but land is scarce. However, there is improvement of farming technology, manure & fertilizers are used to increase production. Irrigation may be used e.g. in Kiambu, Kisii, Vihiga.

Problems of intensive Subsistence Farming

Production potential of land not fully utilized because of constraint of resources and technology.

Population pressure leads to migration to ecologically marginal areas (e.g. semi-arid) with severe degradation of environment.

Increase in population pressure results in increased subdivision of land & settlement. This reduces productive land size.

Low yields are obtained.

Subsistence farmers (SHFs) are resource poor & risk averse.

Farmers are ignorant of new farming technology.

Use simple tools – hoes, pangas, resulting in low yields.

Pests and diseases – damage crops.

Lack storage facilities.

Pastoralism

Refers to subsistence herding of large animals. Pastoralism is a subsistence pattern in which people make their living by tending herds of large animals e.g. cattle (EA), horses (in Mongolia). In the mountainous regions of Southwest Asia, it is mainly sheep and goats.Pastoralism is characterized by extensive grazing – where animals exhaust pasture on a given piece of land and then move to a different location in search of pasture & water. There are essentially two forms of pastoralism, nomadism and transhumance.

(a) Pastoral nomads follow a seasonal migratory pattern that can vary from year to year. The timing and destinations of migrations are determined primarily by the needs of the herd animals for water and fodder (pasture, crop residues).

(b) Transhumance pastoralists follow a cyclical pattern of migrations that usually take them tocool highland valleys in the summer and warmer lowland valleys in the winter. This is seasonal migration between the same two locations in which they have regular encampments or stable villages often with permanent houses.

Nomadic pastoralism

Involves seasonal migration of large numbers of animals – migrating from place to place in search of pasture (or crop residues) & water. Timing & destinations may vary from year to year. Practiced by – Maasai, Turkana, Samburu, Pokot, Boran, Somali, Gabbra among others.

Main features

Migrations conditioned by seasonal availability of grass, water & crop residues

Practiced in ASALS with low, unreliable rainfall

Land owned by community

Cattle kept as sign of wealth

Overgrazing is common due to overstocking

Disadvantages

There is overgrazing which causes degradation of vegetation & land. Soil exhaustion & erosion discourage pasture growth

There is uncontrolled breeding

Deforestation due to overstocking

Cattle diseases & low productivity are common – the quality of pastures is low

Death of cattle due to drought & stock theft common.

There is frequent conflicts over available pastures & water – e.g. cattle rustling between Turkana / Pokot.

Government of Kenya Interventions to Improve Pastoralsim.

Established Kenya Meat Commision (KMC) – to improve marketing & income to pastoralists.

Demonstration ranches – established to sensitize pastoralist on better methods of animal husbandry.

Encourage cross-breeding

Provision of extension services to advice & offer veterinary services

Construction of boreholes & water points to provide water for livestock

The animals herded by pastoralists are rarely killed for family use alone. Fresh meat is distributed throughout the community. This is the most efficient use of their animals because they usually do not have the capability of adequately preserving meat. Not only does it ensure that no spoilage takes place, but it also sets up numerous obligations to reciprocate within the community. It promotes cooperation and solidarity. Often the slaughter of an animal is for a ritual occasion so that its death serves multiple purposes. It feeds both the gods and the people. Most pastoralists also get food from their animals without killing them. Horses, goats, sheep, cattle, and camels are milked. In East Africa, cattle herding societies also bleed their animals. The blood is mixed with fresh milk to make a protein rich drink.

Threats to pastoralists

1. Have little political influence because they inhabit remote areas and are widely dispersed.
2. They have a history of strained relations with central authorities, sometimes leading to outright hostilities.
3. Borders interfering with their traditional migration patterns.
4. Government policies usually favor settled farming and crop production and are implemented at the expense of pastoral existences.
5. Existence threatened by

a. Expansion of crop cultivation as former pastoral grazing grounds are being alienated for crop cultivation.

b. population growth has led to great scarcity of resources in the semi-arid areas that pastoralists inhabit

c. Shortening of the fallow period – previously pastoralists were welcomed by farmers for the fertilizing effect of the manure of the animals

d. Wildlife parks -There are many instances where pastoralists are suddenly prevented from using their traditional pastures, because these are declared nature protection areas or wildlife parks. In Tanzania, the Maasai have been evacuated from the Serengeti plains

e. Privatization of former common pastures -In traditional pastoral societies, land is not owned individually, but represents communal property. In Kenya, a "group ranch" programme was imposed on the Maasai that conferred individual land ownership to groups living together. This has now resulted in most of the land being owned individually and being used for maize cultivation (Galaty, 1992). This change was expected to encourage the Maasai to limit their livestock numbers to match the group ranch resources.

f. Sedentarization projects -Many countries feel that nomadic pastoralism is a backward way of life and that traditional pastoral animal husbandry techniques are inefficient. Under the banner of socioeconomic improvement and modernization, they therefore attempt to settle their nomadic populations by providing them with plots of land and housing. This has happened for instance in Sudan, Israel and Jordan.

g. Food aid and drilling of wells -Often, settlement is also "voluntary" and occurs as a result of processes that have made a pastoral existence impossible. Interventions such as food-aid and drilling of wells, seduce pastoralists to give up their way of life and remain in one spot.

h. Irrigation and hydroelectric projects -Building dams to generate power or to increase agricultural productivity of low rainfall areas is a favorite policy that usually deprives pastoralists of their traditional grazing areas.

COMMERCIAL FARMING

They produce specifically for sale and from the money received; they are able to buy what they need from the market. This large-scale production of a specific agricultural product entirely for sale is called commercial farming. To increase farming efficiency, some commercial farmers have resorted to producing a single crop or livestock. Large-scale cultivation of a single crop is called monoculture. Commercial farming thus mostly involves producing a specific product, (specialisation), thereby leading to increased efficiency. Most often only one product is produced.

Commercial agriculture include the following;- Plantation farming (e.g. monocropping), and Ranching (or intensive livestock farming), commercial dairy farming and specialized horticulture. Commercial mixed farming is also practiced by farmers especially in developed countries. This is a type of farming that is done on a large scale through the use of machinery (mechanization).

Mechanization increases productivity in agriculture. Most farm operations are mechanized. Bulldozers for land clearing, Tractors and tractor mounted implements (ploughs, harrows) are used in tilling the soil, planters, ridgers, spraying machines and combine harvesters are used. In livestock farming, incubators, instead of natural hatching of eggs by fowls, egg graders and milking machines are used. These machines make work in the farm easier, faster and less burdensome.

Major features

Large areas of farmland are required.

Requires the use of machinery like cultivators, tractors, ploughs, harrows, planters, harvesters, etc.

It is very expensive to practice.

Labour is both manual and mechanical and specialists are required, in some stages of production.

The returns are very high at harvest time.

Marketing of products is specialized and sometimes advertisement done.

It involves a lot of research into the various aspects of production.

It requires good record keeping so as to make proper and adequate decisions.

Commercial plantation Farming

Plantations / estates – refers to the cultivation of cash crops on large pieces of land. It was introduced in Kenya by Europeans. The aim is to produce farm products in large quantities for sale, for cash income and export. It involves specialization in production of one crop type, referred to as monocultures, such as sisal plantations (in Marigat, Taita/Taveta), Tea, coffee, sugar, cotton, wheat, maize plantations.

The plantations are large and situated in thinly populated areas.

Advantages

Only one crop is grown –leads to specialization. This leads to increased yields.

Adequate food security for domestic and for export can be achieved. Products are exported & sold in the world market to earn the country foreign exchange.

Employment creation of workers in plantations.

Disadvantages

Capital intensive /high capital requirement.

Soil fertility depletion – soil exhaustion, erosion esp. for row crops. Large quantities of fertilizers are used and this brings environmental concerns – groundwater nitrate pollution.

Risk of huge losses if world prices fall.

Ranching

This refers to extensive commercial grazing, an economically advanced form of pastoral nomadism. Ranching is defined as the commercial keeping of large herds of animals. The difference between the pastoralists and the ranchers is that the ranchers are not subsistence pastoralists but they are businessmen who produce a commodity for national and international markets

Noted: Pastoralists & subsistence farmers own ~95% of beef animals; ranching only 5%. Ranches are found in the Rift valley – Nakuru county, Kajiado, TransNzoia, Coast –Kilifi and Kwale counties.

Main features:

Practiced in confined, enclosed grazing areas.

practiced in drier regions to better utilize poor vegetation & soils

there is improved pastures, disease control, supplementary feeds, water provision

most ranches keep cross-breeds e.g. Zebu, Boran.

Major slaughter houses are in urban centres e.g. Dagoreti, Miritini – Mbsa, Ksm, NKR, Eldoret.

The main meat processing factory – the Kenya Meat Commission (KMC) - at Athi River, which markets & supplies meat for export & local consumption (Individuals with butcheries market own meat products).

Mixed Farming.

Mixed crop-livestock farming systems are characteristics of large areas of Africa. Crops and Animals (livestock) are kept on the same farm at the same time. It is mutually beneficial. Commercial crops & livestock mixed systems are associated with densely populated areas. The animals involved include cattle, goats, sheep, poultry etc and provide ploughing, milk, manure, bride wealth, savings, emergence sale. The crops commonly involved include maize, wheat, Irish potatoes, beans, barley.

Advantages

Animals feed on crop residues - The stover from maize is the source of feed to livestock during the dry season when there is shortage of high quality fodder

Manure from livestock help maintain soil fertility for increased crop yields. Manure is an important source of nutrients, especially N, P and K. increase soil organic matter, water holding capacity, and infiltration rate.

There is effective use of farm resources – land; labour

Farm income (& profitability) is increased through sale of farm products

Security against total loss – if one enterprise fails, for instance due to disease, drought, the farmer gets income from the other alternative enterprises

Total farm productivity is increased.

Livestock is a form of capital to buy inorganic inputs for crop (e.g. maize, wheat) production

Livestock can also provide transport to farmers to carry inputs and outputs from the markets

Environmental impacts of commercial farming

Due to the large-scale nature of this farming method, large tracts of land are often cleared, and at times leading to soil erosion.

Specialised farming has at times led to an accumulation of some pests and / or disease pathogens in the farms and their environs

Monoculture associated with commercial farming has often led to reduced biodiversity, as there is an intense selection among and within varieties.

This method often involves use of chemicals as farm inputs- and some of these are dangerous to the environment. For instance, DDT causes immediate and long-term pollution. Some effects are: biomagnification, bioaccumulation, birth defects (teratogenesis, reduced sexual activity (fertility), nervous disorders, allergy, cancers etc.

AGROFORESTRY

Agroforestry involves raising trees in combination with other agricultural enterprises, crops and/ or livestock. Different species of trees can be planted with many types of crops in a variety of patterns. For example, fast-growing trees can be planted when the land is fallow or they can be grown at the same time as agricultural crops

Components of AF are 3 = Tree or woody perennials, Crops, and Animals

Hedgerow intercropping

Hedgerow intercropping or alley cropping - consists of growing food crops in the alleys / lanes formed by hedgerows of multipurpose trees and shrubs (MPTs) that are usually N2 fixing. The hedgerows are cut back (lopped) and periodically pruned during the cropping season to prevent shading. The prunings (twigs and leaves) provide nutrients when incorporated into the soil as green manure or spread on the soil surface as mulch (Kang et al., 1990).

Trees used – leucaena, Calliandra

In this system, the crop is supplied with nitrogen, and the level of organic matter in the soil is increased. Deep roots capture subsoil N and also reduces nutrient loss through leaching. After the leaves or cuttings fall to the ground, the nutrients once again become available to the crops via decomposition. This effect of woody species is sometimes called a .nutrient pump. The alley cropping system performance is location – specific and greatly influenced by the choice of tree species and type of management adopted. Low yield of hedgerow prunings (2 to 3 t ha-1 yr-1) and competition of hedgerows for nutrients and water are serious limitations causing low crop yields in water-limited areas and in poor soils are the main challenges.

Inadequate water limited the response of crops even though hedgerow intercropping improved soil fertility in certain sites of the semiarid tropics. In favorable environments, hedgerows produced 8 to 12 t ha-1 yr-1 of prunings (dry weight) which increased SOM and ensured adequate supply of nutrients, especially N, to the alleys. In such cases, a biomass transfer (cut- and –carry) system has been proposed as a more beneficial system than alley cropping (Jama et al., 1995). However, this system is very labour intensive. If the trees are neglected (not pruned on time), then crop production will often be lower.

In conclusion, hedgerow intercropping is site-specific and is highly influenced by soil and climatic conditions. Whereas it is a viable technology in humid and sub-humid zones and with the right choice of trees, it does not work in arid and semi-arid regions. In such areas, biomass transfer has been proposed as a more beneficial system.

Biomass transfer

Biomass transfer technology involves ex-situ production of biomass away from the cropping land in designated areas around or within the farm. The biomass is then harvested and applied as mulch or incorporated into the soil before and/or after the crop is planted. In this technology the interactions between the components is only through the nutrient release from applied mulch and the uptake by the crop. The system eliminates competition but often requires substantial labor inputs during biomass transfer activities.

Use of Tithonia diversifolia

Tithonia diversifolia commonly known as Mexican sunflower, is a non-N2-fixing shrub of the family Asteraceae and grows 1 – 3 m in height. Although it is not a N fixing shrub, it accumulates as much N in its leaves as N-fixing legumes, presumably because of its greater root volume and ability to scavenge nutrients from the soil (Gachengo, 1996).

In Kenya, it commonly grows along fields and farm boundaries as well as along roads and it produces large quantities of leaf biomass and tolerates regular pruning (Buresh and Niang, 1997). Early studies on using tithonia were conducted in Western Kenya (Gachango et al., 1996, Bashir 1999) These studies identified green biomass from tithonia-grown ex-situ and transferred into a field to be an effective source of nutrients for maize.

It was as effective in supplying N, P, and K to maize as an equivalent amount of commercial NPK fertilizer. Tithonia applied at 5 Mg ha-1 on a dry weight basis contains about 60 kg N ha-1, 5 kg P ha-1 and 60 kg K ha-1. Tithonia decomposes and releases its nutrients rapidly. Gachengo et al. (1999) reported a half-life of about one week for the disappearance of dry matter in the rainy season in western Kenya. Phosphorus release from tithonia green biomass is rapid, and it supplies plant available P at least as effective as an equivalent amount of P from soluble fertilizer (Nziguheba et al., 1998).

Bashir et al. (1999), reported that green biomass of tithonia is high in nutrients, averaging about 3.5% N, 0.37% P, and 4.1% K on a dry matter basis. Thus, tithonia green biomass has relatively high macronutrient contents as compared to green biomass of other shrubs and trees. The N concentrations in tithonia are comparable to those found in N2-fixing leguminous shrubs and trees, whereas the P and K concentrations are higher than those typically found in shrubs and trees. The sustainability of tithonia use by farmers to recycle nutrients in farming systems is limited by the long-term availability of the plant material and intensive labor involved in biomass collection, processing and application (Bashir et al., 1999; Mutuo et al., 2000).

Use of Leguminous trees for biomass transfer

Calliandra calothysus and Leucaena tricandra are some of the leguminous trees that have widely been used in biomass transfer systems. The properties of leucaena that make it a useful agroforestry tree in improvement of soil fertility include; high biomass production about 10-25 Mg dry matter ha-1 yr-1, high nitrogen fixation of about 100-500 kg N ha-1 yr-1 (Young, 1989), high level of nitrogen in leaves of about 2.5-4.0% (Delve et al., 2000) and thus high rate of return in litter or pruning. It also contains substantial content of other nutrients in leaves; about 0.28% P, 2.5% K, and 1.49% Ca. In addition to providing nitrogen, the application of leucaena biomass greatly increases the availability of phosphorus and exchangeable potassium in the soil.

Calliandra produces high yields of protein rich leaf material that are suitable as green manure (Gutteridge and Shelton, 1994). Fresh pruning of calliandra have been used as green manure. It has the ability to decrease soil bulk density, increase soil porosity and availability of water in soil (Liyanage and Abeysoma, 1996), therefore, the improvement of the soil physical conditions is pronounced with calliandra. IITA (1987/8) estimated the nutrients from calliandra pruning as 169 kg N, 8 kg P, 83 kg K, 423 kg Ca and 26 kg Mg ha-1 yr-1.

The incorporation of calliandra and leucaena biomass into the soil improves fertility and hence significantly increases crop yields. Mugendi et al. (1999) reported that calliandra and leucaena green biomass (with or without fertilizer) had significant increase in maize grain yields as compared to the inorganic fertilizer. With biomass incorporation, an average of 4.1 t ha-1 of maize grain per season was obtained compared to 3.3 t ha-1 obtained from the application of the recommended level of inorganic fertilizer and 3.1 t ha-1 per season for the absolute control.

Improved Fallows

Farmers are using improved fallow technology as an alternative to mineral fertilizers to improve soil productivity. Traditionally, shifting cultivation with long fallow periods was practiced to restore soil fertility. However, with increase in population pressure, fallow periods have reduced hence shifting is no longer effective in maintaining soil fertility.
A natural fallow is land resting from cultivation, usually used for grazing or left to natural vegetation to restore soil fertility lost from growing crops. Improved fallow, on the other hand, refers to targeted use of planted species – usually legumes; with primary purpose of enriching soil within a shorter period compared to natural fallows. This is mainly through N2 fixation when N2 fixing trees are used, deposition of recycled nutrients through litter fall or when biomass harvested at the end of the fallow period is incorporated into the soil (Prinz, 1986; Sanchez, 1999).

Results on an acid soil in western Kenya (pH=5.1, clay=29%) indicated that rotation of maize with Sesbania sesban could lessen losses in maize yield resulting from potassium deficiency (ICRAF, 1998). Control of Striga hermontheca, a parasitic weed of many cereal crops, is an added benefit of the repeated use of improved fallows (ICRAF, 1998). Large-scale adoption of short-term improved fallows by farmers is now taking place in East Africa (Niang et al., 1998; Rao et al., 1998) and other parts of Africa. The main species used are legumes of the genus sesbania, tephrosia, leucaena, mucuna, crotalaria, cajanus, indigofera and mimosa.

Improved fallows can be established during the short rains, when yields are lower and less reliable than during the long rains. Improved fallows often result in low quantity of leaf biomass production in degraded sites (nutrient-depleted and eroded) due to poor growth, especially in dry areas/ those with vertisols

CROPPING SYSTEMS

Definition of Cropping System

The term cropping system is used to describe the pattern in which crops are grown in a given area over a period of time and includes the technical and managerial resources that are utilized. In short, when we talk about the cropping system of a given area, we are not only interested in how those crops are distributed on the field at any given time but also how this distribution changes over time. In addition, the level of management and amount of resource inputs are integral aspects of a cropping system.

Cropping systems are classified based on the following criteria

1. The distribution of crops in time, i.e. whether shifting cultivation, continuous cropping, monoculture, or crop rotation is practised.
2. The distribution of the crops in space on the field, i.e. whether intercropping or sole cropping is practised.
3. The level of management and resources utilized to produce the crop, i.e. whether production is intensive or extensive.
4. The type of crop grown, i.e. whether orchard, arable cropping, pasturing, forestry, etc. is practised.

MONOCROPPING

Monoculture - one crop grown in a field at a time. This is the practice of incessantly cultivating the same type of crop on the same piece of land year after year. For example maize farming in Kitale

Disadvantages of monoculture

1. In monoculture always have their host present, and therefore have the opportunity to build up over the years.
2. Monoculture encourage rapid depletion of soil nutrients and
3. The risk of crop failure is great and ever present.

Advantages of monoculture

The advantages of monocropping e.g. of maize include ease of mechanization, weed control and use of fertilizers. Maize yields up to 10 t /ha have been obtained with inorganic fertilizer inputs of 600 – 1000 kg of nutrients per ha. Thus it gives higher yields, and more efficient. The high yields under maize monoculture is important for food security demands of the nation, and for export markets. It also stimulates economies and provides jobs for the working classes

Some of the disadvantages include

Soil nutrient depletion and erosion; -Nutrient imbalances in soils
destruction of the soil structures.
Encourage weeds, diseases and pests of the particular crops - Monocrops favor disease development
Loss of seed biodiversity and plant genetic resources.

CROP ROTATION

One of the most important aspects of organic farming is the strict avoidance of monoculture, whether annuals or perennials. Besides the proverbial "putting all eggs into one basket", monoculture systems are unhealthy for the ecosystem they are a part of. The prime requirement for any natural ecosystem to thrive and be healthy is diversity.

Crop rotation is the growing of different crops in sequence (season after season) in one plot or field. i.e. two dissimilar type of crops follow each other. A good rotation system needs interchanges or variation of atleast 4 different families / classes of crops such as
Leguminaceae (legumes) – groundnuts, field beans, soya, cowpeas, French beans, chickpeas, etc
Graminae (cereals) - maize, sorghum , millet, wheat
Cucurbitaceae – pumpkins, melon, squash, marrow
Cruciferae – cabbage, kale, spinach, broccoli, cauliflower radish
Others include : Solanaceae – pepper, potatoes, tomatoes, blacknight shade, eggplant
Umberlliferae – carrots, parsley etc; and Liliaceae – Garlic, onions, shallot.

Crop rotation means having times where the fertility of the soil is being built up and times where crops are grown which remove nutrients. Crop rotation also helps a variety of natural predators to survive on the farm by providing diverse habitats and sources of food for them. Include crops with ability to fix nitrogen, ability to break lifecycles of important pests, diseases and noxious weeds - allelopathy. The crops should have different characteristics, including

- Ability to fix Nitrogen – eg legumes
- Different nutrient demand - In a good rotation regime, heavy feeders / nutrient hungry crops eg brassicas, are followed by moderate feeders eg cucurbits. Moderate feeders are followed by light feeders eg carrots which are finally followed by legumes or nutrient enhancing crops to complete the rotation cycles
- Soil cover against the sun, erosion, and humus accumulation
- Deep rooting ability which helps to bring nutrients from lower soil horizons to upper horizons where shallow rooted crops can use them

- Ability to break lifecycles of pests and diseases, hence ease in controlling them
- Ability to deter weed growth and development 9 allelopathy) among others.

An example of a typical 4 year rotation include
maize -beans -Grass or bush fallow (a fallow period where no crops are grown)- and green manure legume crop eg crotalaria (sunhemp)

Example 2: Maize –heavy feeder……….Chillies – moderate ……..Sweet potato – light feeder ….. Legumes – beans

Example 3

1st Season	2nd Season	3rd Season	4th Season
Brassicas (takers)	Cucurbits (moderates)	Roots (Light)	Legumes (givers)
Cabbage	Pumpkins	Sweet potato	Cowpeas
Kales	Melon	Carrot	Beans
Spinach	Marrow	Cassava	Soyabeans
Lettuce	Gourds		

NB: Agroforestry crops / trees in intermittent rows within the field for shade, N-fixation & livestock fodder

In crop rotation, crops should be moved to a different area of land each year, and not returned to the original site for several years. For vegetables a 3 to 4 year rotation is usually recommended as a minimum.

Factors to consider – before starting a rotation program
Crop nutrient requirement - crops such as carrots, garlic, and raddish have very low nutrient requirement; beans, onions, tomatoes, turnip, cucumber have low nutrient requirement;, cauliflower, kales, lettuce, potatoes, pumpkins, spinach, sweet paper and sweet poptatoes have medium nutrient requirement while beet roots, celery and brussel's sprout have high requirement and cabbages broccoli, blacknight shade have very high nutrient requirements.

Rooting system – consider whether deep or shallow rooted. Deep rooted crops have tap roots that go deep into the soil to reach the leached nutrients. These are later brought up to the surface horizons where shallow rooted crops can reach them. Deep rooted crops also help in breaking hard pans.

Nature of the plants – refers to whether the crop is upright, creeping, short or tall, its ability to cover the soil against the sun, erosion and humus accumulation. Examples of upright – kale, cabbage, amarancth, Irish potato. Creeping – vines = yams, sweet potatoes,

pumpkins; tall crops – eg garden peas, broccoli, kales, amaranch, tomatoes, maize, sorghum, millet; and short crops – carrots, cabbage, white onions, irish potatoes etc
The type of soil and its suitability for crop chosen -
Climatic condition – ecological requirements of different crops. Consider rainfall reliability if crops are rainfed.
Market availability – farmers might opt to grow similar crops. He must explore mkt outlets for the crops included in rotation, and which are not staple in his locality.

Intercropping

The practice of growing one crop variety in pure stands on a field is referred to as sole cropping. In this practice, only one crop variety occupies the land at any one time. The alternative practice of growing two or more crops simultaneously on the same field is called intercropping. The various crops in the intercrop do not necessarily have to be sown or harvested at the same time, the main requirement is that they are on the field at the same time for a significant part of their growing periods.

Types of intercropping

1. Row intercropping: this is when the various crops are grown in separate rows.
2. Mixed cropping: this is when the various crops are grown intermingled more or less at random with each other.
3. Relay inter-cropping: this is when a second crop variety is sown between the stands of an existing sole crop just before the first crop is harvested. As such, both the first and second crops spend most of their field lives as sole crop, and grow together on the field for only a brief period.

Factors that determine the crops combination and spatial arrangement

1. Tillage practices: when ridges have been made, the spatial arrangement of the various crops may be determined by particular needs of each crop. For example, yam which requires a deep layer of tilled soil, is planted at the top of the mounds, while rice because of its high moisture requirement, is planted in the lower ground between the mounds. Other crops such as maize, pumpkins and melons are planted at intermediate positions between the rice and the yams.

2. The crop the farmer considers as target crop and which one is considered a subsidiary influence the proportion of the crops.

3. Nature of the crops themselves: a few strands of pumpkin or melon occupy a lot of land and have high economic yield, whereas rice strands, for example, would have to be much more numerous to be meaningful.

Evaluating yield from intercropping

The relative yield of each component crop in an intercropping situation is the yield of that component in the intercropping situation divided by what that crop would have yielded as a sole crop, covering the same area as the intercrop and managed at the same level. Suppose, for example, that a field with a crop combination of maize and cowpeas yields 1.5 tonnes/hectare of maize and 0.25 tonnes/ha. of cowpeas. If the expected sole crop yield of maize is 2.0 t/ha and that of cowpeas is 0.5 t/ha. Then the relative yield of maize is 1.5/2.0 = 0.75 and the relative yield of cowpeas is 0.25/0.50 = 0.50.

The sum of the relative yields of the various component crops in the intercrop is sometimes called the relative yield total. A little reflection of how many times the land area used for intercrop would be required to produced the same yields of the component crops when they are grown as sole crops. The relative yield total is therefore more conventionally referred to as land equivalent ratio (LER).

Mathematically:

LER = relative yield of crop A and relative yield of crop B + relative yield of crop n

In the maize/ cowpeas combination considered above the LER = 0.75 + 0.25 = 1.25. An LER greater than 1.0 implies that for that particular crop combination, intercropping yielded more than growing the same number of stands of each crop as sole crops. An LER of less than 1.0 implies that the intercropping was less beneficial than sole cropping.

LER can also be calculated based on the monetary value of the yield obtained from the various intercrop and sole crop situation and make comparisons on this basis.

Calculating LER by comparing the total energy value of the yield in kilo calories in various situations and compared the values to determine which arrangement was most beneficial.

Advantages of intercropping

1. In a carefully planned intercropping the LER is usually greater than one. This means that there is yield advantage in growing crops together than growing each one separately. This advantage may rise from several sources.

2. The crops may complement one another in their use of field time. The periods of their peak demands for light, water, nutrients and other resources may differ, so that in general

there is a more efficient utilization of the resources available. For example, during the two or three month that yam takes to sprout and establish adequately on the field, a quick intercrop of maize or melon would beneficially utilize the field resources during this period.

3. The component crop may complement each other in their use of space. For example, an intercrop of a deep rooted crop can exploit various horizons of the soil.
4. An intercrop may be able to utilize resources which the main crop may not be able to utilize or which may even be disadvantageous to it.
5. Certain crops may exert specific beneficial effect on others. For example, plantains intercropped with young cocoa seedlings provide shade for the seedlings. Similarly, in an intercrop of a legume with a cereal crop, the cereal would benefit from the nitrogen fixed by the legumes.
6. By having many crops growing simultaneously on the field the farmer is more or less buffered against failure of one of the crop.
7. Intercropping allows for a more uniform distribution of labour throughout the year.
8. When one component of an intercrop combination fails, the other combinations are able to utilize the resources that would have been available to the failed crop, and so yield better than they would have done otherwise. In other words there is yield stability.
9. The spread of diseases and pests is less rapid than in sole cropping. This is probably because the mean distances between the plants of the same component crops are greater. In many instances, the other component crops are not susceptible to the particular disease or pest afflicting one component and may act as physical barriers to the spread of diseases and pest.

Disadvantages of intercropping

1. Since many crops exist together on the field, it is not possible to tailor production practices to the needs of any particular crop.
2. Control of pests and diseases is particularly difficult because pesticides which have been developed to control a disease on one particular component crop may have deleterious effect on other crops in the combination.
3. It is difficult to mechanize operations such as planting, weeding and harvesting.

SOIL AND WATER CONSERVATION

Soil conservation – refers to prevention of soil from deterioration and loss, and using it within its capability and applying conservation practices needed for its protection and improvement. It does not only concern erosion control but also include the wise use of land (optimum use) and maintaining soil fertility through good treatment of the soil.

In sustainable land use, production must be combined with conservation of the resources it depends on. Therefore conservation efforts should

Control erosion sufficiently;

Maintain OM by covering the soil (mulch), leaving the crop residues, green manuring and application of compost or boma manure, among others

Maintain the soil physical properties like structure

Maintain appropriate levels of soil nutrients through rotation, prevention of water runoff, as well as performing required soil amendments.

What is soil erosion?

Soil erosion refers to the removal (detachment) and carrying away (transport) of the top fertile soil by either water or wind. It is common mostly in cultivated soils, steep slopes and where there is intensive rainfall or where the vegetation cover has been removed.

How to recognize soil erosion

There are several ways in which soil erosion can be recognized in a given area.of the signs of erosion include :

Appearance of soil furrows – linear depressions on the bare soil surface seen after rainfall.

Deposition of fine particles – occur where there are conservation trash lines, contour banks and flood plains. Crops may be covered by soil deposited where they are growing

Colour of running water – runoff looks brownish in colour of soil particles

Eventual appearance of gullies – deep trenches are formed which may be impassable, esp after long rains. Roots and left exposed, crops get uprooted and carried away.

Why conserve soil?

The main reasons for soil conservation include

To control soil erosion and prevent loss of valuable nutrients from the soil.

To enable soil fertility to build up in order to improve crop yields

To increase the value of land

To make terraces which make cultivation on steep slopes easier

To provide fodder from grass strips, hedge barriers and terrace embankments

To create employment opportunities for people in soil conservation work

Types of erosion

Raindrop or Splash erosion – occurs when raindrops fall on bare unprotected ground / soil. The impact on the soil splashes away the soil particles and digs a crater. Hence its defined as the detachment and airborne movement of small soil particles caused by the impact of raindrops on soil.

Sheet erosion (by overland flow) – Sheet erosion is the removal of a thin layer of soil from the surface (topsoil) by the force of running water – i.e. is caused by 'overland' flow moving uniformly across the surface and soil is removed more or less uniformly from every part of the slope. Occurs on farms with very gentle slopes / on hillsides.

Rill erosion. Rills are tiny channels. Caused by water runoff concentrated in tiny channels /rills – small linear depressions on the soil surface.

concentrated flow has greater erosive power. The concentration of running water causes rill erosion to be more erosive than sheet erosion. Rills are easily removed during land tillage.

Gulley erosion – Gully erosion occurs when larger quantities of runoff concentrate and create large channels in the landscape. Gullies have deeper and wider than rills. Are obstacles to tillage & cannot be removed by ordinary tillage practices. It is associated with landscape instability. It starts as small depressions on hillsides which concentrate water, and enlarge eventually forming a channel - as a result of localized weakening of the vegetation cover by grazing / or fire

Water conservation

Deals with methods to increase the amount of water stored in the soil profile by trapping or holding rain where it falls, or through harvesting / storage of excess run-off for subsequent use.

In arid lands the careful use of water is as much a part of organic growing as is any other technique.

There are many ways to use water carefully, including:

Trap runoff in dams (damming), use of terracing, pits, rain water basins or catchments or other water reservoirs – slow down running water, store water, increase infiltration & storage.

Contour ditches – help spread water out over a large area.

Maintaining protective cover of vegetation on the soil surface – slow down water & increase infiltration.

Careful irrigation.

The addition of organic matter to the soil to improve its ability to hold water

The use of mulches to hold water in the soil by stopping the soil surface from drying out or becoming too hot

Advantages of water conservation

It makes water available for crops, livestock and improves crop yields

Makes water available for domestic use over a long period of time.
Help control soil erosion, enabling the soil to maintain its fertility for better crop or pasture yields
Increases value of land – terraces makes it easy to cultivate on the steep slopes.

Soil and water conservation measures

Crop rotation – refers to combining plants of different families by growing them in a rotation and ensures 2 continuous yrs of planting a row crop be avoided. Benefits include: reduce soil erosion to tolerable levels, improve soil fertility & increase crop yields. These are achieved by 3 effects of crop rotation, including provision of good ground cover, Improve & maintain OM, soil fertility and by Improving soil structure / stable aggregates

Use of cover crops. Cover crops protect soil from impact of raindrops by interception, thus prevent crusting & increase infiltration. Also prevent blowing of the soil by reducing wind speeds. Cover cropping is normally carried out also with nitrogen-fixing crops that grow fast and require little or no inputs like water or additional manuring. While cover crops can yield some returns, they are mostly used for covering the soil in the fallow months, adding nitrogen to the soil, suppressing weeds, preventing soil erosion and later used as biomass or fodder. Fallowing with suitable cover crops conserves soil water, improves water use efficiency, weed control and soil organic matter. It is the most satisfactory method of building up the organic content of soil. Examples - Velvet bean - used as a fodder crop and biomass generator. Dolichos lablab which is a source of fodder and food. Others are hairy vetch, sweet clover, Lucerne, crotalaria, rye, oats, and peas. Others are mucuna, pueraria etc. Cover crops should give income (grain harvested) e.g. beans, peas, otherwise adoption = difficult. Disadvantage – shade, cost, competition for moisture & nutrients

Mulch Farming: Mulch farming is a system of maintaining a protective cover of vegetative residues such as straw, maize stalks, and stubbles on the soil surface at all times. Mulching is valuable where a satisfactory plant cover cannot be established at the time of year when erosion risk is greatest. The beneficial effects of mulching in erosion control include protection of the soil surface against raindrop impact, decrease in flow velocity by imparting roughness, and improved infiltration capacity. It also enhances burrowing activity of some species of earthworms (e.g. Hyperiodrilus spp. and Eudrilus spp. (Lal 1976)) which improves transmission of water through the soil profile (Aina 1984) and reduces surface crusting and runoff and improves soil moisture storage in the root zone. Apart from increased infiltration, mulch also moderates soil temperatures and also conserves water by reducing evaporation.

Strip Cropping: Row crops & protection effective crops (e.g. grass) are grown in alternating strips aligned at right angles to the direction of natural flow of runoff or perpendicular to the wind. Erosion is limited to the row crop strips while the close-growing strip slows down runoff and filters out soil washed from the land in the row crop strip.

Multiple cropping or intercropping, relay cropping, intercropping offers a two-storey canopy giving higher interception, and reduced soil loss
Agroforesty / Alley Cropping
Trees used preserve fertility & soil structure. Prevent erosion
Examples = leucaena, Prosopis chilensis, Acacia albida, Acacia nilotica
Alley cropping is an agro-forestry system integrating trees and shrubs with annual food crop production. In this system arable crops are grown in the spaces between rows of planted woody shrubs or trees, which are pruned during the cropping season to provide in situ green manure and to prevent shading of crops. The beneficial effects of the system include reducing erosion, surface runoff and soil moisture loss.

Trees / shrubs intercept raindrop impact thus reducing the amount of splash /raindrop erosion. On contours they intercept impede flow of runoff, increasing infiltration. Their roots bind the soil particles improving soil structure. They shade the soil, reducing its temperature and the amount of water evaporation from soil surface. Also break the wind reducing wind erosion. Trees also recycle nutrients from deep in the soil (to the surface), while leguminous trees fix nitrogen used by food crops. Success depends on the proper choice of the protective species. Example - maize with Gliricidia and Leucaena.

Applying organic matter: Regular application of OM to the soil provides the required plant nutrients for vigorous crop growth. This covers the ground quickly, protecting it from erosion and allows water to seep /infiltrate for use by crops leading to better yields. OM also improves soil structure (humus binds soil particles), and water holding capacity.

No-tillage Farming: Sound soil management is essential to the success of soil conservation practices because it ensures good plant cover. No-tillage involves a method of seeding through a crop residue mulch by opening a narrow slot in the soil for seed placement without mechanical or secondary tillage operations. The benefits include soil moisture conservation due to reduction in storm runoff, improved infiltration capacity, enhanced earthworm activity, and reduced evaporation loss. It also reduces soil erosion and maintains organic matter content at high levels.

Contour Farming: Contour farming involves ploughing, planting, and weeding along the contour. i.e. the plant rows and tillage lines are aligned at right angles to normal flow of runoff. It creates detention storage in the soil surface horizon and slows down the runoff thus giving the water time to infiltrate into the soil. The effectiveness of contour farming in water and soil conservation depends not only on the design of the system but also on soil, climate, slope aspect and land use. The beneficial effect is least marked on compact or slowly permeable soils because these become saturated quickly compared to highly permeable soil.

Ridges and Tied ridges: The ridge-furrow system is commonly used physical conservation practice for enhancing infiltration and reducing runoff. Ridge-and-furrow systems when aligned parallel to the contour lines have the dual purpose of erosion control and surface drainage. Ridging is widely practiced and is effective in soil and water conservation, especially contour ridging in the more humid regions and tied riding in the semi-arid and arid regions (for harvesting water & reduce erosion). In higher rainfall areas its used for growing potatoes / drainage. Are suitable for the less steep the terrain and in permeable soils. Ridge-tying is an effective soil and water conservation system especially in arid and semi-arid regions. Tied ridging on clay soils may induce waterlogging which may be followed by mass movement. Tying ridges or mounds to permit more rainwater to infiltrate is an effective system in drier areas (< 1000 mm annual rainfall) on gentle slopes (< 7%) but not in wet years or more humid areas.

Trashlines: Are made by laying crop residues or trash in lines along the contours. They slow down runoff and trap eroded soil, eventually forming terraces.

Contour bunds: Contour bunds are another important physical measure. These are earth banks, 1.5 to 2 m wide, forming buffer strips at 10 to 20 m intervals across the slope. Although little reduction in runoff occurs, there is considerable reduction in soil loss as demonstrated by Roose (1967) in West Africa.

Terracing: Terrace farming involves earthworks made at right angles to the steepest slope consisting of an excavated channel on the uphill side, the spoil from which forms a bank on the downhill side. It converts a steep slope into a series of steps with horizontal or nearly horizontal ledges and vertical walls of stone, brick or timber between ledges.

Stone bunds terraces – are useful in areas with steep slopes and many stones scattered all over the farm and where digging is not possible. The terrace risers are made of stones collected from the land. The terraces themselves can be sloping or level.

Cut-off drains: Cut-off drains are dug across a slope to intercept (excess) surface runoff and carry water safely to an outlet such as a canal, dam or stream. i.e. undesirable runoff is intercepted or diverted from an area above cultivated land and led away safely without causing erosion. This protects cultivated land from uncontrolled rainwater runoff. The runoff may cause gullies if not diverted.

Retention ditches / infiltration ditches : Ditches are dug along the contour. They catch & retain incoming runoff and hold it until it seeps into the ground. They are alternatives to cut-off drains when there is no nearby waterway to discharge the runoff into. Are used in ASALs to harvest water. Infiltration ditches are similar to retention ditches. Water from roadsides are diverted to infiltration ditches to ensure infiltration and retention.

Making check dams: A small check dam can be constructed on a farm where slope is quite prominent. A check dam is usually made along a gully where rainwater flows in large volumes and cuts and erodes the soil. It has a silt trap where sediment in runoff water is trapped. The check dam itself is constructed with easily available local material and the aim is to slow down water flow, reduce erosion and increase percolation.

Water retaining pits: A series of pits are dug into the ground where runoff normally occurs. The pits trap runoff water and encourage infiltration and storage. The soil from the pits is used to make banks around the pits.

Other Water harvesting methods
Micro catchments – semi circular loops
Trapezoidal bunds
Provision of swales: Swales are shallow, circular depressions around trees filled with mulch. They help direct rainwater towards the roots and also aid in reducing irrigation quantities for the same reason. The mulch also helps conserve moisture.

MODULE 5: PRINCIPLES OF CROP PRODUCTION (AGRONOMY)

Main Types Of Crops Produced In Eastern, Southern And Central Africa
Crop Husbandry: Introduction
Some Crops Commonly Produced In East, Central And Southern Africa.
Voulnteer Crops: The Indigenous Vegetables.
Routine Management Activities:
Some Crop Diseases:
Weeds:
Maturity:
Harvesting:
Record Keeping (One Needs Diary, Financial And Production Records
Yield:
Government Legislation On Import Of The Crop
Spinach
Climatic Requirements
Suitable Soil
Fertiliser Requirements
Sowing:
Routine Management Activities
Harvesting:
Marketing:
Pests
Diseases
Record Keeping (One Needs Diary, Financial And Production Records
Wheat
Rice:
Irish Potato (Potato- Solanum Tuberosum)
Domestication And Spread
Economic Importance
Ecological Requirements
Field Operations
Harvesting, Storage And Yield(S)
Growth Habits And Varieties Of Sweet Potatoes
Chemical Composition And Uses Of Sweet Potatoes
Bottlenecks To Sweet Potato Production:
Kenya Government Policy On Sweet Potato Production And Research Priorities
Cassava (Manihot Esculentum) - Manioc - Tapioca
Growth Habits And Importance Of Cassava
Ecological Requirements
Agronomic Requirements Of Tapioca
Box: Grandmother's Testimony On Casava
Possible Reasons For Weed Resistance By Cassava
Harvesting And Yield

MODULE 5: PRINCIPLES OF CROP PRODUCTION (AGRONOMY)

Figures 39: Left: The seeds of some crops. Right: The spinach crop

MAIN TYPES OF CROPS PRODUCED IN EASTERN, SOUTHERN AND CENTRAL AFRICA

CROP HUSBANDRY: INTRODUCTION

Crops are cultivated plants, including what may have germinated by themselves but are recognized by man as useful in some way. Man therefore normally takes care of her / his crops. Crops are useful sources of vitamins (vegetable and fruit crops), energy (cereals and tuber crops), proteins (pulses), etc. Most superpowers in history have built their strength and base on solid food security (mostly food self sufficiency). This is why to date, the Russian steppes, the Canadian prairies, and the corn belt of the USA are intact. These are their prime lands for cereal crop production. Upon this lie their sustained superiority and domination of the world. A country aspiring to dream of being strong and recognised should therefore start by training its people properly in food (cereal) production. Cereals are very useful and if taken in sufficient amounts, can provide almost 95 % of all the body needs, including vitamins and proteins. A student therefore becomes a useful member of a society if he / she has a sound basis for food production. There are broadly two classes of crops that fall in most syllabi. These are:

Cereal crops:

These include maize, sorghum, millet (Finger and bulrush), rice, wheat, barley, oats etc. Some are shown in the pictures below.

Cereal products

Figure 40: The sugarcane plantation

The cereals are also called the grass family, meaning that all grasses fall in this category. It is an important source of energy (Carbohydrate). Many countries in the developed world have spared large chunks of good quality land for them. These include: The Russian steppes; the Canadian Prairies for Wheat, and the Corn belt in the United States of America (USA) for corn / mealies / maize. The same applies to other developed countries, even though the names of the regions they spare for cereal production may not be very well known. It is a strategy the developing countries, especially African countries have to learn from. Without this vital energy source, the country is doomed. Once there is surplus, then it can be channeled to the industry for processing. Increased production will automatically lead to some level of industrialization, even if it is of low class or rudimentary. But all countries developed their now sophisticated technologies from those simple ideas. There surely must be a starting point.

Figure 41: The millets: Left: Finger millet (Eleucine indica) Right: Bullrush millet

Other crops, especially vegetable crops. These fall in many categories such as: (a) Fruit vegetables e.g. tomatoes, green pepper, pumpkins etc (b) Leafy vegetables e.g. spinach, kales / choumolia, rape, etc (c) Root vegetables e.g. potatoes (both Irish and sweet), cassava, carrots, raddish etc (d) Pulse or leguminous vegetables e.g. beans, cowpeas, french beans, etc.

Figure 42 :Left-Bird scaring in a rice field at West kano scheme, kano plains, Kisumu, Kenya; Right: Marketing of cereals

The cereals: Left-Sorghum and Right-maize

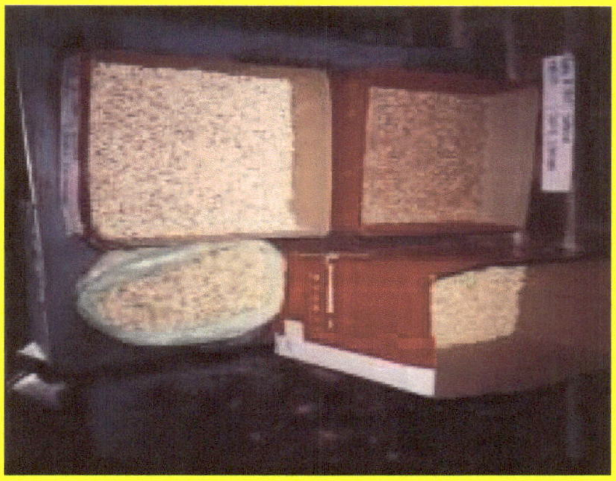

Figures 43: Left- sorghum crop; Right: An assortment of maize seeds The other side of the pumpkin- foe celebrating haloween end of October annually

Fruit products from fruit crops

Figure 44: Left- Tomato crop; Right- The green pepper crop

Figure 45: The barley crop in a field in Mau Narok, Kenya

Figure 46: Left- Spinach crop and assorted other fruits Right: Sweet potato crop in the field

Figure 47: Left: harvested Beetroot crop; Right: The pigeon peas crop in field (Cajanus cajan)

Figure 48: Left Cabbage crop in the field Right: The Green Peas crop in the field.

Figure 49: Healthy cabbage crops

SOME CROPS COMMONLY PRODUCED IN EASTERN, CENTRAL AND SOUTHERN AFRICA.

VOULNTEER CROPS: THE INDIGENOUS VEGETABLES.

There are many plants which volunteer to grow on their own,m and are used by man in many ways, but largely as food. These are indigenous vegetables, hereby refered to as volunteer crops. Traditionally, they have been refered to as weeds in typical agriculture classes. Due to the negative portrayal of them, they have been victims of extinction, since all efforts are put towards eradicating them in farms. Fortunaltely, after repreated failures of attempts to discard them, they are now bieing recognized by some magriculture ministries as foods. This subtopic attempts to bring out some of the nitty grities about indigenous vegetables among the Kenyan communities.

Useful indigenous crops sometimes mistaken as weeds

Spider plant

The nightshades

Most African societies have been very unique in their utilization of the many types of wild plants for different purposes. Kenyan's diverse ethnic strength should help it identify these edible wild plants (branded as weeds by modern agriculture) and utilize them for long term food security. The Luos, a Kenyan Nilotic group living majorly along the Lake Victoria basin, are a group with a strong traditional affiliation and cultural beliefs. These have helped it to be quite conservative, especially with regards to adopting some modern habits. May be good in the short rains, but quite sustainable. However, this culture as not been utilized very well partly because of the misplaced priorities and policies of the agricultural sector, and also due to Kenya's false start at the independence with a focus on the exotic crops.

The aim of intensifying studies of traditional vegetables is to broaden the base for agricultural development. The objectives include:
To stress the importance of indigenous vegetables in ensuring food security in the Lake Victoria basin of Kenya.
To stress the importance of indigenous knowledge of traditional vegetables in environmental protection.
To emphasize the importance of intermarrying traditional to modern agricultural and feeding in order to ascertain sustainability.

These are based on the premise that the more the number of plants species identified as important in agricultural development the less the risks involved in this effort. This is because more species offer wider genetic diversity which, in turn restrains man from branding literally any odd plant a weed. It should be remembered that a weed is a plant growing where it is not wanted (i.e. an odd plant, or a plant out of place). To put in simple terms, a weed is an undesired plant in modern agriculture. It is high time this notion is removed.

The fact that a weed is considered an hindrance or barrier to getting economic yields as gone a long way shaping man's attitude towards these "odd plants." This is because the immediate response to man towards them is irreversible, fatal i.e. killing those plants. A variety of methods are available for controlling plants out of the place. The most popular ones are cultural, chemical, biological and physical (mechanical). The physical method involves some extent of soil manipulation, therefore destroying the soil physical properties especially if heavy machinery are used when the soil is in a lubrication (liquid) phase. The extent of destruction is largely puddling, and is more or less permanent.

Chemical control of weeds is by use of herbicides (weedicides), chemical biocides that are applied on the plants. Their action is majorly by suppressing photosynthetic or respiration reactions, the most ….. Cellular metabolisms which are the backbone of life. Other herbicides simply kill the plant tissues due to their dehydrative (desiccating) nature. Sometimes the application of these herbicides is done by using heavy machinery such as tractors hitched traided to sprayers. The effects of chemical weed control are therefore two fold. First, the use of heavy machinery especially on saturated or poorly drained wet soils is bound to destroy the soil structure permanently, also the chemicals have a lot of undesirable effects on the biosphere, herbicides being biocides, are bound to kill the non- targeted flora and fauna, including man's economic crops. They have done this, and they will continue doing it. For one organism's food is another's poison, and a drug to yet others. There difference is their size and physiological differences which cause different dose, response, relationships and therefore response.

No wonder some of the organic pesticides, of which herbicides are a subset, are broken down as a source of carbon by some bacteria and macro flora. The same process requires that, if the herbicide has a high C/N ratio of more than soil, it requires a temporary immobilization of soil N to complete the decomposition. They create a N demand from the soil. This means that application of the chemical herbicides of high C/N ratio, especially those which are les volatile (resistant to solar radiation, heat and light) need a commensurate application of nitrogenous fertilizers to the crops being protected from the weeds. The volatile ones on their part are major atmosphere pollutants.

However, if it were actively growing leguminous crops being thus treated, the efforts would be counter productive. This is because at aslope such crops have their nodules heavily infected by the symbiotic Rhizoba sep whose activity is greatly suppressed by application of exogenous nitrogen. For short-term economic grains, food would be desirable, but definitely food is the interest of the long –term user/...... of such a soil with bacterial activity suppressed on unwarranted chemical applications, the soil health is greatly at risk.

Besides the above chemical impact on the soil upon herbicide application, there is the effect of no-biodegradation of some herbicides, as well as biomaphication among other organism. Non biodegradable of a chemical means its ability to stay intact in the environment for along time, or its derivatives (partial decomposition products). This is the chemical half-life is rather long. This is especially dangerous if the said materials have toxic active ingredients, for they will continue killing often organism as long as it is intact. Since different chemicals are more dangerous to some organisms than others, there is bound to be selective killing of flora and faun at class, genera, species, subspecies or variety. These inevitable changes the ecological balance of the environment botanically or zoologically, and/ or kills the environment altogether in the long runs .... To the food- web dynamics. This has obvious repercussion on balance of nature.

Bio-amplification (magnification) and bio-accumulation on the other hand, mean the build up of the chemical along the food web, and accumulation of herbicides in the living organisms respectively. For inorganic herbicides (and other pesticides) the accumulation is most prevalent in internal body organs especially the liver, kidney, lungs and the nervous system. Once in a while the body runs short of nutrients/ sugar in the blood (upon prolonged starvation) and the reserves stored on these tissues have to be mobilized. This gives the toxins, there in stored, a chance to get into the blood, and in turn into the cells where they eventually take their toll. Their effects are seen as malfunction of the most delicate tissues especially haemopoietic, erythropoietic and such processes. Obvious symptoms are then observed which in the long run result in death. These bio-processes involve majorly the fauna.

Cultural methods of weed control have a diverse traditional base. Mixed cropping, crop rotation, and agroforestry among other cultural practices long have been practiced the many tribes in the tropical world (i.e.Luos). They still have a big role to play in the weed control. Thus ... effects have been traditionally reduced. This applies to other types of weeds.

Biological control of weeds has been part and parcel of agriculture and social life of Luos, literally each volunteer plant had to use. These were especially for horticultural (either as fruits or vegetable), medicinal, or insecticidal purpose. Like the Bukhusu community, different age-group and garden command varied knowledge and use of different plants considered as needs by modern agriculturalists. This is unfair since only up to 10% of the local volunteer plants have unidentified uses. (Juma, 1998).

In the traditional Luo society there was a clear-cut division of labour among different age-groups and gender. The young men were responsible for taking care (cattle, sheep, and goats) and milking the animals. The young boys (under 10 years old) could accompany the young men to the grazing and watering points. The (young ones) particularly liked this job of being free riders, since they were actually engaged in gathering wild fruits and sometimes hunting. The variety of wild fruits fed on included Chwa, Ambidi, |Osangla, Akado, Olemo, Ochwoga, Oxalis, Cactus e.t.c. one could probably dismiss this as uncivilized and a mere remnant of development, since hunting and gathering are merely at the base of each developing agricultural economy. But we have to be real, call a spade a spade, therefore we have to accept the fact that we are basically agricultural state, with a majority of agricultural communities at the base of agricultural economy. Luos are not an exception; survival is largely by subsistence agriculture and to large extent hunting and gathering play vital roles.

Note that Luos are conservative lot and do not want to adopt new agricultural technology. Far from it! Luos have not been understood by modern agriculturalists, who rights at the onset are told that Luos occupy high potential areas of the country. So for their extension services/roles, the agricultural workers bring to the community wrong modern, inappropriate to changes such as hybrid maize, tea, coffee, e.t.c. these wrong packages are mainly attributed to colonial mentalities that Kenyan and West African countries were left with upon getting independence. The base of the agricultural sector is built upon exotic plant and animal resources, emphasizing large scale cash crop economy, beyond reach of the relatively helpless villagers. It is upon this background that most villagers in Luo community have always considered the modern agricultural technologies as inappropriate, as they tend to be biased, only emphasizing wrong crops. Most agricultural workers and administrators have misunderstood this as lack of innovativeness in the community. Indeed, it is the credibility of the workers that the community has always had questions about.

The other active groups, adult men engage majorly in building houses, granaries, cowshed fences e.t.c. due to this, they are very competent in identification of different tree species. From there experience of gathering wild fruits while younger, they still have rich knowledge of wild fruit species.

Among the plants used for building purposes are Dwele, Sisal, Pow, Osangla, and Ollalo

Those used for making furniture and shelves include Ober Albisia coriana, Chwa T. indica

Those used for fencing include Ojuok, Kwogo, Dwele, Okwato, and Cactus.

The other active age-group include teenage girls and young women who were trained on the types of the edible vegetables by elderly women. Among such vegetables, wild plant/Volunteer crops included Apoth, Osuga, Akeyo (Dek), Hariadho, Ododo, Bo, Nyabodhok, Ondhundho, Nyamtukru, Atipa, Okuro, Amara, e.t.c. In the lake belt of Kenya, these are some of the most prevalent field weeds. These groups of volunteers are important because, they are annuals or perennials therefore the most 'dangerous' weeds experienced by farmers. These are known to be some of the most serious volunteer crops, what modern agriculturalists call weeds. Indeed, they grow in large numbers, and vigorously so in Nyanza. Instead of bothering much to weed them/eradicate them, the Luos for centuries have used them as foods. This in turn has help reduced the intensity of weeding demanded by the planted crops, and has also increased the biological and economic yields of the forms, it has been a form of mixed cropping albeit at low scale.

Looked at in the modern sense of agriculture, it is a method of biological weed control. This writer proposes to call it biological control of volunteer crops. Since the so called weeds are simple volunteer crops with specific uses, a lot of economic losses have been incurred in trying to eradicate them. This has seriously resulted in reduction of farm returns.

In typical pasture, one would find mixed species of grasses and legumes. Some of the grass species such as Bothriochwa, Hyperrhania (Hatch grass), Wandi setaria (setaria spp), Guinea grass (Panicum maximum) e.t.c. which were used for thatching houses. Other grass species such as Panicum maximum, Love grass (Seteria vercilata),Star grass, Couch grass, themeda triandra, e.t.c. were favorites food for cattle; the shrubs were favoured by browsing goats. The harvesting process from atypical pasture would therefore be so orderly and complementary that it was considered a very convenient weed control process. Thatch grass would be harvest first, then animals largely goats and cattle are released to graze there. This could be done immediately after the rainy seasons. Nandi setana (setana spp) and Wise grass… remain in the field

Past studies of traditional agriculture of Luos

Not many studies have been done on the history of the Luo agriculture, the colonialist drew up their own plan of what crops they wanted grown in specific regions for Nyanza, they coerced the cultivation of crops and maize. To date this h seriously affected the agricultural history of the region.

Juma (1998) studied the indigenous vegetable and fruits of the Bukhusu community of Western province. He identified the area as a good center of diversity for a variety of these horticultural crops up to genera and species of indigenous vegetables were identified. This was largely due to the fact feeding is being closely related to tradition, is held in high esteem due to the religious and nutritional relational between some weeds and the culture of the community. No studies of traditional vegetables have been undertaken. AAS (1988) has undertaken studies among different Kenyan ethnic groups with no mention of feeding habits. No studies have been done on the possible uses of these weeds as indigenous vegetables.

Indigenous Traditional Vegetables And Their Processing/Cooking Techniques

Indigenous vegetables
Black nightshade (S. nigrium)
Spider weed-Sakia
Pig weeds (Amaranthis Spp)
Pigweed- Chepkarta
Commelna Banghalensis (Wandering Jew)
Amaranthus-Chepkarta
Stinging nettle (Urtica massaica)
Tomatoes
Leonotis Lepetifolia (Lion ear)
Altenathera pungers (Khaki weed)
Galiun soga parviflora (Macdonald eye)
Potatoes
Pumpkins leaves
Mexican marigold (tagetes minuta)
Oxalis latifolia
Tarchonanthus camphorates
Devil's horse whip
Spider weed
Oleraceac-kipche
Sweet potato leaves mapera
Chepkarta-Amaranthus (Pig weed)
Kisoyo-Black nightshade
Chepkit-Wandering Jew
Spider weed-Sagai
Tarchonanthus camphorates-Leleshwa
Black bean weed-Chepkut?
Chemitrit-Oxalis latifolia
Macdonald eye-Chepsitaki

KISOYO (Black mgutshade)/Osuga Solarium negrium

Is a traditional vegetables found in the farms during the rainy season. It is a better kind of vegetable. First it is boiled to remove the green bitter water from it then mix with other vegetables to make it taste sweet eaten with ugali cooked from millet.

It is harvested, cut and dried… sunlight for late consumption, or consumed.

It is also used as medicine to heal infected ears. Also used as livestock feed.

.. by the old people.

Cooked in a pot for about 1 hour, washed to remove the bitterness. Then it is ready for eating.

It is so tasteful than any other vegetable.

Grows well in a new cleared piece of land and in places where there is plenty of humus.

It is boiled or sold fresh.

A narrow leaved plant, growing well in the land tilled for the first time, and in the cowsheds.

Harvested by plucking its leaves.

Because it is bitter, it is cooked and pressed after boiling so as to remove the bitter taste.

It is eaten with ugali.

Used for appetizing ugali especially when taken also with meat.

Suitable for harvesting when has borne fruit

Dried in the sun for some days. Then on any place for some weeks for future use.

A good appetizer of food especially when a little milk is added. It is harvested when still green.

Grow mostly during dry season, and rarely during dry conditions. The surplus is harvested and sun dried, then put into sacks for storage. When it is to be re-used again, it is boiled for softness to be regain/recovered. It is mixed and cooked with meat. The disadvantage of drying is that it deprive the vegetable of its nutritional values; it also become less appetizing, sometimes only mixed with meat after being cooked.

Separately it gives a good taste.

They are sometimes mixed with other vegetables to reduce their bitterness, increase their taste and nutritional values.

After cooking, milk is added to reduce the bitterness, other people eat it raw (uncooked).

It can sometimes be mixed with other vegetables to make it soft (if it is harvested late).

It is used to prevent stomachache.

Boil in a lot of water, pour out the water and squeeze the boiled vegetable, then fry the product. It needs very little time from this staple to finish up its cooking some salt is added to improve taste. Contains a lot of vitamins.

Sometimes just boiled in water, add fat and a little milk and salt, and then boil for a few minutes. It is then ready for eating.

It is a wild growing crop especially within cultivated land, eaten with ugali.

Boil the vegetable in the pot, until most of the bitterness is removed. Press the bitter water away, cook briefly again, and add cream, milk and salt.

It needs moderately climatic conditions, fertile soil and enough rain. The plant is a volunteer crop.

After boiling for about 25-30 minutes, add salt and it is ready for consumption. If one wants to remove the bitter soup, pour the water out, add more water, and put the sufuria back to the fire place for a few minutes.

When the fruits open they become yellowish to orange in color. It is used for home consumption, and sometimes for local markets. The vegetable is washed and cooked for 30 minutes.

A traditional vegetable with small fruits and dark green in color. When the fruits ripen, children eat them, they are very delicious. It is boiled for 30 minutes.

Sometimes mixed with food such as gidheri. Boil squeeze out the liquid, add milk or butter fat and fry.

Can be mixed with other vegetables. Naturally grows in farms during rainy season; sometimes also during dry season. It is a bitter kind of vegetables. Boil to remove the green bitter water, and then add some vegetables to make it taste well, and then eat with millet ugali.

It adds taste to other vegetables; it is also medicinal plant sometimes sold commercially.

Are essential plants of meals, the old prefer not to eat if there is no such a vegetables because it has a lot of taste. Taken with ugali. Needs a specialist to cook, as it is very bitter, this bitterness is reduced by pouring out the first water used in boiling it and reboil again with fresh water.

It is dried in the sun for about 2 days then kept. It can even stay for one month depending on quality. When to be re-used, it boiled in warm water and thereafter eaten with ugali.

A drought resistant vegetable. After initial boiling, it is fried with tomatoes and other spices to make to taste delicious.

Kipchikin / Kipchigin / Kipchiken

A traditional vegetable available during dry season and during the rains. It is also a medicinal plant. A drought resistant vegetable, doing well in hot areas, it is found growing wild in the bushes and leaves used as vegetables.

It is the crop mainly in the Kerio valley and arrow regions. It is eaten with millet ugali and taste sweet today, it is also sold in markets. It is boiled, then cream added and then ready for eating

A drought resistant crop/vegetable. It has good pleasant taste. Sometimes eaten raw due its good taste. Exchanged with other foods during famine (these are food that are not locally available, i.e....trade). The turgen exchange the milk with people of the markwet who have the vegetable. This is because the turgen land is drier and do not have this vegetables when it is too dry.

A traditional vegetable found in the desert places. It is basically used for human consumption and sold in local markets; the very soft green leaves are picked and boiled for about 20 minutes.

A traditional vegetable found in the semi-arid areas, used for home consumption, and sold in the local markets. Harvested, cut, and the boiled for hours.

A traditional vegetable found in dry seasons and rainy seasons. It is a medicinal vegetable. It is dried for some days. After drying it is taken piecemeal and cooked with meat. It makes a delicious soup. It is normally prepared with animal... (head, legs e.t.c.). It stays for along time if cooked after drying.

Chepkaerta (Amaranthus spp)

Pig weed- Started to be eaten in 1970 among the turgens, but popular among the neighboring communities. Commonly found in the cowshed or fertile soils. It is considered a very dangerous weed. It is not drought resistant, but immediately the first rain start, it takes only a week or 2 to be ready for consumption (march/April). The vegetable takes a short time to cook, about 15 minutes to be ready. It is sometimes mixed with black nightshade to neutralize it, tomatoes and onion are also added. It can sometimes be sun-dried and stored in sacks to preserve it, later it is removed piecemeal and cooked, and eaten with ugali.

It is used as a livestock feed and home consumption. Found during rainy season mostly. It is not good to eat them alone if possible, but also eat with ugali. It is rich in vitamins. Used as vegetable to neutralize other vegetables or soften tough vegetables. It is boiled with very little water, it take very short time to get cooked. It is a soft vegetable that does not require high temperature boiling as it can be overcooked. Chepkaerta (Uhali weed) has medium size leaves; it grows in rainy season, and is in highland vegetable. It is boiled for as less as 5 minutes, then fried for 2 minutes, of small leaves, boiled for a short time (10 minutes) when used with or. Requires fertile soil, moderately cool and warm climate, and enough rain.

Used as vegetable or to neutralize acidic and bitter vegetables, picked from the shamba, wash the leaves, boil for about 10 minutes, pour off the water and add more water and salt –then boil again for 5 minutes. The vegetable is ready!. A common Kalejin vegetable, they are common weeds pick, wash, fry then apply milk or cream plus salt to be ready. Found in plenty during rainy seasons. Therefore either cooked with other vegetables or alone. It is washed, boiled in a pot, add a little water, and then fry.
The vegetable can be cooked in a little water, or use fat only. Add some salt after cooking the vegetable is cooked for a short time thereafter. It is used at times to feed rabbits; as well as soften other vegetables. Grow in areas of moderate humidity. It is picked when still very young, as it becomes less tasty with maturity/age. It is not stored (partly because when it is in season it is abundant). It is normally cooked with water during the day for only 15 minutes and salt added then kept to stay overnight in the morning or midday hours, milk is added or cream and salt, then warmed while stirring. It is then consumed.

A narrow leaved vegetable, growing both in highlands and well in animal sheds, and the land that have been tilled the previous year, leaves are edible. It flowers easily and producing a lot of seeds. When the plant eventually dries, a lot seeds fall to the ground. It is normally good when cooked with the vegetable, black nightshade to make it sweeter. It is normally soft when cooked, and some animals also eat it. It is taken with ugali.

Is a vegetable eaten with ugali. It is boiled, fried and eaten or cut and dried. A traditional vegetable that grows naturally in fields, regarded as a weed. It s not found commonly in the dry season. When rains begin, it germinates and grows very fast as a weed. It needs very little cooking time, it also requires only a little as too much makes it tasteless. It is mixed with other vegetables to make it tasty. It is highly popular among old mothers, and they are specialist in making it. It is not eaten raw by human beings, it is also an animal feed; mixed with pasture. It is a wild plant, growing very vigorously, and… it has invaded a place it easily spreads.

Found majorly during the rainy season. It is preferably eaten with other vegetables to make it tasty. It has a lot of mineral salts, and good for eating with ugali, but can also be eaten alone if necessary. It grows locally in very fertile areas especially where the livestock have vacated e.g. cattle boma.

It is a traditional vegetable, a weed of wet areas growing naturally in maize and other crop fields. It is cooked with a little water as it takes short time to cook (10 minutes). It is treated with onion, tomatoes, and cream. It is mixed with other vegetables to make it tasty and delicious.

Locally growing as a weed in ploughed lands. It is harvested, cut, dry and prepared for eating. It is used as a vegetable, used to neutralize bitter vegetables. Used also to soften tough vegetables, and also mixed with other vegetables to increase there taste.

Used as vegetable, as well as animal feed (to feed rabbits). It is used to neutralize bitter and acidic vegetables. It is boiled with water, stop the boiling for sometime, remove the water, add tomatoes onion and salt, and then mix with other vegetables. Since it is bitter it should be thoroughly washed before mixing with other types.

Needs cool climate and moderate rainfall; grows mostly in highland areas, it is a good rabbit feed and also human food; it is used to neutralize bitter and acidic vegetables. It is harvested boiled washed, boiled for 5 minutes, remove the water, add more water and let boil for another 5 minutes. Add salt, tomatoes, and onions, boil it lonely, remove from fire and serve.

MAIZE

Maize is a field crop. Other names are: Zea mays; semanga, mahindi, oduma, Bando, etc. Is a cereal crop produced for its cob which contains seeds rich in carbohydrates and a number of vitamins (especially vitamin B).

Varieties: Kalahari top cross, Kalahari early pearl, among others.

It is a high-potential-land crop from which to expect reasonable yields. Extremely low yielding varieties survive well in dryland areas (ASALS). Needs 800 - 1545 mm pa; 18 - 270 C and very fertile soils. In Kenya, it is produced in the Rift Valley and parts of central Kenya - Nakuru, Bungoma, Kakamega, Trans Nzoia, Nandi, Uasin Gishu and Marakwet districts. Maize yields here are the highest in the country.

The production of this crop in Kenya increased by an average 4.4% pa between 1963 and 1973. This was the first decade of Kenya's independence. Since then, production has greatly dwindled. Between 1980 and 1990, there was only a marginal increase in maize production. However, the situation worsened since 1991, with the import bill increasing annually. By 1994, crop forecast and food supply showed the lowest grain production in 10 years.

Figure 52: The maize cropin a field, and thoe mature crop being processed

This coincided with previous records of 1993/94 as being the worst drought and food insecurity period Kenya had ever faced in 50 years. Thus, the maize imports shot from 128,740 tons to 650,387 tons within the 1993-94 period. An equally high amount of wheat and rice were imported, with the previous imports being almost trebled in each case. Nobody even thought of alternative crops and foods of the dryland. Within this same critical period, the highest sugar, skimmed milk powder, cooking fat and oils imports were recorded. Maize consumption in Kenya is projected to grow to 40 million bags (3.6 million tons) by the year 2000. Many Kenyans (southerners) go for it. If it is not available, then normally the government declares a food crisis (famine) period and an import signal is switched on. India produces the crop very intensively to suffice its needs. In the 1994/5 period, it produced 192 million tons, enough to feed its over 870 million inhabitants. This was attributed to heavy chemical inputs, using even those globally declared environment health hazards decades ago. The current food crisis in many Sub-Saharan African countries is attributed to the narrow subsistence sector and sharp taste for maize. It is a high potential crop whose production in export crop producing countries has greatly dwindled. It has been relegated to the periphery, compared to cash crops. In the 1960's and 1970's maize production growth rate was 2.5 and 4.7% (global) and 2.4% and 3.1% (Kenya) respectively. For Kenya, these valves were far below the population growth rate of 3.3 -3.9%.

Climatic requirements:
Hot and sunny, with enough moisture is best. Temperatures of 10-30oC is the optimal range, depending on variety. Rainfall should be well distributed and reliable especially during flowering and grain filling stage (7-14 weeks after planting). Susceptible to frosts, and grows very slowly in winter. Rainfall should be at least 450 mm; yield increases with total rainfall availability.

Suitable soil:
Can do in many soil types, but should be fertile, well drained and clay loam, sandy loam to loam, with pH of 5.5-7.0. Needs soils rich in organic matter; does not withstand water-logging. Sandy soils need more irrigation and fertilizers

Land preparation:
Clearing, primary cultivation, secondary cultivation necessary with 1-4 weeks intervals in between. Maize needs medium tilth and free from weeds, pests and diseases. Flat to slight slope preferred as the crop is amenable to soil erosion.

Fertiliser requirements:

Needs basal dressing with any phosphate-rich fertilizer such as DSP, TSP, SSP, 2:3:2 or rock phosphate. 80 g/m2 of 2:3:2 or TSP is applied before planting, or 200 kg/ ha 2:3:2 before planting. 5-10 kg kraal manure / m2 is helpful (even in raw form). It is better to use composted organic matter; Chicken manure should not be used in raw form. Topdressing with SOA, CAN, 100 kg LAN/ha; 50 kg urea/ ha; etc important at 10-20 g/ plant at 6 weeks after emergence when crops almost 30 cm tall. May be repeated if necessary, depending on the variety, soil type, and OM content.

Table 6: Some locally available crops.

ASPECTS	CASH CROP	LEGUME	VEGETABLE
Crop	Sunflower	Cowpeas	Spinach
Cultivars	Russian no4	Black eye beans	Ford hook giant
Basal dressing	Super phosphate 170kg/ha	Super phosphate 250kg/ha	2:3:2 80g/m2every 2 weeks
	2:3:2180kg/ha		
Top dressing			2:3:2 70g/m2 every 2 weeks
Sowing method	Row planting	Row planting	Row planting
Spacing			
Between rows	75cm	30cm	35ccm
Depth	35cm	15-20cm	15-20cm
Common pests	3-4cm	1-2cm	2-3cm
Control/prev		Beetle larvae	Aphids
	Rust	Harvest before seed pods	Systemic pesticides
Common disease	Avoid wet weather	Open	
Control /prev	Flower turns brown	CAMV	Leaf spot
Signs of maturity	Back turns green to yellow	Control aphids/systemic	Avoid wet weather
	Manual or combine harvest	Pods light brown &dry	Leaves large &green
Harvesting	Dry flowers & remove seed		
Storage	Dry seeds <12%moisture	Remove pods,	Cut mature leaves with
Yields			Knife 2cm above soil
Use	1000kg –2000kg/ha	Coat with oil, dry & bag	Stand in water or wrap with plastic
By-product	Cooking oil		Human consumption
	Animal feed	Human consumption	Green manure
		Animal feed	

Sowing: Direct sowing is done at onset of rains at 45-75 cm apart, with 2-3 seeds per hole at 2-5 cm deep. These can be later thinned to 2 plants per hole. Gapping should be done when necessary. Can also plant during dry season where irrigation is possible.

Spacing: Depend on variety, soil type and climate. Generally range between 45- 90 cm X 45-90 cm. Closer spacing for smaller varieties in less fertile and dry areas; wider spacing for larger varieties in rich soils and good rainfall. Thinning normally done at first weeding.

ROUTINE MANAGEMENT ACTIVITIES:
Spinach is maintained by watering, disease and pest control, topdressing, weeding and mulching.

Harvesting:
Spinach is ready / mature when leaves start bending hind wards, normally at 30- 45 cm long. This is when the crop is 2-4 months old. The crop can continue producing for up to even 1 year depending on care. Harvesting should be done early in the morning using a sharp knife immediately above the soil surface.

Marketing:
Leaves tied up in bundles for marketing and kept in plastic bags to prevent wilting. Can also be kept in a cold room if operating on large scale, or at least a fridge to keep the harvest fresh for long. T is better to only harvest the little that will sell on the same day.

Pests:
Stalk borer (Buseola fusca) (Seboko sa letlhaka) digs through the funnel and weakens the stem, leading to crop fall. Apply Endosulfan when crop 20-30 cm tall. Also destroy crop residues.
American bollworm (these days called African bollworm) (Heliothus armigera)(Sebokwana) bores through the cob and causes it to rot, leading to Fusarium cob rot to infect the crop). Apply Endosulfan.

Armyworms-
African army-worm and the lesser army worms (Spodoptera exigua).
Eat the whole crop, starting with the leaves. Apply Malathion, carbaryl or Endosulfan (DDT was banned many years ago).

Armoured cricket / Corn cricket- (Acantholus speisei) (Setotojane):
Feeds on all stages of the crop, leading to total loss. Control is by weeding and clearing the fence so that no wild grass / plants are close to the crops. Can also construct deep ditches (20-30 cm deep along the boundary; they are so lazy thet they cannot climb these, and die. Can also use food baits e.g. carbaryl mixed with any cereal product.

Sudan diorch (Quelea Quelea) (Thaga, osogo):
These flock and feed on the grains at grain-filling stage. Are very destructive. Control by birds scaring. Can also stook once the crop has attained physiological maturity to avoid further losses.

Fgure 47: Sudan diorch (Quelea quelea), a common bird pest among cereals

SOME CROP DISEASES:

Leaf blight: This is viral disease which results of from aphid infection.

Fusarium cob rot: Is a fungal disease; pink- brown rot on the cobs; poisonous to man; (use dithane M 45, crop rotation; field sanitation).

Maize streak virus disease: Leaves of crop establish yellow- green strips; a viral disease transmitted by the leaf hopper (Cicadulina mbila). Control the leafhopper using any pesticide e.g. malathion, pyrethrin etc.

WEEDS: Striga spp – witch weed is a parasitic weed.

MATURITY: Maize is ready when the cobs become hard (non-milky), dry up, and leaves, silk and tassel dry up .

HARVESTING:
Is done either by stooking (30-40% moisture content), or by directly detaching the cobs from the stalk, then shelling and threshing. Ready after 3-4 months. Stored at 12% moisture content to prevent rotting. Store should be clean, dry and well ventilated.

Figure 50: Striga spp in one of the maize fields in Kasau Village, Usigu division, Bondo District, Kenya.

RECORD KEEPING (ONE NEEDS DIARY, FINANCIAL AND PRODUCTION RECORDS

Farm diary: record all major daily activities and observations

Financial record: used to estimate all expected expenses and returns- budget,

Production records: record of all physical quantities, sources and description of all inputs and outputs relevant to the enterprise.

Uses of the crop

Seeds boiled ant eaten; Seeds ground and used to make porridge; Seeds can be used as poultry feed; Cobs and stalks can be used as livestock feed

YIELD:

Maize yields depend on climate, soils, variety and management. May vary between 100 kg and 6 tonnes.

GOVERNMENT LEGISLATION ON IMPORT OF THE CROP

Both Kenya and Botswana governemnts have laws governing import and export of any maize material. The seeds, live plants etc have to be inspected first, and a certificate issued before import or export. This is to control diseases, weeds and pests.

SPINACH

Spinach is a leafy vegetable crop whose leaves are rich in minerals and vitamin C.

Figure 51: The spinach crop

Varieties recommended in many East and central African countries:

Ford Hook giant

CLIMATIC REQUIREMENTS

Temperatures of 10-30oC are best. It can resist light frosts. Grows best during winter due to less heat stress.

SUITABLE SOIL

Can do in many soil types, but should be fertile, well drained and sandy loam to loam, with pH of 5.8-7.

FERTILISER REQUIREMENTS

Needs basal dressing with any phosphate-rich fertilizer such as DSP, TSP, SSP, 2:3:2 or rock phosphate. 80 g/m2 of 2:3:2 or TSP applied before planting. 5-10 kg kraal manure /

m2 is helpful (even in raw form). Better to use composted organic matter; Chicken manure should not be used in raw form as it generates a lot of heat as the worms use the rich manure. Topdressing with SOA, CAN, LAN urea etc important at 10-20 g/ plant at 3-4 weeks intervals. This leads to large size of leaves.

SOWING:

Can be direct or indirect. Direct is done at onset of rains at 45-60 cm apart, with 2-3 seeds per hole. These can be later thinned to 2 plants per hole. Indirect planting involves use of seedlings and is recommended at onset of rains when plants 10-20 cm tall. One plant per hole okay, and gapping should be done when necessary.

ROUTINE MANAGEMENT ACTIVITIES

Spinach is maintained by watering, disease and pest control, topdressing, weeding and mulching.

HARVESTING:

Spinach is ready / mature when leaves start bending hind wards, normally at 30- 45 cm long. This is when the crop is 2-4 months old. The crops can continue producing for up to even 1 year depending on care. Harvesting should be done early in the morning using a sharp knife immediately above the soil surface.

MARKETING:

Leaves tied up in bundles for marketing and kept in plastic bags to prevent wilting. Can also be kept in a cold room if operating on large scale, or at least a fridge to keep the harvest fresh for long. T is better to only harvest the little that will sell on the same day.

PESTS

Cabbage moth (a caterpillar with light stripes on body; feed on underside of young leaves; use pyrethrin to control)

Aphids (under leaves of crop, dehydrate leaves) (use Dimethoate to control)

Cutworms (in soil, cut the stems of young seedlings; apply furadan in soil before or at planting).

DISEASES

Leaf curl (a viral disease which results of from aphid infection)

Leaf spot- a fungal disease; makes circular spots with dark edges on leaves, thereby lowers quality of marketable leaves; (use dithane M 45).

RECORD KEEPING (ONE NEEDS DIARY, FINANCIAL AND PRODUCTION RECORDS

Farm diary: record all major daily activities and observations

Financial record: used to estimate all expected expenses and returns- budget,

Production records: record of all physical quantities, sources and description of all inputs and outputs relevant to the enterprise.

Uses of the crop: Leaves cooked and eaten, rich in vitamin C and minerals.

By-products : Prunes leaves and uprooted stumps used as livestock feed.

Government Legislation on import of the crop. The governments within the Eastern, Southern and central African countires have laws governing import and export of plant materials, including spinach material. The entry points for most of these counties have plant inspection organizations, which would also issue importation or exportation certificates in case of approval. The seeds, live plants etc have to be inspected first, and a certificate issued before import or export. This is to control diseases, weeds and pests.

WHEAT

This grows best in temperate regions of the world. It is produced in large quantities and with high efficiency in Australia, former Soviet Union and United States. The economies of scale greatly enhance their efficiency. In Kenya, 1 ton is produced at 15,560/= while in US, it is produced at 12,537/= and in Argentina it is 9,625/=. India also produces wheat efficiently albeit of what is feared to be lower quality. It is affected at times by Karnal bunt fungus which causes yield losses and affect grain quality. However, they still proudly feed on what they produce.

Wheat crop

Canada currently produces 20 million tons of wheat in 10 million hectares. The Northern Africa wheat yields average 670 kg/ha, while Western Europe yields 5 tons/ha. Kenya's yield is around 1 ton/ha. There seems to be no prospects of improving Kenya's production since it is associated with large-scale plantation farming and high capital outlay. Thus, it remains a preserve of the rich, who have an access to a combine harvester. By 1989, the first and senior most Kenyan National University, the university of Nairobi had only one "stalled" combine harvester, yet in the same period, the private Narok, Mau Narok, Keiyo and Uasin Gishu wheat and barely farms had the same machines ranging to tens - live and operational. Thus the monopolistic nature of cultivation of this crop renders its market in Kenya very inefficient. Kenya only produces Durum wheat, which is also grown in Ethiopia, India and Egypt. This variety grows in relatively dryland regions; have higher straw; heat resistant, but susceptible to rust. It is soft wheat used for making paste, baking cakes and other soft wheat products.

The other regions, USA, Canada, Argentina, South Africa, Australia, Portugal, USSR etc. produce the hard wheat of the variety Triticale. It has short vegetative growth, long reproductive growth, high quality and high yield. The hard wheat does not crack on baking cakes and other soft wheat products. Since Kenya's wheat is almost entirely soft, it has to source the hard - wheat requirements from the North. Its annual wheat production ranges between 250,000 and 300,000t, against an annual demand of over 700,000t. The short fall is made good by imports. The crop is hardy, requiring 300 -1015 mm pa of precipitation and 6 - 200C. It is a hardy crop sown in well prepared lands in June and harvested in October. In North Africa, it is cultivated in the more temperate Morocco and Algeria. The bulk is produced in Russia, USA, Canada, France and India (In that order).

Wheat is the greatest cereal in global output with 387 million tons produced annually. Rice is the next with 367 million tons annual output, although it is the most widely utilized (fed-on) cereal. In the 1961/66 period, Kenya imported 0.5 million Kenya pounds worth of wheat. This amounted to 1% of the country's total foreign earnings. The trend has remained to date, but on a worsening side. More and more of the forex is being used to meet import requirements of selected cereals. The high proportion of these cereals imported in the total food imports of the African countries tends to reflect not a relative failure of indigenous staple food production, but a partial shift from locally produced foodstuffs. In the 1960's and 70's global wheat production growth rate was 2.7 and 3.5% respectively. Kenya's production was negligible on global scale.

RICE:

Originated in Asia. This is a water-loving crop that needs 1145 -1780 mm pa for good yields. There are, however dryland rice varieties that need less water. It is also a heat-loving crop performing best at 20-270 C. It is an indigenous tropical crop having originated in either India or China, both in Asia. 30% of China's land is under rice and this produces 40% of the world's output. China produces 36% of all rice of the world. China and India produces rice for local consumption whereas Thailand exports a lot. In Thailand about 50% of the land is under rice. Other chief producing countries include Indonesia, Bangladesh and Japan. Tanzania produces 75% of East African rice. In Kenya, it is produced largely in the great Kano plains in Nyanza province under irrigation. Currently only 52,000 ha are being utilized whereas the total irrigable land in Kenya is over 540,000 ha. There is also a clean, reliable water source Lake Victoria. Besides, the Kano plain is a large stretch of flat land, which can conveniently serve for irrigation use. This is only a wish, which can only be operational with political and economic will.

The production of rice in Kenya is coordinated by the National Irrigation Board (NIB), a government parastatal. The production inefficiencies are just but unbearable. Thus, the NIB tenants get less financial returns than their freelance rice-producing neighbors do. Secondly, the current production of rice in the two main Kano plain schemes - West Kano and Ahero is unsustainable due to the pollution of the water sources and the soil. There is too much fertilizer input which has greatly impoverished the irrigated soils. The concerns about environmental sustainability, conservation and management of natural resources have become a first priority. Unfortunately, the world is increasingly becoming complacent to them, the increased awareness notwithstanding.

Rice is one of the most important crops of the world. It gives a higher production of food per hectare than any other crop and can be grown every year in the same field without decline in yields. About 50% of the world's population, especially in Asia, depends on rice as the main food. China and India produce more than 50% of world rice. No wonder, they are most populous sub-continents, yet still able to feed themselves-unassisted.

Figure 53a-f: Rice crop in the field: Seeds broadcasted in a nursery; seedbed ready for transplanting; crop in the field; bird scaring in mature crop; harvesting and making of rice heaps

Figure 54: Rice crop: harvesting and making of rice heaps

ROOT AND TUBER CROPS

Root crops

IRISH POTATO (POTATO- Solanum tuberosum)

A crop of the rift valley, western and central parts of Kenya. In 1995, Nakuru district alone produced 89,000 tons and 1994 about 93,000 tons. Nakuru, Narok, Uasin Gishu are the main producers of the crop nationally. 17 districts in the rift valley produced in 1993, 1994 and 1995 a total of 156,000t, 20300t and 290,000t respectively. The average yield is 12 -15 t/ha. The main hindrance to increased production are bacterial diseases (soft rot), poorly available seed, poor producer prices, fungal disease etc. ADC (Agricultural Development Corporation) is the only seed producer in the country, but only produces 10% of farmers requirements. Very frequently, bacterial diseases destroy seeds in the Molo store, yet this prides to have the best potato storage facility in Africa.

The crop has great nutrient demands from the soil. Heavy fertilizer application is therefore inevitable. It is a cool season crop which needs 15 -200 C, and at least 1000 mm pa of rainfall uniformly distributed over the 4 month growing period of the crop. Unfortunately, Kenyan's taste has preferred this potato so much to the neglect and expense of other root and tuber crops. There is an urgent need to broaden the feeding habit so that the dryland root and tuber crops can also have their own niche in the food market of Kenya. These include cassava and sweet potatoes. Already there is competition for land between this crop and others which need high potential prime land for good yields. An emphasis should be shifted and urgently laid to crops that can produce well in dryland zones which is 80% of Kenya. The El Nino and Lanina phenomena are already too clear to Kenyans.

The Irish potato crop

SWEET POTATOES (IPOMOEA BATATAS)

Sweet potatoes, Ipomoea batatas, originated in Central America west of the Andes Mountains and parts of Latin America. These were also the centres of diversity.

Figure 55: The different varieties of sweet potato crop

DOMESTICATION AND SPREAD

It is believed that Ipomoea batatas was domesticated at the centres of origin. During the 16th Century, the new crops, one of which the crop is, were spreading throughout the tropics. By the Precambrian times, it had reached Polynesia. The sweet potato capsules are thought to have crossed the Pacific Ocean by floating from the Western Coast of the Americas to Asia. Human migrations intensified the crop introduction to other parts of the world, especially as part of trade.

In the Eastern African region, the origins look mysterious. The Luo of Kenya inhabiting Nyanza Province, believe that the crop was introduced at the Nyanza Lake shores by a type of Hippo called Nyaundi. This Hippo is believed to have deposited the crop capsules with its stool. The `seeds' then germinated into attractive crops that were very much liked by goats. Human beings living along the lake then became curious, and coupled with craze for food gathering, they decided to start uprooting the sweet potatoes. To their dismay, it had well developed tubers underground. They started using the tubers in different forms, and finally fully took it as part of their food. They started cultivating it. From the Pacific Ocean and Lake Victoria associations of sweet potatoes, it is clear that the crop has been widely spread via water-bodies. The crop is now a stable crop among many African and Asian communities. Only the two continents commercially produce the crop, with Africa producing 4% and Asia 92% of the total global production.

ECONOMIC IMPORTANCE

Sweet potato is the only economically important species of the more than 400 tropical members of the Ipomoea genus. It is of convolvulaceae family. Its tubers are a staple food among many in the sub-Saharan Africa, as well as Asia. The vines are used as livestock feed in many parts of the highland and temperate regions. Its tubers are also a source of raw material for many industries. The leaves are an important source of vitamins and minerals when utilized as vegetables. Once established, the crop needs minimum attention.

ECOLOGICAL REQUIREMENTS

Ipomoea batatas requires a warm climate of at least 20oC throughout the season. This is best available in areas of altitudes less than 1500 m above sea level. It does best as a tuber crop within the tropical region 30oN and 30oS of the equator. The crop requires between 750mm and 1000mm pa of precipitation. Of this, at least 500mm should be available during the active growing period. Sweet potatoes withstand drought, but low yields are experienced if drought strikes within the first 6 weeks of planting. This period is critical because tuberization occurs between the 4th and the 7th week of planting. The rest of the period is simply for tuber enlargement. Roots, which normally grow to 20cm - 2m deep, develop into tubers as a result of secondary growth. There are very few roots within the top 20-25cm.

Sweet potato does well in a wide range of soils - from sandy to sandy loam to clay loam. However, it is very susceptible to waterlogging, thereby rendering clay soils unsuitable for its production. The crop is a short-day plant requiring at most 13.5 hours of daylight in order to flower. This is typical of most tropical crops, most of which are either short-day or day-neutral. Tuberization promotes vine development, a quality also promoted by low temperatures. Since the latter is typical of highland or temperate environments, sweet potatoes for vine yield is very common in such cool, high altitude areas.

Day-length of at most 11 hours promote flowering. Since the critical yield component is the tuber, flowering is an insignificant phenomenon. Vines also act as planting material thereby rendering seed production and therefore reproductive growth merely a luxury. However, for production of new varieties with desirable traits, seed production in selected environments such as research institutions and seed multiplication centres can still be encouraged.

CROP ESTABLISHMENT

After thorough primary and secondary cultivation, the field is ready for sweet potato planting. Vines are used. The crop is unusual in the ease with which it can be propagated successfully. Apical cuttings of vine are used, and planted with 50-75% of their length buried. Crops establish whether their basal or apical ends of the cuttings are buried. The cuttings, usually 23-90cm long, are planted on flat land, ridges or mounds. Vines are placed 30-60cm apart in ridges 90-150cm apart, or 4 or 5 vines placed in mounds 1m apart.

The decision on whether to plant the sweet potato vines on mounds, ridges or flat land depends on many factors. These include soil depth, water table, drainage and cropping system. Sweet potato needs deep soils. Therefore, any soil that is less than 2m deep needs establishment of either mounds or ridges to avail enough pliable soil easily penetrated by the roots. The main function of plant roots is moisture absorption. The drier an environment is, the less the soil moisture and therefore the need for more elaborate root systems. A plant in a waterlogged soil therefore is liable to produce very weak root system. Since sweet potato tubers are simply enlarged roots, the roots need to have a sub-optimal soil moisture content to establish in large numbers. The stage that needs more moisture is tuberization - the enlargement or secondary thickening of the already established roots. Thus water logging is a hindrance to good tuber establishment if high tuber yield is the desired goal.

FIELD OPERATIONS

Application of farmyard manure (FYM) is the main known field operation. It improves yields tremendously. The fact that no inorganic fertilizer is recommended for application in a sweet potato field implies that it is very likely that such a research has never been conducted. Being largely a tropical and therefore a poor-man's crop, sweet potato seems to have had minimum research attention both nationally and internationally. The next essential field operation is weed control. Any form of wild or volunteer growth typical of tropical countries is found in a potato field. These are better controlled within the first one month of planting. The second month is very useful as far as tuberization is concerned. Thus the developing roots should be left intact from the second month. Sweet potato is very sensitive to weeding shocks it gets from the second month. After the second month, however, there is no need bothering about weeds. This is because sweet potato is a very effective fast growing cover crop which fast smothers any weeds. It is therefore hard to come by weeds in a well-established and timely wed sweet potato field by the third month.

Traditionally, the crop was established as an under-crop in a cereal field. It was planted after weeding the main crop. Cultivation of sweet potatoes is relatively labour saving. No other known field operation is undertaken other than weeding and FYM application. However, some light earthing up can be done to ensure that any tubers getting exposed to the soil surface due to soil displacement is covered. This saves them from being preyed on by man or rodents. Earthing up also reduces chances of the tubers becoming fibrous. Fibres develop in tubers if the soil temperature and therefore, moisture content is not stable. Direct heat from the sun exposes a tuber to a false stimulus of drought, thereby making it fibrous. This is an undesirable quality. Lastly, earthing up is done to ease tuber expansion.

HARVESTING, STORAGE AND YIELD(S)

Africa used to produce 5% and Asia 95% of the total global sweet potato production in 1970s. However, in 1980s Africa dropped to 4% and Asia 96%, produced in 11% and 85% of the total land areas respectively. In Asia, Korea is the major and most efficient sweet potato producer. The yields there average 16.3 tons/ha, while the global yield is 8.3 ton/ha. Africa's yield is a limping 3.9 tons/ha. Among other producing countries, sweet potato field yield range between 2.5 and 50 tons/ha. The high yield potential, coupled with high calorific density, render sweet potatoes a great famine - reserve crop. It is no wonder; therefore that it is popular to many.

Being sweet, it has good organoleptic acceptability. It also has disease, pest and drought resistance once established. This has made it popular globally. The advantages should be used positively to promote its production and yield. Harvesting is done by either uprooting the entire crop or removing individual tubers piece-meal. The crop is ready when the soil is seen cracking, showing the pressure exerted to the soil by a mature or fast expanding tuber. By opening up the cracks a little, the tubers can be either harvested or re-buried, depending on their sizes. Due to their succulent nature, sweet potato tubers have a short shelf life. However, they can keep for long if exposed to a temperature of 27-29.5o C and relative humidity of 85-90% for between 4-7days, then stored at 13-16oC at 85-90% Relative humidity. Drying and chopping can also improve their storage qualities.

GROWTH HABITS AND VARIETIES OF SWEET POTATOES

There are numerous sweet potato varieties, a majority of which are only traditionally known. Those with extensive shoot systems even in soils considered suitable for tuber establishment normally have poor tuber yields. They therefore best serve as livestock feeds especially the vines. Then there is a second more popular type among communities that consider sweet potato a staple food. This type has less shoot but elaborate root system from which tubers develop. Generally, the crop is a trailing vine with sweet, succulent tubers. The tubers are of different shapes, sizes and colours depending on the variety. Known varieties include Sifwembe, Nyakarika, Nyamayare, Nyaguta, Nyamilambo, Olombo jopidi. These varieties can again be classified on basis of their nature or colour of flesh and / or skin. The skin colour can be white, brown, yellow or red. The flesh colour, on the other hand, can be purple, white, yellow or a mixture. The sweet potato varieties can further be classified on basis of nature of the tuber flesh. The first group has a firm, dry mealy flesh after cooking. The second group has soft moist gelatinous flesh after cooking, while the third group has very course tuber only suitable for industrial use or livestock feed.

CHEMICAL COMPOSITION AND USES OF SWEET POTATOES

Sweet potato tubers have the following composition: (mg/100g fresh weight)
Ether extract (fats and oils) - 1.8-6.4mg; Moisture content - 50-81mg; Starch - 8-29mg (25% amylose, 75% amylopectin); Protein - 0.95-2.4mg; at times up to 39% crude proteins. (Protein comprises 2/3globulin has no tryptophan and sulphur amino acids)
Reducing sugars - 0.5-2.5mg; Calcium - 200mg; Iron - 3.0mg; Thiamine - 0.1mg; Nicotinic acid - 0.9mg; Carotene - 1-1.2mg; Riboflavin - 0.06-0.17mg; Ascorbic acid - 29-40mg.

Provides 7.4 million calories per hectare, with 96 kg protein/ha.
When boiled, sweet potato tubers become sweeter. This is because much of the available starch is converted into sucrose, making it sugary. This happens readily because much of the starch is in the easily hydrolysed amylopectin form and only 1/4 in the amylose form. The sweetness of the tubers and vines expose them to pest attack. Thus, their shelf lives are relatively short. To cope with these weaknesses while capitalising on the sweetness of the tubers, a number of processing techniques have been introduced. These techniques are closely related to the uses to which the tubers are put.

The tubers are often washed and boiled or chopped and boiled. It becomes soft and sweeter, a form in which it is readily used for breakfast with liquids such as tea or porridge. It can be used also for major meals such as lunch and supper. At times, traditionally, the Luo of Kenya used to chop, dry and grind the sweet potato tubers. The flour could then be mixed with that of other grains such as millet, sorghum and/or cassava and used for making any form of gruel or traditional brew. Any form of drink thus made from sweet potato powder was called 'AMARA' by the Luo.

When harvested and not to be put into immediate use, the sweet potatoes traditionally used to be preserved in different other ways. The intact uninjured tubers could be put into holes dug about 1m deep and covered with the soil. In that state, the tubers could keep for long, and only removed piece-meal for use. This method could at times be necessary during famine when it could be deemed necessary to harvest the entire crop from the field to save it from being stolen. Otherwise, field storage was a popular method among the Luo. In this method, harvesting was only done piece-meal as need arose. Yet another method of keeping harvested tubers involved washing, chopping and drying. In this form, the tubers could be soaked and boiled or ground depending on the intended use. These sweet potato processing technologies and uses are known by both Luhya and Luo communities of Western Kenya, some Uganda and Tanzania communities bordering Kenya. The crop thus seems to have deep association with Lake Victoria basin. The vines of sweet potatoes are used both as planting material and/or livestock feed.

BOTTLENECKS TO SWEET POTATO PRODUCTION:

i) The main pests of sweet potatoes include:

a) Squirrels and monkeys - eat tubers

b) Sweet potato weevil (Cylas spp) produces a vine and tuber-burrowing larvae, causing bitterness and odd colours of tubers. They also open up wounds for secondary infection by other pests and pathogens.

ii) The main diseases include:

a) Virus - B disease is transmitted by whiteflies (Bemissia spp). This causes distortions of the leaves, especially at the shoot apex, greatly reducing photosynthetic surface area and yield.

There are virus resistant potato varieties that can be effectively used against virus-B disease. In addition, control of whiteflies using any means is essential. There are many varieties resistant both to the disease and potato weevil. The knowledge of these remains a preserve of the traditional sweet potato producing and consuming communities.

iii) Besides the pests and disease, perishability is the next main problem of sweet potato production. There are varieties that cannot keep in the soil for long once they are ready. They either rot or become prone to Cylas Spp attack. Some tubers also crack, a sign of either irregular moisture availability and/or over maturity.

iv) The last main bottleneck to sweet potato production is lack of appropriate planting material. The crop is drought escaping as it has a very short productive life, therefore maturing very fast once planted on time. In the short run, therefore, there is sufficient tuber production. In Kenya, sweet potato production as a staple crop is practiced in the lowland zones very prone to environmental excesses such as droughts and floods. Thus, long droughts can easily render useless all the vines leading to an acute shortage of planting material in the next season. The weevil-infested vines are even more prone to drought.

At the onset of rains, it takes time to assemble sufficient planting material, frequently delaying planting in such areas. This tends to render ineffective integrated pest management (IPM) methods as the most environmentally friendly component - the cultural method is already compromised. The time lapse therefore significantly reduces crop yields, as the birch effect does not benefit them, besides other advantages of early planting. These problems call for research in a number of areas.

KENYA GOVERNMENT POLICY ON SWEET POTATO PRODUCTION AND RESEARCH PRIORITIES

Sweet potato, like other root crops such as cassava is listed as a famine reserve crop in Kenya Government documents. These include the National Development plans (1989 - 93, 94 - 96 and 97 - 2001) and Sessional Paper No (1986). It is described as a drought-escaping crop that suits the Kenya ASALS. It has of late in the 1990's been an official policy of Kenya Government to emphasis the utilisation of ASALS, which comprise at least 75% of the total land area of the country. This in itself speaks volumes about the importance attached to sweet potatoes. However there seems to be a contradiction in the government's approach to implementing the policies. KARI, the main government agricultural research body, rates sweet potato 37th out of 52 crops on research prioritization. This was in a 1990 - 2000 crops research prioritization in Kenya. Again, unlike other temperate crops, KARI has no formal research station for sweet potatoes (at least up to 1997). The fact that sweet potato is described as a subsistence crop with good production potentials means that it is not given a lot of importance.

Kenya has 50 sweet potato accessions, of which 32 (64%) are local. Thus, exotic accessions are only 36%, rendering East Africa a probable centre of origin of sweet potato besides the tropical America. Thus, there seems to be a lot of indigenous/folk knowledge of agronomic and processing packages of the crop. Thus, the existing yield gap 2.5T and 50T per hectare means that there is a lot in stock for agricultural extentionists. The yield component can be improved while also attending to the previously stated production hindrances. Since many traditional farmers seem to have internalised the cultivation and use of this crop, they should be involved in a collaborative research. This should aim at maximizing yields while increasing chemical quality, organoleptic acceptability, industrial use and processing techniques. Drought, pest and disease resistance, coupled with long shelf life, should be top in the sweet potato research.

It is due to some of the stated problems and poor attention by researchers that the yields and production in Africa remain low. The 1996 target of 556,220 tons has never been achieved to date. The 1992 production was 504,220 tons has against a target of over 550,00 tons.

CASSAVA (MANIHOT ESCULENTUM) - MANIOC - TAPIOCA

Cassava, otherwise known as Tapioca or manioc, is believed to have originated in North Eastern Brazil. However the centre of diversity is Latin America east of the Andes Mountains. It was the third crop developed by the ancient American farmers. Another possible centre of origin is Central America. The European exploration and expedition of the 13th Century helped in the introduction of the crop to Africa. This was reinforced by long periods of European settlement in the Tropical Africa in the 16th and 17th centuries. In the 16th Century, new world crops including cassava had reached West Africa. Like other root crops, it is believed to have evolved between the tropical forest and savanna zones of West Africa from where it was taken by the Portuguese to East Africa. Since the introduction, it has been virtually a peasant crop in the dryland. The most recent development in the dispersal of tropical crops has been associated with expansion of tropical research in the tropics coupled with international cooperation, which has included the exchange of seeds. This effort culminated in the establishment of IBPGR (International Board for Plant Genetic Resources) in 1970's. This collects, conserves and disperses plant genetic resources on global cases. This, however, conserves a negligible proportion of genetic resources since it is too urban - biased and urban - based.

GROWTH HABITS AND IMPORTANCE OF CASSAVA

Cassava is a famine reserve crop in many parts of the tropical world. It stores food in swollen underground structures whose yield is often thrice that of cereals. The starchy roots provide more food energy per unit land than any other known root crop. It provides 8.2 million calories per hectare, the second highest among all known food crops. It is second to sugarcane in productivity. It is only followed by sweet potatoes, which provide 7.4 million calories per hectare. The roots/tubers can normally store in the soil for as long as they are not immediately required for use. Thus even when mature, the crop can be left to stay in the field to await needy famine seasons. It is due to these that it is called a famine reserve crop. Since famine is more prevalent in drier environments, where Cassava also best performs, it even serves the purpose of famine reserve better. These qualities render cassava the most widely grown root crop globally and are one of the most dominant food crops of Africa especially in West and Central Africa. Since the tubers grow deep into the soil, it helps remove leached nutrients from those deeper soil zones. Thus it is an important nutrient recycling crop worth considering in a crop rotation cycle. The leaves of cassava are also a rich source of vitamins and minerals when utilized as a vegetable. It is a delicacy among many communities in Africa and Asia.

Figure 56: The cassava crop

ECOLOGICAL REQUIREMENTS

Cassava is a dryland crop capable of performing impressively even in highly impoverished soils. It is very drought resistant, rendering it a marginal and small-scale crop. Cassava best does at altitudes below 1500m above sea level. It requires about 1050 - 1500 mm of precipitation in very hot environments for maximum yield. However it is a drought resistant and does reasonably well even with 500 - 800 mm pa. It only needs good rains to establish. The temperature where manioc best performs is 25 - 290 C. Going by these climatic requirements, cassava is a typical tropical crop. It best performs in sunny weather in the tropical regions lying between 150N and 150S of the equator. It is very popular in the lake Victoria basin.

As far as edaphic conditions are concerned, tapioca requires light soils. They can be either sandy, sandy clay or sandy loam. Very compact soils of pure clay suppress tuber formation and development. Clay soils are also susceptible to water logging due to poor drainage, a condition to which cassava responds very negatively. It needs soft, pliable, well-drained soils of medium fertility. Soils that are fertile or generally have a lot of nitrogen are inappropriate for cassava production. High amounts of nitrogen tend to increase shoot growth at the expense of tuber growth resulting in low yields. Besides high nitrogen in the soil increases the concentration of prussic acid in both the leaves and the tubers. The nutritional significance of prussic acid will be discussed later. The main component of the crop normally utilized almost universally wherever cassava is cultivated are the tubers. These are products of secondary thickening of the previously laid down fibrous roots. Thus tuberisation is an increase in the girth of a root. It is affected by day length, rendering it a photoperiodic phenomenon. Under short day conditions, tuberization occurs readily. The process is delayed and low yields result when days lengths exceed 10 - 12 hours.

AGRONOMIC REQUIREMENTS OF TAPIOCA

The crop is propagated vegetatively by use of stem cuttings. Once a field is well prepared, free from weeds either ridges or holes are dug. Ridges are necessary where there is frequent water logging and/or soils are relatively shallow. Holes 40 - 50 cm long and slanting (i.e. deeper at one end than the other) are made uniformly 50 cm - 1 m apart. Cuttings from mature stems are cut each 30 - 45 cm long. They should have at least 4 node points with root and shoot initials. They are planted at the onset of rains at a slanting angle of at most 45o. The stem is then half buried by the soil. With right conditions, the cuttings soon develop roots and shoots within 1 -2 weeks. Water logging at this stage normally gives a very poor start, with roots only poorly developed. At worst, it causes rotting of the cuttings. This is why choice of land is important in successful establishment of cassava.

Response to fertilizers is said to be poor. However, positive contribution of Phosphatic fertilizers should be expected since this ensures proper and elaborate root establishment. It is only that formal researchers have never seen the need of applying inorganic fertilizers to cassava fields, one of the African crops openly neglected. The negative effects of nitrogenous fertilizers look possible and convincing. The prussic acid content on the tubers is proportional to soil nitrogen content and inversely proportional to soil potassium content. The roots develop much earlier in the life of the crop. Tuberization follows later with the established of roots enlarging. Tubers normally even force their way through rocks thus acquiring differently distorted shapes. Otherwise the normal shape of a cassava tuber is either spherical or cylindrical. Obviously, water logging renders useless the role of roots, resulting in their poor development. Thus where waterlogging is likely, proper drainage should be ensured.

It takes six to nine months for a cassava crop to be ready. Thus early weeding is essential to ensure proper shoot growth and utilization of available nutrients. Weeds intensify competition for light, minerals and water. Thus early weeding ensures that both the shoot and the root have no competition. In root crops, it is believed that with proper soil conditions, shoot biomass is directly proportional to tuber yield. The weeding aspect remains controversial in many tropical African countries. Many traditional cassava producing and consuming communities believe that weeding does not affect yield as long as the crop is cultivated as a famine reserve crop. It is not uncommon to come by heavily weed infested cassava fields in many rural environments. Such crops when given enough time to mature, normally to the tune of 1 year, produce handsomely. An old mother in Kasau village of Bondo District, Kenya had this experience (see box below). Weeding of cassava thus looks only advantageous from the ease of harvesting point of view. From yield point of view, it may play an insignificant role as long as the crop is left long enough to mature.

BOX: GRANDMOTHER'S TESTIMONY ON CASAVA

"My husband has 12 heads of cattle to take care of. You know, these days he tethers them. He used to lightly graze the animals in that lower field. However in April last year, I planted cassava in that lower field so he started bringing the cattle to my other weed infested paddock that still had some mature cassava. Both the goats and even the cattle were left free to graze here, untethered. Within no time, they had eaten up all the shoots of the crop. I abandoned the field. The livestock used that plot for eight good months. They left it for dead. I did not bother about the field again. However when the rains came this year, some cassava plants started coming up from that wasted overgrazed plot. Out of curiosity I went to dig wherever these young plants were sprouting. To my dismay from one site I harvested 10 tubers, each the size of the head of a mature human being."

POSSIBLE REASONS FOR WEED RESISTANCE BY CASSAVA

Cassava seems to be weed resistant. Weeds intensify competition for water and minerals from the soil. Thus for a plant to out - compete others in such a competitive environment, it must produce a very elaborate root system. Thus weeds intensify cassava root development which enables it to effectively absorb sufficient nutrients and water even in relatively marginal environments. If the soils are infertile, this is even more to the advantage of cassava at the expense of its competitors. This is so because cassava does well even in soils deemed too poor for other crops. The latter will show deficiency symptoms, the most common and immediate of which are stunting and chlorosis. With reduced photosynthetic efficiency and growth rate, cassava readily smothers the weeds. It is thus healthy even in relatively unhealthy soils. The physiological and biochemical explanations remain a mystery. However, the crop has an obvious advantage over others (weeds included) in relatively poor soils.

Once it smothers the weeds with well-established root system, the roots soon start undergoing secondary thickening (i.e. tuberization). It tuberises by actively utilising any rains to photosynthesis and storing the photosynthates in the well-established sinks, the roots. It thus utilizes the competition - free environment to maximize productivity and translocation. The roots thus enlarge uninhibited. It looks clear that as far as yields are concerned, weeding of cassava plays minimum, if any, role. However from management point of view, weeding is as important as any other operation. However, it should be done very early before active root establishment and enlargement sets in. Should late weeding of a cassava field be necessary, it should be very light and shallow. It should be accomplished by simply slashing. Alternatively a cover crop such as cowpeas or beans can be planted in a cassava field as an under crop. These can effectively smother weeds. Late deep weeding in cassava fields account for at least 10% of yield reduction. This is because there are risks of cutting, bruising and creating wounds on established tubers. When thus treated, the tubers rot or become watery. The wounds also become entries for secondary infections, thereby resulting in a disease infected crop.

HARVESTING AND YIELD

Depending on variety and ecological conditions, the crop matures in 4 - 9 months. When mature, cracks are often seen on the soil surface overlying the tubers. Harvesting can be done with either the entire plant removed or individual tubers harvested. Individual tubers can be located from the positions of the soil cracks from where they are dug out. Therefore if only a few tubers are to be harvested, even a crop with only a few soil cracks around its roots can be harvested from. However, if the whole crop is to be uprooted it is better only plants with at least 3 large soil cracks are uprooted. This is because it can be assumed that the plant has had most of its roots fully developed into tubers. The ones with a few cracks most likely still have many underdeveloped tubers. Thus uprooting everything at that stage substantially compromises the potential yield of the crop and the whole field.

It is therefore important to keep a cassava field clean to ensure efficient harvesting. Soft, well-pulverized soils tremendously ease harvesting since the whole crop can be pulled out with minimum, if any, loses. Almost all tubers come out attached to the stem cutting from which they developed. Then they can be individually removed. Whereas this method saves time, labour and produce, it is only possible in humid environments where soils are also moist most of the time. It happens that cassava is not a crop for which medium and high potential environments are spared. It is largely a dryland crop. Thus the use of hand tools and equipment such as jembes and hoes in cassava harvesting is very common. In many traditional rural settings, a field in which cassava is established is permanently a cassava field. This is because any single cassava plant harvested normally sees a new cutting planted at the exact spot from where the old crop has been uprooted. A stage thus reaches when in a single field, cassava at different growth and maturity stages all exist. The field is only regularly slashed, or lightly cultivated for an under crop to be established. Occasionally, it is burnt when it is too bushy.

YIELD

Cassava yields are as varied as the number of communities cultivating it, agro-ecological zones, varieties, prevailing food situation and cropping system. The inhabitants of Kambe and Ribe locations of Kilifi District, Kenya, plant the stem cuttings vertically. Thus the resulting yields are very low with only one large tuber from a single plant. The Luo of Homabay district have several cassava varieties, each with a distinct use for which it is produced. The Luo of Siaya district on the other hand has very few types and varieties of cassava. Thus the cassava cultivated in this area serves all purposes - boiling, grinding etc. Thus harvesting is often done even when crops are not mature due to the many competing uses. Again the cooking types of cassava yield much less than the processing types.

Similarly, in marginal areas such as Siaya, Bondo and Kitui districts famine is a very common occurrence. Therefore cassava is at times harvested long before maturity. This also happens in areas with many cassava thieves. These strike a field and harvest everything at night when there are any signs of mature crops. Whatever the production and harvesting circumstances, cassava yields are never a disappointment. It ranges between 2.5 and 25 tons/ha. In 1974, global and Africa's cassava yields were 9t/ha and 7.2 t/ha respectively. Root tubers have a relatively high harvesting index. This is the proportion of the biomass that is edible. They have 70% while cereals have between 20% and 60%. Thus not any photosynthetic time of a root crop is wasted; all is converted into edible matter. Cassava alone provides 8.2 million calories/ha, the second highest among all crops. It is second to sugarcane, a known C4 plant.

CHEMICAL COMPOSITION OF CASSAVA LEAVES AND TUBER

Both tubers and leaves of cassava are utilized as food. The cassava peel comprises 10 - 20% of the entire tuber yield. The cork layers on the other hand, comprise 0.5 - 2% of the total tuber weight. The edible fleshy portion is 80 - 90% with 20 - 25% starch. It has 2% digestible crude protein content and a gross crude protein level of 3.5%.

Leaves are rich in vitamin C (35 mg/100g);

The tuber is rich in Arginine but low in methionine, Phenylalanine, tyrosine, lysine and Tryptophan. These are amino acids.

Has 35g crude protein/kg fresh matter of tuber.

Contains 5.29 mg/kg total oxalate

Has 63% moisture content and provides 12.8 mJ /kg DM

Has prussic acid content of

a) 10 -490 mg/kg in fresh tuber depending on variety

b) 200 mg/kg in fresh leaves

Because it belongs to the Euphorbiaceae family, it contains laticifers, thus produces latex. This white fluid comes out of tuber peels, leaves or stems. This chemical reinforces its drought resistance and heat tolerance.

NUTRITIONAL SIGNIFICANCE OF PRUSSIC ACID IN CASSAVA

Cassava contains 2 major cyanogenic glucoside, linamarin and lotaustralin. The former is synthesized from Valine, an amino acid, while the latter is synthesized from Isoleucine. Both are highly water-soluble and decompose when heated to temperatures exceeding 1500C. Under the influence of an enzyme, linamarase (linase), also present in the cassava plant, both glucosides are broken down to prussic (Hydrocyanic) acid - in the intact plant.

RELATIONSHIP WITH PROTEIN SYNTHESIS AND RELATIVE CONCENTRATIONS:

The prussic acid produced by the hydrolysis may be used as a building block for the amino acid, asparagine. The concentration of prussic acid range between 10 and 490 mg/kg of tuber. The leaves, however, have a concentration of 200 mg/kg of fresh leaves. Leaves tend to contain more of the glucosides than tubers. The leaves also contain at least 100 times more linase than the tubers. Therefore there is a tendency for more prussic acid to be in leaves than in tubers. If 0.1 mg of the glucoside is taken in one meal, it is lethal. The concentration of prussic acid tends to vary with the level of soil fertility and the soil moisture content. Cassava growing in soils low in potassium or high in nitrogen has more prussic acid. Similarly, cassava grown in wet/moist regions has higher prussic acid than those in drier ecozones. Thus prussic acid concentration increases with moisture content of tubers. It is for this reason that cassava harvested in dry season is sweet, while the same cassava becomes bitter when harvested in wet season. In the Luo Nyanza, cassava tends to be sweet from September to April, while they tend to be bitter between May and August. The bitter taste in cassava is imparted by prussic acid.

PHYSIOLOGICAL EFFECTS OF PRUSSIC ACID, AND ITS DETOXICATION:
Prussic acid/hydrocyanic acid (HCN) is a respiratory poison. Coupled with the bitter taste it imparts into foods, most cassava processing techniques aim at reducing its concentration. It is also on basis of prussic acid concentration that some cassava are eaten fresh or after boiling, while others are only used after thorough drying. In South Nyanza, there is clear demarcation among varieties either used as food directly or thoroughly processed prior to use. Whereas a stranger would not dare feed on a cassava meal, those used to producing and utilising cassava in their homes even know what varieties and which part of cassava to eat raw. They know that 'selele' variety is the sweetest in both raw and boiled form since it has least prussic acid content. Even for those with relatively higher acid content, the rural folks know that the peel and the outermost layer of the peeled tuber have most of the acid. Thus they peel, and then scrape the outermost layer before eating such tubers raw. Otherwise one has stomach upsets and some temporary restlessness. This is common village knowledge. Boiling and drying is sufficient treatment, however.

VARIETIES AND KINDS OF CASSAVA
Traditional cassava producing and utilising communities know the varieties for eating and those for processing. The processing types are those that have to be chopped, dried and ground before use. The eating types, on the other hand, are used in their fresh form, raw or roasted. The processing type of tapioca tends to be bitter, seemingly because of high prussic acid content, with relatively higher yields. One plant can yield 50 - 100 t/ha. Some also grow into tall shrubs of even 4 - 5m. The eating type however tend to be sweeter, lower yielding, and between 1 - 2 m tall. Among known varieties are:
Nyandhaga (Cannabis Sativa (Njaga) like in leaf structure)
Otuga (otugo diep); Nyauganda; Kamisi; Selele - sweet type; tubers whitish in their coat, lower yields.

PROCESSING AND USE OF CASSAVA

There are numerous processing and cooking techniques of cassava. Each method depends on the intended use of the product. The fresh tuber has moisture content of up to 60%. This renders it perishable once harvested, as it becomes very prone to fungal attack. Many processing techniques therefore strive to either reduce the moisture content and/or the prussic acid content.

Boiled Cassava: Since prussic acid is not thermo-stable, it is greatly broken down by boiling. At a temperature of 730 C, the concentration of prussic acid in cassava tubers greatly reduce to non - lethal levels. The better if the tuber is boiled at this temperature until it becomes soft. Thus whatever the variety of cassava, boiling renders it safe for direct consumption. This is what will be referred to thereafter as boiled cassava. Prior to heat treatment, the fresh tuber is peeled, halved, the pith removed and divided into convenient sizes to enhance cooking. Smaller pieces cook faster.

b) Chips: Alternatively the cooking cassava can be peeled, halved, de-pithed and broken into finger like pieces called chips. These are washed, salted, left to rest for some minutes, then either deep-fried or half boiled before being deep-fried. This produces chips.

c) Crisps: A third method of processing the cooking cassava is the making of crisps. Once the tuber is peeled halved and de-pithed, thin paper like platy pieces are sliced out, dried and deep-fried. The result, called crisps, is very common in coastal towns in Kenya.

d) Marsh: A fourth form of processing of the eating cassava is done like this. After the obvious first three stages of processing, the tuber is sliced into conveniently small pieces, boiled thoroughly then cooked. This produces a mash.

e) Roast: The eating cassava can also be roasted and eaten as a roast. This is common among the Luo, but is an adored processing method among the Marakwet, a sub tribe of the kalenjin of Rift Valley of Kenya. The producers of this crop hail from the drier, hotter Kerio Valley - areas such as Tot and Aror. They simply roast the cassava and take them even to far markets such as Kapsowar, Cheptongei or Chebiemit for sale.

f) Cassava Flour: For the processing type of cassava, there is relatively higher prussic acid content besides the tubers being much larger. Therefore it undergoes similar processing pathways such as peeling, halving, chopping, drying and grinding. Before chopping, de-pithing can either be done or omitted depending on the community in question.

The resulting flour is normally very fine, with some sticky wheat like qualities when moistened. The flour can be used for making any form of gruel alone or mixed with one of the following, sorghum, millet, maize and sweet potato flour. Cassava can also be used in the industry to produce starch. Starch is used as a filler material or as a laboratory chemical.

USES OF PROCESSED CASSAVA.

The boiled cassava makes a good cassava meal. The chips are popular, very fast replacing the traditionally known Irish potato chips in many urban kiosks. Crisps are popular snacks in Coast province towns as well as main upcountry hotels as part of either snack or breakfast. The roast remains largely a home - based technique except among a few Kenyan communities eg the the Marakwet who take it to the public markets for sale. The flour can be used to make any form of gruel thick or thin (i.e. porridge or Ugali). It is also used in some ratio with wheat and/or sorghum, millet flour for baking.

For the leaves, they are rich source of vitamins and minerals when prepared as vegetables. However, first the prussic acid content has to be reduced to a convenient and safe level. This is done by first sun drying, then chopping or crushing and boiling the leaves for a few minutes. The resulting water is discarded as it contains the bulk of the acid. The resulting leaves are cooked in the normal way any vegetable is treated such as frying. It makes a cheap nourishing meal. After elucidating these important aspects of cassava, it is important at this stage to discuss the spatial (local and global) production and utilisation.

OIL CROPS: SIMSIM AND GROUNDNUTS SIMSIM (Sesamum indicum); Sesame

Background

Simsim is a wild plant in Africa. Better varieties seem to have been grown by African farmers for a very long time before its spread to Asia. It certainly seems to be a very old farm crop in Africa because of the ceremonies used for ensuring good yields in which it was involved. In Botswana, there are some wild varieties of Simsim, depicting it as a pure dryland crop. Indeed in Botswana, wild varieties of simsim exist in large numbers, showing that simsim might have originated from there, among other African countries.

Use of Simsim in cultural activities including songs was common. In an Acoli traditional poem, translated by Okot P' Bitek, line 4 reads: " Her teeth are white like dry season Simsim". This statement tells a lot about the African knowledge and use of Simsim. In Arthur Thomas's book (1967) entitled Farming in Hot climates, it is stated that in one African country the seed would not germinate in one experimental plot until some old ladies of the country were asked to grow it. Certainly, Africa seems to be the centre of origin (diversity) and domestication of Simsim. The knowledge of its production processing and use remains a preserve of Africans.

Figure 57:The simsim crop (Sesamum indicum)

Figure 58: The wild simsim in southern Africa

ECONOMIC IMPORTANCE

Simsim is an oil, proteinous crop traditionally grown in Africa for many cultural uses. It is grown for confectionery purposes. It is one of the richest oilseeds known, with a concentration of 45%. The protein content is also very high being 20% and 1.5 % Calcium. Thus in terms of proteins, energy and minerals it rates very highly. It is of great industrial potential for extraction of oil. Considering the dangerous levels of Protein Energy Malnutrition (PEM) and desertification in sub - Saharan Africa, Simsim can be a very important compromise crop. It is drought resistant and drought escaping due to a short maturing period, and a good source of essential fatty acids. It can give sesame cakes and oils thereby acting as a source of food and feed. Like its groundnut counterpart Simsim provides essential fatty acids such as linoleic, linolenic, arachidoic and arachidonic acids. Coupled with the high energy density, it is a crop worth producing in bulk.

ECOLOGICAL REQUIREMENTS

Simsim is an extremely drought resistant crop. It needs 400 mm pa during the growing season, though it grows best with 750 mm pa. It requires warm climates of temperatures at least 20 C. This is most common in attitudes less than 1500 m ASL. It is intolerant to water logging. Due to the attitude and temperatures requirements, Simsim best does in

Coast and Nyanza provinces and to some extent Western province of Kenya. Simsim can do reasonably well even in impoverished soils. This advantage, coupled with drought resistance and short maturity period, makes it a crop of the semi-arid parts of Kenya, which is at least 75% of the total area. Since aridity is fast catching up even with the hitherto moist environments, it is a crop of the future.

AGRONOMIC REQUIREMENTS

The seeds of Simsim are very tiny. This requires that the seedbed meant for the crop be very fine. However, when it is extremely fine, winds tend to blow away many seeds leading to losses. In addition, the seeds become prone to pests, as it is a delicacy to many kinds of birds. Therefore, a relatively rough seedbed is needed. Care needs to be taken to ensure that the tilth is not too rough as the clods tend to cover many seeds. This leads to lots of losses as well as seedling abortion. This results when the seed gets the right germination conditions but due to excess soil cover fails to emerge. The failure to emerge arises from exhaustion of the stored nutrients, which at some stage needs to be supplemented with photosynthates. If the soil/clod cover exceeds the penetrating power of the seedling, no light reaches the seedling resulting in its abortion/death. Therefore a rough seedbed of moderate tilth is needed.

The crop, with very small seeds, is hard to sow by hill placement. Therefore broadcasting is the most common and indeed the oldest sowing method. The seed size has made the mechanizations of sesame production hard. It is therefore still entirely produced manually. The seeds are sown early in the rainy season at a rate of 5.5 - 9 kg/ha. The sowing time should be such that the crop gets 1½ - 2 months of rain, followed by a dry season for seed maturity, ripening and drying. It can be weeded twice during its production season. However it is relatively tolerant to weed infestation due to its fast growth and maturity.

HARVESTING AND YIELD

When ready, some leaves start yellowing and falling. Timing should be proper because a lot of losses can be incurred. The whole plant is uprooted or cut at the lower end with a Panga. They are then hung by tying up on a rack so that as the capsules dry up and crack, seeds may fall out. After 2 - 4 weeks, depending on the stage of harvesting, all the capsules have dried up. Some may even have started opening up. Therefore a mat has to be spread below the rack where the mature plants have been hung.

Threshing is done by beating (hitting) the hung plants with a stick, causing the capsules to break open and release their contents. These fall on the mat. Further processing involves winnowing to remove any chaff from the rest of the seeds. These are then dried and stored. If well taken care of, up to 1.5 tons of seeds per hectare can be yielded. However, the most common yield is 500 kg/ha. This is terribly low. The low sesame yields can partly be attributed to pests and partly to the sub-optimal seed rate. The latter arises from the fact that the crop is largely produced under mixed crop system. At times the area covered by rocks and tree stumps are never considered.\

STORAGE

Simsim seeds are very small and platy. Therefore they dry up very readily and store very well. Their small size is not conducive for many storage pests. Traditionally Simsim used to be kept in pit or cave. This is why in the tale of "Ali Baba and the forty thieves" the password was 'open sesame'. Sesame and Gingerly are other names of Simsim. It can also be kept in gourds and/or earthen pots.

PROCESSING

The Simsim seeds can be roasted. The result can be

mixed with roasted groundnuts and sugar to make a snack called Simsim common in Bondo town and the Line Saba stage in Kibera.

finely ground/pound to make Odi. Odi is either used directly as a salad with Ugali or added to vegetables.

CHEMICAL COMPOSITION OF SESAME

This is on per 100g DM basis of edible food:

- 2400 KJ (2.4 MJ)
- 20 g proteins. The proteins are rich in Arginine, Leucine and methionine but low in lysine.
- 1500 mg calcium (15%)
- 15 mg Iron (liver has 12 mg iron per 100 g).
- 0.4 mg Thiamin
- 45 g oils

Sesame seeds are therefore important energy, protein and mineral sources and to a less extent, vitamins. These qualities make it a good crop for cottage industries for oil extraction and simple processing, as it is very rich in energy.

PRODUCTION LIMITATIONS

A number of factors hinder large scale and efficient production of sesame. These are:

i) Pests

a) Simsim webworm (Antigastria catalaunalis). This affects the terminal leaves, tender shoots and pods, which are damaged. The terminal leaves are webbed together with silk, greatly retarding growth of the terminal bud. In effect, this reduces pod production and development. The caterpillars reach 13 mm long, are green bodied with black head.

b) Fleas beetles (Aphthoria bimaculata) eat the foliage.

ii) Diseases

Leaf spot - a fungal disease caused by Cercospora sesame and Alternaria spp

Marketing:

There is no formal marketing institution of Simsim in Kenya and many countries. Its production therefore largely remains at subsistence level.

Research:

No much research is done on Simsim productivity. The agronomic packages largely remain traditional. KARI rates Simsim in the 27th position among crops research prioritization 1990 - 2000. It is thus not considered important to the economy of the country.

GROUNDNUT (ARACHIS HYPOGEA)

Groundnuts originated in South America East of the Andes Mountains. This is the same centre of origin for cassava, cocoa, rubber and pineapple. At least 90 wild species of groundnuts exist here. his crop was domesticated in South America. The specific areas include South Western Brazil, Northern Argentina, Paraguay, Bolivia, Eastern Peru and the neighboring districts. Out of the 90 species, only 15 are described formally.

SPREAD

Groundnut was never known in the new world in the pre-Colombian times. It was taken from Brazil to West Africa by the Portuguese and from Peru to the Philippines by the Spanish. Since then, groundnuts spread throughout the tropics and sub tropics to latitude 40oN and 40oS of equator where rain, during the growing season, is more than 500 mm pa. In the 16th century, groundnut value was recognized by the Portuguese who introduced it into Africa. In Kenya it was introduced by the British in 1920's and 1930's. In the latter years, the colonialists were providing free seeds to Kenyan indigenous communities. Ordinary groundnut was brought from America to West Africa over 380 years ago. It is currently grown in the drier parts of most hot countries like USA, which have long hot summers. Other producers are West Africa, India, China, East Indies and Coromandel Coast.

SPATIAL DISTRIBUTION

Asia and Africa produce the bulk of the crop on global scale. In the 1970's India was producing at least 25% of the total groundnut. It is still a main producer. West Africa produces the bulk of the crop in Africa. It is the main export crop in Liberia. Nigeria's export contributions of groundnuts also rates at 20% per annum. The country produces 7% of global groundnut produce; Sudan contributes about 3% and Senegal 4%. The production in Africa has not been very steady because of quality restrictions in the importing countries in Europe. However, it is the most important Africa oil seed and form the chief crop of Senegal and Gambia.

GROWTH HABITS

Groundnut is a warm season leguminous cover crop that takes 4 - 5 months to mature. It is relatively drought resistant with a reflective coefficient (albedo) of 0.17. It produces flowers after 2 months. Upon fertilization, the pods containing fertilized ovules bend downwards and penetrate the soil with their sharp pegged ends. This develop into mature fruits. The number of edible nuts per pod range between 1 and 6.

ECOLOGICAL REQUIREMENTS

Arachis hypogea needs good rainfall, being a warm season tropical crop of humid environments. The best amount of rainfall is 600 - 1200 mm pa, with a good amount falling during the active growing period (500 mm). It only needs 3 good months of rainfall followed by at least 3 weeks of drought to enhance crop maturity and drying. Reliable rainfall especially during flowering is important to ease pegging, a process that largely relies on softness and moistness of the soil. Groundnut needs a warm climate throughout its growth period. Temperature requirements therefore range between 20 - 30oC. However, the higher the temperature, the higher the transpiration losses, necessitating the availability of a lot of reliable rainfall.

Light soils are most preferred by groundnuts because this enhances pegging. This is the penetration of the soil surface by a groundnut peg. A peg is a young conical shaped pod resulting from fertilization of the ovary and with a sharp peg like tip to enhance its penetration. Heavy clay soils depress pegging and therefore maturity of fruits. Thus sandy, sandy loam or loamy soils are the best. They need to be well pulverized, well drained and fertile. Waterlogging is a hindrance to groundnut production. It depresses both the growth and pegging of the crop. Thus the need for ridges in soils of poor drainage.

AGRONOMIC REQUIREMENTS

Arachis hypogea needs a rough tilth since the seeds are large. The seedbed needs to be well prepared with both primary and secondary cultivation well done. The soil should have rested enough between secondary cultivation and planting to ensure that soil borne pests and pathogens are exposed to the harsh sunshine. This reduces their numbers. This is an important cultural method of pest and disease control. Tertiary cultivation such as ridging is essential if the soil is prone to waterlogging. Planting is done by making holes, 10.0 cm apart on both sides of the ridges. If ridging is not necessary, holes are dug on flat land of a spacing of 15 cm x 9 cm. if mixed cropping is desired. If mono-cropping a pure stand of groundnuts is planned, then a spacing of 60 cm x 15 cm or 45 cm x 15 cm is appropriate. 2 - 3 seeds are placed in each hole, covered and the soil compacted over the planting holes. This is essential to ensure proper seed coverage by the soil as well as good contact between the soil and the seed. The seed soil contact enhances germination since the seed soon starts exerting osmotic force and therefore absorbing moisture. The erect type needs closer spacing.

The seeds germinate within 5 - 10 days. Weed control should be ensured from the earliest time possible. Up to 3 weedings are normally done in many Nyanza groundnut-producing areas. This is necessary to enforce minimum crop - weed competition and to ease harvesting. Clean field at harvesting time ensures maximum retrieval of pods, and therefore minimum crop loss. This maximizes yields. Harvesting is done when 50% of the crops have had their foliage drying up. At this stage, a good proportion of pods of matured and very few, if any, have started detaching themselves from the mother plants. The detached pods often remain in the soil and re-grow in the following season as if another crop has been planted. Timely harvesting greatly reduces these over - maturity losses. The whole plant is uprooted using jembes, hoes or any other implement. They are spread on the soil surface to dry up for 2 - 3 days. This is why it is recommended that harvesting be done during the dry sunny months. Then individual pods are plucked, further dried, shelled and the seeds stored. The yield depends on variety and management, and averages 1 ton/ha.

POST HARVEST PROCESSING, STORAGE AND USES.

Once the nuts are removed from the pods, they are dried to moisture content of 12 - 15 %. It takes a long time to fully dry up the nuts because of the high oil content. However, dry seeds store better and for a long time. The nuts can be put to any of the following processing procedures:

1) Roasting: The nuts are sorted, moistened put in a heating pan and regularly stirred (mixed). When patches of the nuts have turned dark, they are ready. This is the roasted groundnut.

2) Pound Ground Nut: The nuts are first roasted before being pound by a pestle and mortar. It acquires pasty foam as the peanut butter. This is the pound ground nut or peanut butter.

3) Rough Pounding: The roasted nuts are roughly pounded, taking a rough pasty but with clodded form.

4) Groundnut Flour: The nuts are roasted and ground or milled. In grinding, the nuts are used wholly without any parts removed. In milling, the testa is first removed after roasting before being ground. The resulting powder is either rough or fine, depending on whether it is produced by grinding or milling.

USES

When processed to the different forms 1 - 4 above, each can be put to many uses. The roasted groundnut is good snack. The pounded groundnut or peanut butter is used to paste bread. It is also used for cooking many kinds of vegetables. Once a vegetable is boiled, and some water removed, the pounded groundnut paste is added as well as salt. This is a complete meal. The peanut butter is traditionally called 'Odi' by the Luo. Odi was a popular dish taken with Ugali (thick gruel) i.e. as a sauce. Roughly pounded groundnut is traditionally used for preparing vegetable dishes. After boiling the vegetables, the pounded groundnut is added. This compensates for frying. Groundnut flour is used to make a dish called 'Dengu'. The flour is mixed with milk and boiled gently but thoroughly. When prepared in a cooking pot, it is very delicious. It is a popular vegetable in South Nyanza and Bondo districts, especially Uyoma.

From the industrial point of view, both oils and groundnut cakes can be extracted from the nuts. The oil contains many essential fatty acids needed by man, while the cake is popular animal concentrate feed. It is highly energy packed. The unshelled immature pods can also be boiled and eaten. Shelled, dried nuts can also be eaten raw. However, some people are allergic to the raw nuts, suffering from stomach upsets upon use. The uses depend on varieties. Different varieties have varying levels of oils, proteins and other essential fatty and amino acids.

STORAGE

The most important reason for keenly observing storage conditions of groundnuts is to keep them from fungal infections. This can take the form of pod or nut rotting to aflatoxin accumulation. The nutritional significance of aflatoxin is discussed later. Groundnut is an oil rich leguminous crop. It is for these two purposes - oil and protein content that it is produced. The details of chemical constitution is as follows on 100g dry matter basis:

Table 8: Chemical composition of groundnuts (per 100 g dry matter)

2.6 g proteins (i.e. 26%)	17.5 mg nicotinic acid	0.8 mg thiamine
46 g lipids (oils i.e. 46%)	0.2 g riboflavin	50 mg calcium
2.34 MJ		

The proteins have sub optimal amounts of cystine and methionine both being sulphur amino acids. The first limiting amino acid is lysine. The groundnut oils are unique in that they supply essential fatty acids such as:

(1) Linoleic C17H31COOOH, melting point - 5oC and

(2) Arachidonic acids.

Other essential fatty acids that other seed-oils/oil crops provide in large quantities are arachidoic and linolenic. Groundnuts supply only a small quantity of these. In as much as groundnuts are important in supplying essential fatty acids, the acids are largely an unsaturated type that induce vitamin E deficiency/shortage if consumed in excess. Groundnut cake is a good livestock concentrate feed. However, it should not form more than 25% of the diet in pigs, as it tends to produce soft body parts and a laxative action. The latter is largely because the proportion of roughage (indigestible matter) is very low. Thus if heavily relied on, it weakens the GIT functioning.

Groundnut contents also have a growth factor and an anti-trypsin factor. Thus some individuals feel stomach upsets after feeding on groundnut meal, especially when raw. The anti-trypsin factor also has an anti-plasmin activity. Thus it shortens bleeding time (i.e. enhances blood clotting). The factor is destroyed by heating, thereby rendering roasted or cooked groundnuts safe. Those allergic to raw groundnuts should first roast them before use.

AFLATOXIN:

Groundnuts also contain a poison called aflatoxin if infected by a fungus Aflatoxin flavus. The infection can take place in the field or in the store. Aflatoxin is classified as Aflatoxin B1, G1, B2 and G2. Of these, Aflatoxin B1 is the most potent. It causes liver damage with marked bile duct proliferation. It also causes hepatic tumors and liver necrosis. The latter is the drying of liver cells. Aflatoxin is also a very active carcinogen. A metabolite called Aflatoxin M has been shown to be present in milk of cows fed on toxic meals. Aflatoxin M also causes liver damage to ducklings. It is thermostable, rendering aflatoxin-infected nuts permanently toxic. Oils and fats (lipids) produce more energy than their carbohydrate counterparts do. 1 gram of carbohydrate provides 16.8 Joules while a similar quantity of lipid provides 38 Joules. This qualifies it as a high energy density food.

LIMITATIONS TO GROUNDNUT PRODUCTION

Going by the production and the yield data, it is clear that the production is dwindling. This is attributable to a number of reasons being:

a) Pests

Groundnut hopper - Hilda patruellis: causes artificial drought symptoms. Ants attach to base of plant stems. Small brown hoping bugs up to 3 mm long in soil around roots.

Systates weevil

Groundnut aphid (Aphis crassivora) is a vector of the groundnut rosette virus, which distorts plant leaves.

Bark eating termites - Macrotermes spp. Feeds on the shells of the nuts, making them rough. They weaken the shells so that they easily release the nuts, which easily germinate. They also detach the pods from the mother plant, thereby resulting in a lot of nut loses. The wet wood termite Coptotermes formosanus, also causes similar damage.

Cotton bud thrips: Frankliniella schulzei distorts the leaves of groundnuts thus reducing photosynthetic surface area and area exposed to the sun.

Bean flower thrips: Taeniothrips sjostedti affects the flowers of groundnuts, seriously reducing yield.

b) Diseases

Aflatoxin flavus: infects the developing nuts inside pods. Produces a toxin, aflatoxin, that is carcinogenic and a potent poison. The fungus-based toxin is encouraged by moist conditions.

Groundnut Rosette virus (GRV): GRV is transmitted by a pest Aphis crassivora and distorts plant leaves. This greatly reduces photosynthetic surface area and the yield.
Leaf Spot Blight - caused by a fungus, Sclerotium rolfsi

c) Marketing:
There are no formalized marketing institutions of groundnuts in Kenya and many other African countries. It remains subsistence crop because of poor market channels. The industries that could consume the product such as cooking oil and soap industries prefer to import from other countries their raw materials. Thus to date, oil seed imports constitute a large proportion of total annual imports of Kenya.

Termite mounds are common features in Africa

RESEARCH NEEDS FOR GROUNDNUTS
The immediate research needs of groundnuts include:
Breeding for high yields and high methionine, cystine and cysteine amino acid - these being sulphur -based essential amino acids.
Breeding for Aflatoxin flavus resistant varieties
Breeding for reduced ant-nutritional factors such as anti-trypsin factor.
Increased drought resistance since many varieties need a minimum of 600 mm pa during active growing season. This condition puts off many otherwise climatically suitable areas in Kenya thereby lowering production.

Other industrial oil crops

Sunflower

The coconut fruit

COWPEAS (VIGNA UNGUICULATA)

HISTORICAL BACKGROUND

Cowpea has its centre of origin in Ethiopian region, the same centre of origin of a wide range of tropical African crops. However, its wild forms are widely distributed in Africa. Vigna is a tropical group of plants, a genus of about 170 species. It has the largest number of species endemic to Africa. Only 4 species, V. unguiculata, V. radiata and V. sinensis and V. mungo are economically important. Cowpea was first domesticated in Ethiopian region. Others also think that the crop was domesticated in West Africa, or perhaps widely throughout the African savannah zone, more than 4000 years ago (i.e. before 2000BC). It is thought to have evolved together with millets and sorghum with which it has, since time immemorial, been intercropped. The wild forms of Vigna unguiculata occur only in Africa.

SPREAD

Cowpea was introduced to America from Africa during the slave trade and given the name cowpea. This is because it was a popular plant for livestock (cow) feed. The earliest cultivars of cowpeas in Africa were the spreading, photosensitive, short day types of subspecies unguiculata. Cowpea probably reached India together with millet and sorghum from East Africa around 1500 BC.

IMPORTANCE OF COWPEAS:

The crop is widely cultivated for their leaves and pods which produce pulse. It is rich in even the often-deficient sulphur amino acids such as cystine and methionine, among others. It contains reasonably high concentrations of minerals such as iron and calcium. Besides, it is so rich in vitamins such as B2 (Riboflavin) and B1 (Thiamine). The leaves are rich in vitamin A. It also supplies a reasonably impressive energy quantity. Cowpea is a pulse crop with the highest potential of all food legumes in the semi and sub humid tropical lowlands. It has the advantages of quick, early growth, short maturity, wide ecological adaptation, drought resistance, grain quality, acceptability, broad range of genetic diversity and ease of hybridization. If production can be promoted, it can greatly help reduce the food insecurity inherent in the dryland South.

ECOLOGICAL REQUIREMENTS

Cowpea is warm climate lowland crop of the tropics. It requires an annual precipitation of 800 mm pa. However, even with 400 - 500 mm pa well distributed over the growing season, it performs modestly.

The best temperature for cowpea is anything above 18oC. For this it does best in lowland areas of altitude less than 1500 m above sea level. However, if cultivated for leaf production, it does well in much higher altitude, even 3000m. The low temperatures associated with high altitudes greatly suppress pod production and development. Therefore if cultivated as a pulse crop, it should be restricted to areas below 1500m ASL. Cowpea is a leguminous crop that can fix atmospheric nitrogen. Thus it does not need very fertile soils if meant for pulse. However, if meant for leaf production, it needs a reasonably fertile soil. The soils need to be of good drainage not prone to waterlogging. The crop thus needs an agro-ecological zone of lower midlands 4 - 5 or lowlands 3 - 4. The upright determinate cultivars of the cylindrical subspecies are more drought resistant than their unguiculata counterparts.

AGRONOMIC REQUIREMENTS

Cowpea evolved together with millet and sorghum in the dryland Africa. Its production requirements are therefore known best by these traditional farmers. In the traditional cereal farming system of West Africa, the spreading, short - day cowpeas were inter-planted with sorghum or pearl millet about six weeks after the cereals were sown in a well prepared land. Broadcasting was the most popular planting method. Up to 20kg of seed was needed per hectare as planting material. However, where only foliage is required seeds were broadcasted at the rate of 100 kg ha-1. Modern agriculture has come up with single/pure stand or monoculture cropping systems that have been applied to almost all crops. Under such circumstances, seeds are sown directly at a rate of 66 kg seed/ha, and with a spacing of 75 cm x 5 - 8 cm. The seeds are placed 4 inches deep. Early weeding is advocated for since the crop provides almost full soil cover within 4 weeks. At that stage, they completely smother any other under - growth. This is why traditionally they were sown at least a month after the cereal crops.

If the crop is meant for seed/pod production, the terminal bud is pruned after 2 - 3 weeks of sowing. This encourages the growth of lateral buds from where most pods later develop. If topping is not done, the fields become too bushy with a lot of leafy growth. The pruned terminal buds are used as vegetables.

If produced for leaf, harvesting begins from the third week. Harvesting can then be done on a regular basis, the duration depending on the extent of each. However, one week is a fair enough time lapse between leaf harvestings.

HARVESTING AND YIELD

Leaf harvesting begins from the third week of sowing. This can continue for even 4 months depending on the prevailing ecological conditions. Up to 2 tons can be harvested from a field set entirely for leaf production. In the traditional African farming system, cowpea was cultivated both for leaves and pod (seed) yields. In such circumstances, an extrapolated seed and leaf yields could be 750 Kg and 1 ton respectively. In some farming systems cowpea is produced purposely for seed yield. The yield depends on the variety. Thus the more heat and drought tolerant cylindrical subspecies yield much higher than their unguiculata counterparts do. Seed yield from an entirely pod purpose single stand of cowpeas yield between 500 kg ha -1 and 6 t/ha. The latter has been achieved in humid research stations. The former is assumed to have been recorded from a traditional mixed crop and multi purpose cowpea field, with clearly significant under-use of the land. In Kenya, Migori district records 2 t/ha. There is no corresponding national yield. Seeds can yield 1.5 t/ha within 70 days of sowing and 3 t/ha within 95 days. Harvesting of seeds continues for 4 - 5 months.

CHEMICAL COMPOSITION

Both leaves and pods are utilized as food. They have the following based on 100 g dry matter:

Total Protein - 24 - 25 g; Calcium - 90 mg (Leaves); Thiamine (both) - 0.9 mg;
Riboflavin (both) - 0.14 mg; Iron - 4 mg (leaves); and Calcium - 65 mg (seeds)
The % composition of amino acids within the cowpea protein is as follows - Range (average) : Lysine 5.7 - 9.6 (6.6); Cystine 0.7 - 1.7 (0.9); Methionine 0.7 - 1.6 (0.9); Histidine 2.7 - 4.0 (3.3); and Threonine 3.4 - 5.3 (4)

PROCESSING AND USES OF COWPEAS

Both cowpea leaves and seeds are eaten. Being a traditional crop, only traditional methods of Africa are known as far as processing is concerned. Some of these techniques are:

Leaves: Are plucked, cut into smaller pieces boiled for at least 30 minutes then fried. Either Odi from groundnuts or butter from churned milk could be added. This was a delicacy among the Luo, Luhya and the Mji Kenda of Kenya. The cowpea could be prepared alone or in combination of a number of traditional vegetable normally branded 'weeds' by agriculturalists.

Leaves: These could be washed, cut into pieces and then sun dried. After being packed in bags, these could store for weeks and even months. When to be cooked, they could be soaked in water for some time or just boiled and treated as in (a) above.

Pulses: The seeds could be used for making Nyoyo by the Luo. The traditional Nyoyo was boiled to softness mixture of sorghum and cowpea seeds. Later, maize replaced sorghum, and beans slowly replaced cowpeas when the production and popularity of the original two crops started dwindling.

Pulses: The seeds could be used for making Dengu. This was thoroughly boiled and fried seeds of cowpeas. The dish could then be eaten with Ugali.

PRODUCTION STATISTICS/DATA

Africa has been producing at least 60% of the total global cowpeas. In 1970, for instance it produced 93%. The total global production then was around 1,222,220 tons. Of this total production, Nigeria produced the bulk 60.2%, Niger 12.7%, Upper Volta 8.1% and Uganda 5.9%. The rest of Africa produced about 6.1% and other parts outside Africa produced 7%. There are however a lot of doubts as to the accuracy of the African data. The crop has traditionally been both a multi-purpose and an inter-crop. Thus whereas the yields in West Africa is estimated to be 375 kg/ha, this can be gross underestimation. There are also lots of unreported plantings where the crop is produced largely for subsistence. This is a case in point in Eastern Africa - Kenya, Sudan, Ethiopia, and Tanzania etc where the crop originated. In West Africa, the reporting could be attributed to the fact that the crop is right now commercially produced. It is obvious that there has been a general bias towards commercial crops at the expense of subsistence crops. Hence the discrepancies in reporting. Kenya produced 79,000t in 1992 and set a target production of 87,030 t for 1996. The global production in 1990 has been about 10,000,000 tons.

RESEARCH AND REQUIREMENTS OF COWPEAS

Kenya's KARI places cowpeas as the 29th crop out of 52 in order of research prioritization. It is surprising that with the great untested and unimproved nutritional and climatic potentials of cowpeas, it can be rated so poorly. It is known that Africa gets the bulk if any at all, its agricultural research funds from outside - the North. The source of the funds dictates the nature of research to be undertaken. Thus a lot of agricultural research is not to the interest of Africans themselves even though their own highly trained personnel are deployed there. There is a need to refocus the research priorities to consider as a matter of urgency the traditional foods that remain largely neglected and despised. For cowpeas, there is need to increase its production and consumption and improving its distribution and purchasing power of many Kenyans who remain below poverty line. The high energy and calorific density of cowpea, beans and groundnuts should put them much at the top agenda in agricultural research in this country.
There is therefore an immediate need for cowpea research in the following areas;

Processing techniques of the hard leaves, which in most cases are wasted.
Keeping techniques for both leaves and pulses so that they can reach a wider market.
Increasing digestibility of the cowpea leaves
Processing techniques for seeds so as to reach a wider market.
It will be hard for Africa to develop as long as it does not give production and research priority to its own indigenous crops, foods and resources.

SOME CASH(?) CROPS

COFFEE

Is widely believed to have originated in Kaffa Province of S.W. Ethiopia. Coffee is produced for its berries which are rich in caffeine, a beverage active ingredient and a stimulant. It is a popular drink among the Europeans and Americans. It is not of any known nutritional significance other than being a source of Vitamin B (Thiamine and Riboflavin). These two nutrients, vitamins B1 and B2 are also found in bulk in meats, pulses and cereals. The two vitamins are needed on daily basis to the tune of 2mg by adults, with children requiring less. Coffee requires an altitude of 600 - 1800 m ASL, a temperature of 14 - 260C and an annual precipitation of 1000 - 2000 mm. These wide ecological ranges cover the two main species, Robusta and Arabica. The Robusta need hotter environment, more rainfall and lower altitude, while the Arabica need cooler and higher altitude zones. Fertile, deep volcanic soils are necessary. These ecological requirements of coffee very well fit to be classified as high potential areas. Coffee therefore does well in highland and upper midland humid zones (UH1-2, LH1-2 and UM 1-2.)

The coffee plant

Coffee is produced in:

1. Latin America - Brazil, Columbia, Mexico, EL Salvador, Guatemala, and Costa Rica
2. Africa - Cote D'Ivore, Ethiopia, Kenya, Malagasy, Cameroon, Uganda and Tanzania, and
3. Asia - Indonesia and India

Arabica is largely produced in Brazil and East Africa, while Robusta is produced in East Africa and West Africa. The main coffee growing zones in Kenya are Kiambu, Embu, Thika, Meru, Murang'a, Nyeri, Kisii, Subukia, Kirinyaga and Bungoma districts. The crop is exported to Europe and North America. In 1986, Kenya produced 84,500 tons of coffee. It is a protected crop which is extremely capital intensive to manage.

TEA

A crop of Asian origin which was introduced into Europe in the 16th Century. This is a beverage crop whose leaves are rich in nicotine, tannin and caffeine. Like coffee, it has no known nutritional significance other than being a good source of vitamins B1 and B2. It is also a stimulant that is very popular in Europe and North America to activate cells during winter - when it is too cold. The Assam (Indian and Sri Lanka originated) tea and Chinese (China originated) tea are produced in Africa (Kenya) and Asia (India, Bangladesh, Sri Lanka and China). Kenya tea is said to be the best in the world and constitutes 50% of Africa's tea exports. Nearly 98% of Kenya's is used for blending and improving other teas from the rest of the world. In 1986, Kenya produced 164,00 tons and earned US dollars 155 million. This boiled down to KSh. 30/kg of tea. Kenya is the third largest tea producer and exporter after India and Sri Lanka. Kenya tea constitutes

12% of the world's exports. The tea industry supports 5% rural workforce. Tea growing in Africa started in Malawi, now the second largest producer after Kenya. Tea provides 70% of export earnings to Malawi. Tea was introduced into Kenya in 1903. The agro ecological requirements of tea are as follows:

1. Precipitation of at least 1500 mm pa
2. Altitude of 1000 - 1700 m ASL.
3. Temperatures of 18 - 21oC, with the range exceeding 21o C not exceeding 8 months.

These correspond with agro ecological zones, lower highlands and upper midland, (LH1-2 and UM1-2). These are found in the Kenyan highlands East and West of the Rift Valley such as Kericho, Kisii, Mau Range, Nandi, Kakamega, Vihiga, Bungoma, Sotik, Trans Nzoia, Meru, Nyeri, Nyamira, Nyambene, Murang'a, Kiambu, Thika and Embu. These are all high potential areas of Kenya. Between 1976 and 1978, East African production record of coffee and tea was as given below (in %).

Tea plantation in Kyenjojo, Uganda

Table 9: % production of coffee and tea by Kenya, Uganda and Tanzania (1976-78) Source: GOK, 1994):

Crop	Kenya	Uganda	Tanzania	TOTAL (%)
Tea	72	12	16	100
Coffee	29	55	16	100

COCOA

Another beverage crop brought to West Africa form America. Currently, Ghana, Nigeria, Cote D'Ivoire and Cameroon produce more than 75% of all World cocoa. The rest is produced in East Indies and Central America. 5% of cocoa is produced by Togo and Equatorial Guinea. Brazil produces 20%. Cocoa's dry beans are bagged and transported overseas. Almost all is exported, as there are no processing facilities in the producing countries. Ghana earns over 60% of export earnings from the crop, which it produces in over 1.2 million hectares. It is exported to Japan, Europe and America. Other cocoa production data (1976 - 78) is as follows:

Table 10: Cocoa production in Africa (x 1,000 tons) (Source: Lester B, 1987)

	1976	1977	1978
Ghana	324	280	270
Cote D' Ivoire	230	240	297
Nigeria	165	202	160
Cameroon	82	105	100

This is a lowland crop that grows at 0 - 300 mm ASL. Cocoa requires at least 1500mm pa of rainfall, low relative humidity, temperatures 23 - 28oC and no long droughts. It is drought prone. It is often mixed with banana to provide shading. It is largely an equatorial crop, growing well within 10oN and 10oS of the equator. It requires fertile soils rich in iron and Potassium, but non acidic. It is exported to West Europe and North America. Since 1970, cocoa production has not been steady due to bad weather, pathogens and neglect at the peak of the oil boom.

RUBBER

It is produced in S.E Asia, Liberia, Nigeria, Ivory Coast, Zaire, Cameroon and Sierra Leone. It was first noticed in Amazon forest in Latin America, and introduced into West Africa. Its product is used for making vehicle tyres and carpets. It is produced in 70% of Malaysia land, more than 120,000 ha of Liberia's land as well as in Zaire. In Liberia, the rubber industry employs at least 85% of all Liberians. Rubber exports contribute 90% of Liberia's forex. The produce is exported to USA. Rubber requires at least 1500 mm pa of precipitation, and more than 21oC of temperature, and fertile soils.

OIL PALM

This originated in West Africa and is right now produced in Nigeria, Ivory Coast (more than 80,000 ha), Ghana and Zaire. Nigeria is the centre of origin for palm oil. At least 50% of the produced oil in Nigeria is used at home, and the rest exported. Nigeria is a net importer of foods. Oil palm requires an annual precipitation not less than 2030 mm, relatively hot environments (27oC) and high relative humidity. These are typical equatorial climates. It is a low altitude crop and needs a small range of temperature (<5^{0}C). It is exported to UK, USA and Italy.

PYRETHRUM

This is an insecticidal crop whose leaves are rich in pyrethrene. Kenya is the main producer especially from Meru, Nyandarua, Nyeri, Kisii and Nakuru. The 1991 world demand was 15,000 metric tons of which Kenya supplied 10,500 metric tons. The produce is exported to Europe and USA. The crop needs 1000 - 1500 mm of rainfall annually, an altitude of at least 1800 M ASL, rich volcanic soils and very cool places especially highland areas. The crop needs at least 10 days of cool weather (15oC or less). It covers 30,000 ha in Kenya. Kenya supplies 90% of global pyrethrum. USA and Austria are importers. Pyrethrum originated in the Palmatian Coast of Yugoslavia. It was introduced into Kenya in 1929.

The pyrethrum plant

TOBACCO

This originated in America. It needs 17 - 26 frost-free weeks during its production period. It needs at least 18oC and a lot of rainfall, not less than 1000 mm pa. It is a crop of lower midland and lowlands (LM1-3 and L1-2) zones. These exist in Kitui, Meru, Tharaka, Mitunguu, Migori and Busia. Tobacco is produced for its broad leaves that are rich in nicotine. It is used for making cigarettes largely in America and Europe. Tobacco smoking is a serious health hazard. The largest producers are USA, China, India and Russia. In Africa, Zimbabwe tops, followed by Kenya, Uganda and Malawi. It accounts for 24% of Zimbabwe's export earnings, greatly relieving it of over-depending on mining.

MODULE 6: PLANT / CROP PROTECTION

General Principles:
General Methods Of Crop Protection
Quarantine
Background And Application Of Quarantine
Cultural Methods
C) Mulching
Classification Of Weeds
Weed Control Methods
Cultural:
Mechanical / Physical:
Biological Weed Control:
Chemical Weed Control
Classification Of Herbicides
Spectrum Of Action:
Mode Of Action:
Time Of Application
Striga Spp (Witch Weed). As A Special Weed In East Africa
Crop Protection: Pests
Classification.
Biting And Chewing Pests
Grasshoppers:
Piercing And Sucking Pest
Effects Of Piercing And Sucking Pests On Plants:
Fungal Diseases
Sorghum Downy Mildew
Mode Of Infection:
Harmful Effects:

Physical Methods
Biological Control
Crop Protection: Weeds (Makwekwe Or Mfero)
Negative Effects Of Weeds
Positive Effects Of Weeds

Boring Pests
Maize Stalk Borer
Damage Caused By Nematodes.
Control Of Nematodes
Sorghum Shootfly (Atherigona Indica):
Maize Stalk Borer (Buseola Fusca)
Sudan Dioch (Quelea Quelea):
Rodents
General Method Of Pest Control
Crop Protection: Diseases
Introduction:
1. Bacterial Diseases:
Soft Rot Of Carrots
Bacterial Wilt Of Tomatoes
Mode Of Infection:
Harmful Effects On Crop
Prevention & Control:

Prevention & Control:
Viral Diseases
Groundnut Rosette Virus
Maize Streak Virus Disease

MODULE 6: PLANT / CROP PROTECTION

GENERAL PRINCIPLES:

A pest is any biological organism that negatively affects growth and development of a plant, hence its yield. It can be an insect, pathogen, weed, rodent or a bird. Various pests attack crops through the country while others are only important in localized areas. Damage to the crop or yield reduction as a result of pest activity can be small or large. A decision to control a given pest must be made after taking into consideration the following:

Potential yield; Potential yield loss to the pest; Cost of controlling the pest (labour and pesticides); and Availability of appropriate implements and pesticides.

The most important investment is therefore to increase the yield potential by following good crop management practices.

GENERAL METHODS OF CROP PROTECTION

There are five main methods of pest control: quarantine (regulatory or preventative), cultural physical (mechanical), biological and chemical. The choice of a suitable method largely depends on the local experience and the availability of resources.

QUARANTINE

This is used to prevent the movement of pests to areas where they do not already exist. The method is effective for pests which are spread through the movement of infected plant material by people (especially travelers). Internationally, this is done through individual counrty's legislation which prohibits or restricts the introduction of plants or plants products into the country. Nationally, this is done though certification of planting material. In Botswana, this service is provided by the department of Agricultural Research.

BACKGROUND AND APPLICATION OF QUARANTINE

This is a legislative (legal) method of controlling pests, diseases and other wanton organisms. The word is borrowed from a Greek word- "QUARANTUM" meaning, keeping in isolation (at bay for 40 days when ships came from voyage). It was assumed that if the ships contained deadly disease, the lives in that ship would die within these 40 days, hence do not contaminate the rest.

It is often recommended by most governments that before importing a produce into a region or country, one has to get a sanitation certificate. This is to reduce chances of introducing weeds, pests and diseases indigen of exporting countries. It is out of negligence that cassava green mite, cassava mealy bugs (Prostephanus truncatus), both from latin America, reached the East African countries such as Kenya. Others were pine wooly aphid (Pineus pini), imported from Australia, for which a bug was imported to control it. A parasite of coffee berry borer (CBB), Prorops nasuta, was bred and taken to various coffee growing countries and it is proving to be effective. Quarantine can be either regulatory or exclusive. Exclusive quarantine is the total embargo of a particular crop or animal product against entry. For Kenya, such include the Basberry plant. Regulatory quarantine means that certain animals, crops or their products can only be imported subject to certain conditions such as:

Should be thoroughly inspected in the exporting country or province

Should be provided with a sanitatory certificate by an authority after inspection and certifying freedom from diseases and dangerous organisms;

After depot, the product should be inspected again, and fumigated at the point of entry (e.g. airport);

Should be grown in an isolated site to certify its purity.

CULTURAL METHODS

This involves the use of good crop management practices that reduce or prevent crop infestation by pests or increase crop tolerance to pest infestation. Such activities include crop rotation, phyto-sanitation, mulching and the use of trap crops.

a) Crop Rotation

This is achieved by growing different types of crops which are not susceptible to the same pest in succession in the same field. The pest population is reduced in the absence of a suitable host. This method is effective against pests which survive on plant debris and in the soil and attacks specific plant species. It is important to know the survival period for the particular pest before recommending a rotation scheme.

b) Phyto- Sanitation

This involves the application of good crop management practices which are essential for producing a vigorous crop that can withstand pest infestation. The management practices are also aimed at preventing the spread of the pest as well as eliminating its source. Such activities include;

Proper fertilizer application to promote healthy and vigorous
plant growth.

Timely planting for the crop to escape pest infestation.

Burning of crop residues that harbor pests.

Deep ploughing to bury plant debris at the end of season

Rouging of infected plants. This is more effective against virus diseases.

c) MULCHING

This involves covering the soil surface with a layer of material such as cut grass, stover or a black plastic sheet. The mulch excludes light and inhibits the growth of weed seedlings. It is applied after crop emergence. Mulching is therefore effective in controlling weds and not other pests. It is also used as a means of covering soil moisture.

d) Trap crops

A trap crop is used to stimulate the germination of the parasite but does not support its development. Therefore it reduces the source of the parasite in the soil. This method is effective in controlling parasitic weeds. Trap crops should be grown for at least three years before there can be substantial reduction in parasitic weed seeds.

e) Crop support

This includes cultural practices such as staking, training, or propping of heavy crops which are likely to fall down, or lean on the ground, thereby acquiring some pests and / or diseases from there. This is shown below.

Figure 59: A Mulched field of Kales

Heat: This can be done by using steam to kill soil pathogens or through solarization, and also with hot water to kill pathogens in seeds or vegetative material.

BIOLOGICAL CONTROL: This is the use of host plant resistance and natural enemies to kill or reduce the population of a pest.

CROP PROTECTION: WEEDS (MAKWEKWE OR MFERO)

These are plants growing where they are not wanted, i.e., plants out of place.

Examples of weeds are: Witch weed (Striga spp), Mexican poppy, Salvinia molesta (a water weed in Kenyan rift valley lakes), Upright Star bur (Acanthospernum hispidum), Black jack (Bidens pilosa), Pig weed (Amaranthus spp), Water hyacinth (Eichhornia crassipes) (water weed in Lake Victoria), Devil's horse whip (Achyranthes aspera), Star grass (Cynodon dactylon), Couch grass (Digitaria scalarum), Lion's ear (Leonotis spp), Wandering Jew (Commelina benghalensis), Thorn apple (Datura stramonium), Sodom

Do you agree that weeds are special volunteer crops in fields?

Think about the following as you attempt to answer the above question:

a: Thepe (Pigweed) – Amaranthus spp

b: Hariadho

c: Wandering jew (Commelina benghalensis)

d: Delele/ Apoth

e: Aboka

f: Osuga(Black nightshade / Solanum nigrium)

NEGATIVE EFFECTS OF WEEDS

Reduce the value of a landscape- so that the owner gets less if selling a weed- infested plot;

Some interfere with farm operations e.g. double thorn (mosetho/ mosetlho), upright star bur etc

Some are parasitic e.g. Striga (matebele, molelwane- witchweed)

Some reduce the quality of marketable produce e.g. black jack and upright star bur may make the wool from sheep fetch less money in the market.

Some contaminate farm produce- e.g. weeds in the onion family impart their taste to milk if lactating livestock feed on them; thorn apple and black jack fruits also stick on the wool of sheep.

Some make the land less attractive e.g. thorn apple.

Most compete with crops for essential nutrients (water and minerals) and sunlight

Some harbour diseases pathogens

Some harbour pests of crops e.g. Phoka (Urochloa spp) habours the leafhopper which transmits maize streak virus to maize

Some are poisonous e.g. thorn apple (Datura spp)

Increase cost of production

POSITIVE EFFECTS OF WEEDS

Some make the land beautiful e.g. witch weed

Some reduce soil erosion by providing cover to the soil surface e.g. couch grass

Some are livestock feed e.g. star grass, couch grass etc

Some are vegetables / morogo e.g. amaranthus (thepe), delele,

CLASSIFICATION OF WEEDS

Weeds can be classified in 3 different ways:

1: Morphology- i.e. nature of plant; some are broad leaved (dicots), others are narrow leaved (monocots / grasses)

Length of growing period / growth habit:

Some are (i) annuals (finish their entire life in a year or season), while others are (ii) perennials (finish their entire life in more than 2 years or seasons); OR (iii) Biennials-finish their entire life in 2 years.

For biennials and perennials, the first year is normally for vegetative growth, and the second year (and beyond) for reproductive growth (flowering and fruit / seed production).

Figure 61: The castor oil plant (Ricinus communis). Right: The Mexican poppy (a thorny weed common in southern Africa)

Figure 62: Comelina bengalensis (The wandering jew) Figure 63: Striga spp (witchweed, in Western Kenya)

NB:Comelina bengalensis (The wandering jew) Is a common weed, but a very important indigenous vegetable among many African communities.

WEED CONTROL METHODS

Due to the many negative impacts of weeds, they need to be controlled to avoid the farmer incurring economic losses from them. The different methods of weed control are: Cultural (2) Chemical (3) Biological (4) Mechanical / physical

Table 11: Classification of weeds.

Class of weed	Annual	Perennial		
Morphology	Narrow leaved	Broad-leaved	Narrow leaved	Broad-leaved
Difference	Finishes life in 1 year or season	Finishes life in more than 2 years or seasons		
	A monocot	A dicot	A monocot	A dicot
Examples	Couch grass (Digitaria scalarum) (Motlho)	Black jack (Bidens pilosa)	Star grass	Castor oil plant (Ricinus communis)
Examples	Eragrostis spp	Witch weed (striga spp) (molelwane)	Nut sedge	Hibiscus spp (Mmabesete, kenaf, moku)
Examples	Small carrot seed grass	Thorn apple (Datura spp-mokhure)	Quack grass	
		Upright starbur (Acanthospernum hispidum) - khonkhorose, mmalenakana	Urochloa mosambicensis (Phoka)	

CULTURAL:

Involves using any mechanism which gives the crop an advantage over the weed e.g. by ensuring a healthy start and proper establishment of the crop, so that it can out-do or smother weeds. Specific methods include: Early planting, use of trap crops, correct plant population, crop rotation, good husbandry practices, mixed cropping, use of cover crops, use of clean certified seeds, flooding (is in the case of rice), burning, and use of cover crops (e.g. sweet potatoes as shown below).

Advantages

Inexpensive; and Has no environmental pollution effects

Disadvantages; Cannot be effective where the weed is already established; and Cannot be used on its own

Figure 64: The sweet potato crop can be used as a cover crop to control weeds

MECHANICAL / PHYSICAL:

This is any energy- expending method of removing weeds e.g. clearing, weeding, uprooting.

Advantages:
Is quick and simple; Effective at any stage in crop production; Simple forms such as uprooting may be very effective in young weeds which are very close to the crop; and Very effective if well timed, properly done and proper tools used

Disadvantage
A lot of energy involves (i.e., tiring); Too frequent cultivations may lead to breakdown of soil structure; There may be extensive root damage e.g. if hoe used (dangerous for root crops suck as carrots, cassava, sweet potatoes); Cultivation enhances water loss by evaporation; and May improve conditions for germination of some weed seeds

BIOLOGICAL WEED CONTROL:
Involves the use of a biological agent to control the weed. The agent may be another plant, animal (e.g. predator), pest, parasite, or pathogen. The agent acts on the weed by reducing its activity, and at times completely killing it. The weevil has been used to control the water hyacinth in Lake Victoria in the recent past.

Figures 65 and 66: The water weeds: Right- the water hyacinth (Eichhornia crassipes).

Advantages:
No pollution of the environment; and Weed is not completely eradicated, but only its population and

Disadvantages:
It takes time to be effective (shows results after a long time); There may be a problem identifying the right biological agent; It is at times difficult to control the agent once it

has played its part (some become pests and parasites of the crop); and Difficult to multiply the right biological agent.

CHEMICAL WEED CONTROL

This is a method whereby chemicals called herbicides are used to manage weeds. The herbicides may either directly cause death of the plant by making it dry out, or may interfere with one or more physiological processes within the plant. The processes include photosynthesis, respiration, protein synthesis, etc.

ADVANTAGES:

No root damage (ii) helps in conserving soil moisture in dry areas since either minimum or zero tillage can be used (iii) Conveniently used in all types of soils, even where the nature of soil may hinder cultivation (iv) Can be used at any time - before and during the growth of a crop (pre- or post-emergence) (v) Systemic herbicides can be used to control persistent perennial weeds.

DISADVANTAGES

(i) Pollute the environment (ii) May poison human beings, including the operator (iii) Expensive (iv) Need skills to use and manage (v) May kill some useful organisms (vi) May kill some or all crops even from a little wind effect from the neighbouring farm

CLASSIFICATION OF HERBICIDES

Herbicides can be classified based on 3 systems

(A)SPECTRUM OF ACTION:

Some herbicides are selective (kill some plants and leave some), others are non-selective or broad spectrum in action (i.e., kill all plants). Chemicals may not kill some plants because:

Leaves and stems have some protection e.g. leaves are slippery

Leaves slanted at an angle so that chemicals drain immediately

Leaves of some plants too small, thus they absorb less chemical not enough to kill them;

Low concentration of chemical (weak) thus can only kill young plants but spare older ones;

Some plants are able to detoxify some herbicides and convert them into useful chemicals.

Some plants have deep roots so that chemicals meant to be absorbed by roots may not easily affect them, but will kill shallow rooted plants

Selective herbicides will kill one category of plants (crops and weeds alike); it is upon the person applying to know whether it will affect the crop or not. The chemical cannot decide to spare your crop- it can be killed just like the weed. Selective herbicides include 2, 4-D, 2, 4, 5-T, MCPA, Dalapon, while Non- selective herbicides include paraquat.

(B)MODE OF ACTION:

Some weeds are only killed by contact with herbicide, while some herbicides have to be absorbed by the plant first before action. The former are called contact herbicides (eg sulphuric acid), while the latter are called systemic or translocated herbicides (eg 2,4-D, 2,4,5-T etc). Contact herbicides are not economical, as the whole plant has to be flooded with the chemical for it to die. Even then, the root may survive. Only the parts of the plant which comes in contact die-, the rest is spared. On the other hand, systemic herbicides have a hormonal action. They are more economical than contact herbicides because once absorbed by any part of the plant, the chemical distributes (translocates, transports) itself throughout the system of the plant, and kills the whole plant. It is useful in controlling very persistent, deep-rooted perennial weeds such as couch grass.

© TIME OF APPLICATION

Some herbicides are applied before the crop emerges, and are called pre-emergence herbicides. They can be used to clear land instead of the normal mechanical methods, and are useful agents for conservation (zero) tillage. They are usually non- selective e.g. atrazine. They kill all weeds before the crop emerges so that once the crop appears; there is a very beautiful uniform crop cover. The post- emergence herbicides are those applied after the crop emerges, and therefore they have to be selective- as they have to spare the crop. Examples are 2, 4-D.

STRIGA SPP (WITCH WEED). AS A SPECIAL WEED IN EAST AFRICA

Witch weed is the main weed affecting sorghum, and indeed most grass crops. It is a parasitic weed which produces lots of seeds even in harsh environments. The seeds can stay in the field for even up to 70 years, still intact and viable. This renders even close season and crop rotation as futile as means of controlling the weed. Even if live plants are killed, many thousands of viable seeds are produced and remain intact in the soil.

Witch weed (Striga spp) causes economic losse4s due to their imbalanced trophic relationship with plants of the gramineae family, especially large-stemmed ones such as maize and sorghum. These crops produce unique types of organic acids at their root tips which diffuse into the soil. These induce the germination of Striga seeds. So any time there is maize or sorghum plant, there would be a problem of witch weed since the two seem to have evolved together. Once the weed germinates, mostly very close to the crop plant, it would send its roots to penetrate that of the crop. It would start to absorb nutrients from the crop, making it fully reliant on the crop. This is a parasitic trophic relationship. There would, however, be some level of autotrophism by the weed since it has functional green leaves.

Due to the difficulties faced by scientists in controlling witch weed, it has become necessary to strike a deal with the weed. At the moment, ICRAF is very busy trying to look at possibilities of biological control of the weed followed by cultural methods. This involves planting maize or sorghum as a trap crop which will induce the germination of the weed. Thereafter, the whole field is cultivated deeply, with both the crop and the weed incorporated into the soil. This has been effective in reducing the infestation of Striga. However, with any new maize or sorghum crop, Striga has always continued to be the major weed even after the involving process of trapping it. The method, however, has a future since it helps reduce the number of viable seeds still kept in the soil. The above method is expensive since it involves using too many crop seeds just to help trap the weed. The common farmer is not that endowed with resources that (s)he can continue wasting like that. The amount of seed involved, as well as the time and the expensive farm operations, all boil down to expenses to the farmer.

Figure 67: Striga spp (witchweed), a parasitic weed in Western Kenya)

Thus the method is not practical. However, it can be done successfully by the government research institutions, Universities and private and multi-national companies who are stakeholders in agriculture. Indeed, all these should be interested because they would help produce more food for sale at reasonable prices. This would go a long way in reducing food insecurity in many African countries. It is important here to challenge the industries to venture into farming since they are well endowed with resources and security to even lease land for food and / or cash crop production, as well as even for research. After all, anything they do to boost the food security in their host countries would not go to waste. The positive effects would soon spill over to them in the form of increased sales of their produce resulting from increased savings the farmer will have due to cheap foods.

A second Method ICRAF has researched on is the use of trap plants. It has been observed that a plant like Sesbania sesban, a local agro-forestry shrub, induces similar germination from of Striga just as does maize or sorghum. However, Sesban does not provide the necessary sustenance that Striga needs to survive as a parasite. Unlike the grass crops which support Striga full time with them acting as the weed's lifeline, the sesbania only offers the germination inducement, and thereafter the crop has to survive on its own. This method can be used to control the number of Striga seeds since Sesbania plant can be planted in a field suspected to be seriously infested. Then another crop is planted alongside Sesbania. After the weed germinates, it will be forced into an abortion since there is no cereal to serve as its host. All germinated Striga plants would therefore die. This is a more farmer-friendly method of controlling Striga in farmers' fields. In the long run, the numbers of seeds of Striga can be substantially reduced in even soils suspected to be heavily infested. It may prove to be a very sustainable method.

Farmers are advised to plant as many sesbania sesban plants as possible in their farms and continue planting the usual crops in a rotation. It will eventually pay dividends if the tree crop is well maintained. If it becomes too big, as the adults normally are, then they can be cut for use in other activities such as building material. In this way, the farmers will be getting income from two sources (indeed the main reason for researchers recommending agro forestry). Lastly, still on Striga control, there may be a lot more trap plants and other methods of Striga control perhaps still only known by the farmers themselves. Farmers even know which plants (weeds) are allelopathic, and which ones attract what type of pest or disease more. These ideas can be used to improve research findings.

Caliandra spp is one of the famous trees for agroforestry

Researchers should be encouraged to come out of their cocoon and integrate farmers in their research. Only collaborative research will solve the farmers' problems, while at the same time enabling researchers to cut down on research costs by benefiting from time-tested and age-old traditional technologies. This is what Robert Chambers in his master-piece called "Putting the last first" calls- a third culture. At the moment, the researchers believe they are the sole source of knowledge because they think they are modern. Yet on the ground, they have nothing to show and they also remain mere observers in markets as prices of foods sky-rocket.

They only complain about the government being responsible for increased food prices without looking at their contribution to it all. The rural person (farmer) has researched on different crops for all the years, and his methods are time- tested. These should be incorporated into our national research programs and collaborative research initiated with the local having a department in which they make their contributions inform of verbal information or skills. These can be recorded and used to improve the lot of everybody.

There are traditionally intercropped crops whose returns were always fair to good. Monoculture has simply worsened the risks from major pests and weeds. Other weeds of sorghum include Bidens pilosa, Conyza spp, Commelina benghalensis, Solanum nigrium, Solanum incanum, Amaranthus spp, Mexican marigold, etc. In Botswana, the common weeds are Mexican poppy, upright starbur, Amaranthus spp (Thepe, ododo, omboga), Double thorn (Mosetho, Okuro), Thorn apple, Apoth (Delele) etc.

CROP PROTECTION: PESTS

Pests are living organisms which cause anger and annoyance to man due to the economic losses they cause directly or indirectly in a field or store. Can be large animals (vertebrates such as rats, birds, etc (or small animals/ invertebrates (insects, worms, etc). This syllabus focuses on the invertebrate pests.

CLASSIFICATION.

This is based on the feeding habit, which is directly related to the kind of mouthpart. They can be (i) biting and chewing pests (ii) Piercing and sucking pests (iii) Boring pests (iv) nematodes/ eelworms.

BITING AND CHEWING PESTS

These include leaf miners, termites, locusts, and beetles. Eat different parts of the plant.

Effects:

Eat leaves, thus reducing the leaf area available for photosynthesis- thus reduces yields;

Stem eaters make the plant weak and can fall over-and break the stem/die

Crop loses shape and becomes less attractive- this leads to less marketable yield.

GRASSHOPPERS:

Life cycle: Is a type that moves in swarms. Undergoes incomplete metamorphosis (Egg-Nymph-Adult). Below is the simplified life cycle.

Adult Female, 10 cm long, digs a small hole in the ground and lays her eggs. These have a frothy secretion cover which hardens.

After 2 weeks, these hatch and produce nymphs, which emerge at soil surface. At this stage they are called hoppers

The hoppers undergo 4-5 moults and an adult locust emerges with fully developed wings

The adults fly away, reach sexual maturity and mate. Fertilization occurs inside the female

The female lays several batches of eggs during her lifetime.

Control:

Spray with dipterex; if the field is small pick by hand. On large scale/ commercial farms, use aerial sprays

PIERCING AND SUCKING PEST

Piercing and sucking pests are those with a sharp needle-like mouthpart (proboscis) which first penetrates through the host surface (pierce), sends its proboscis to a tissue with fluids, and starts taking in the fluids (sucking) from the host. For it to penetrate the host surface, some pests produce chemicals to soften and lubricate the surface. These chemicals may be contained in the saliva. At times, these lubricants contain viruses. Examples of piercing and sucking pests include aphids (whose life cycle is described below), Bagrada bug, mealy bugs, red spider mites, whiteflies, Scale insects etc. Below is the life cycle of an aphid.

Figure 68: The whitefly- infested tomato plant

Figure 69: The aphid on cowpeas leaves

EFFECTS OF PIERCING AND SUCKING PESTS ON PLANTS:

Transmit viral diseases e.g. tobacco mosaic virus, groundnut rosette virus disease, and

Make holes as they pierce, thereby creating wounds used by bacteria and fungi for secondary infection of the plant;

Suck fluids from the plant, causing artificial drought and wilting of the host plant;

May make the plant be deficient in some minerals lost through the sucking process;

Some like the aphids leave behind some sugary stuff that attracts fungi, thus exposing the host plant to other diseases.

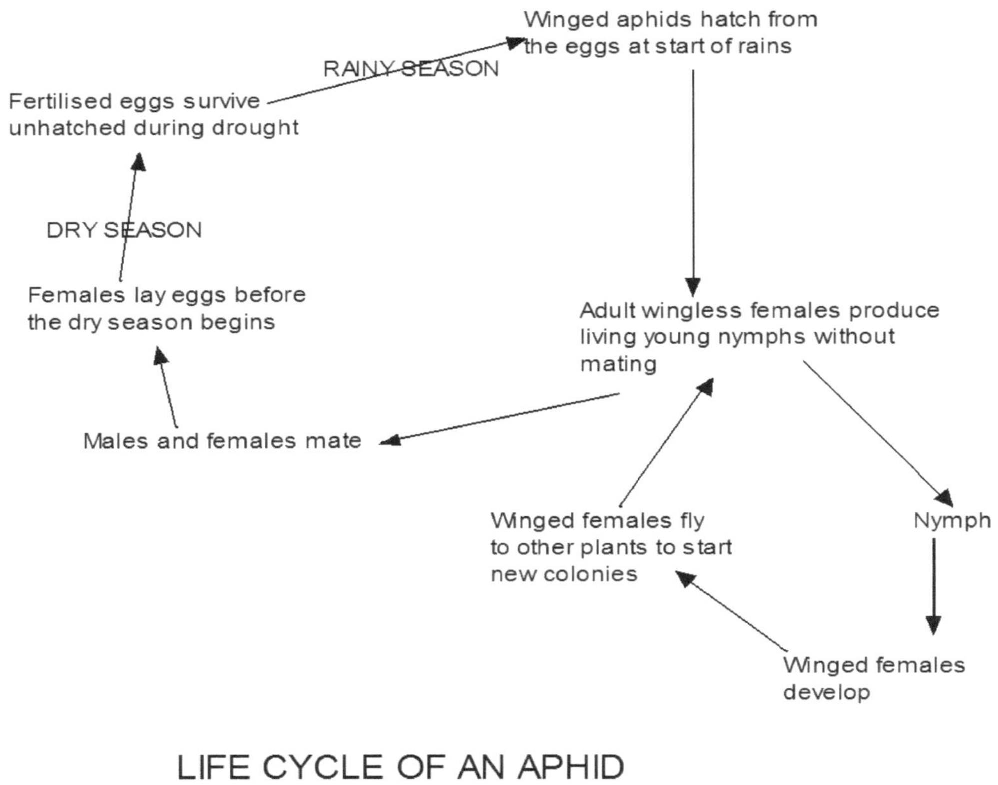

Figure 70: Life cycle of an aphid.

BORING PESTS

These include the maize stalk borer, American / African bollworm (maize, tomato), the bruchids (pulses e.g. bean and peas), and the weevil (cereals)

MAIZE STALK BORER

Freshly hatched larva migrates towards the central shoot where they feed on tender leaves for 2-3 days. They make circular holes on these leaves and later bore into top internodes and move downwards. They stay in the stem, boring vertically downwards making a hole throughout. Dead heart results. This meant that the central shoot dries up. The plant then may be unable to grow further, or its growth stunted. However, the stem remains weak, and can break even with gentle breeze. The transport of water and food within the plant is also interfered with.

DAMAGE CAUSED BY NEMATODES.

Nematodes are small tiny (microscopic) worms which often invade different parts of plants, especially the roots. They have the following effects:

Distortion of plant roots e.g. swellings- caused by the Meloidogyne spp (root knot nematodes);
May render the roots useless, thereby causing artificial drought to the plant, which ends up wilting.
Makes holes on the plant as it pierces to enter, thereby causing wounds which act as points for secondary infection of the plant;
Produce chemicals which may be toxic to the plant;

CONTROL OF NEMATODES

Nematodes can be controlled by treating the soil using chemicals called fumigants. These produce vapours which choke the pest and it dies. Other methods of control include crop rotation, sanitation etc.

OTHER PESTS:

SORGHUM SHOOTFLY (Atherigona indica):

This attacks sorghum when they are 1-2 inches (3-5 cm) tall. The fly lays eggs on the abaxial side of young leaves. The lava enter the growing shoot tip, killing it. The crop, however, can make compensatory growth by tillering.

MAIZE STALK BORER (Buseola fusca)

These attack the sweet tasting stems of sorghum. They create wounds on the stem and at the centre, causing secondary infections to occur (ie infections by opportunistic pathogens such as fungi and bacteria). Also, they create hollow centres of the plant, leading to dead hearts. Lastly, they seriously weaken the stem of the plant, thereby exposing it to lodging (ie it becomes more susceptible to lodging).

SPOTTED STALK BORER (Chilo partellus):

This also affects sorghum stems in low latitudes, causing lodging and rotting

SUDAN DIOCH (Quelea quelea):

These eat young grains in their milk stage. They make their nests hanging above points almost beyond reach of man. When they invade a crop, they move in hundreds. Children are usually used to scare them away in small fields where economic losses are easy to incur. Larger farms are self-protected because the effects of the bird are insignificant. This is why traditionally, farmers in given villages used to plan earlier and plant almost together. There was an order of planting, starting from the eldest person and/ or family in a homestead or village. Others would follow without delay, resulting in almost uniform crop growth and maturity. This ensured that the pests had a share from each field, resulting in almost no significant loss to any one individual. Only those who planted maize and light coloured sorghum were more affected from the effects of quelea than the rest.

RODENTS

Rodents are animals with gnawing teeth. Their teeth tend to out-grow the mouth, and to control this, they always want to "gnaw"- i.e. keep biting solid objects. This is part of the reason why they like digging the ground, chewing hard substances such as pieces of wood etc. Examples are rabbits, mice, rats, squirrels, etc.

The rodents

Effects / Destruction:

Dig and destroy root crops such as cassava, sweet potatoes, etc.

Dig and uproot young seedlings;

Dig out seeds before they germinate or emerge

Small mammals, such as the common house mouse, Mus musculus, pictured here, generally have more rapid heart rates than large mammals. A mouse's resting heart rate is 500 to 600 beats per minute. By contrast, an adult human's resting heart rate is about 70 beats per minute. http://encarta.msn.com/media_461556012_761572608_-1_1/House_Mice.html

Control:

Can be controlled using dogs, traps and food bait.

GENERAL METHOD OF PEST CONTROL

Use of mixed cropping as shown in in the figure below.

CROP PROTECTION: DISEASES

INTRODUCTION:

Diseases are abnormal conditions in living organisms such as crops, often caused by living organisms (pathogens) such as bacteria, viruses or fungi.

1. BACTERIAL DISEASES:

Examples: Bacterial blight of cowpeas, black rot of cabbage, soft rot of carrots, bacterial wilt of tomatoes, soft rot of cabbage etc.

Figure 71: Mixed cropping as a method of pest control

SOFT ROT OF CARROTS

This is a bacterial disease, caused by Ewirnia carotovora. It infects the taproot of carrot, making it soft. In this state, the tissues become too soft and other opportuniastic pathogens easily penetrate. The disease is believed to be soil and seed borne, and thus disease free seeds should be selected and used. Crop rotation should be practiced, coupled with crop hygiene to ensure the crop is not affected.

The carrot plant

BACTERIAL WILT OF TOMATOES

Also affects potatoes, and other crops in the solanacea family. It is caused by bacteria called Pseudomonas spp.

MODE OF INFECTION:

Is seed and soil- borne; strains of the bacteria can live in the soil and in the crop indefinitely. Is transmitted to a crop via rain-splash, irrigation, or by cutting a planting material using a knife that is already infected (i.e., has been used to cut an infected potato tuber). It enters through wounds created on the crop by pests or by any accidental injury to the crop during spraying or weeding; or as a result of pruning.

HARMFUL EFFECTS ON CROP

(i) Leaves and stem (foliage) wilt even when it is not short of water. To confirm it is not nematode infection, pull up an infected plant and inspect the roots for knots. If absent, squeeze the dying stems. A white milky sap exudes, indicating a bacterial infection. A whole crop can be lost by wilting. (ii) Masses of bacteria block the xylem tissue and disrupt water and mineral absorption and transport into and within the crop

PREVENTION & CONTROL:

(i) Use disease resistant varieties; (ii)Dress the seeds/ use certified seed; (iii) Use clean
(ii) cutting knife in potatoes; and (iv) Burn affected crops/ practice field sanitation

FUNGAL DISEASES

Examples: The mildews of vegetables, Sorghum downy mildew, sorghum smut, maize smut, leaf blight of sorghum, leaf blight of maize, Early or late leaf spots of groundnuts, charcoal rot of sorghum or maize.

SORGHUM DOWNY MILDEW

MODE OF INFECTION:

The fungal spores are carried from an infected leaf/ crop residues to a healthy plant by wind, rain splash, etc. They enter the crop via stomata and / or wounds created by pests or farm operations. It then spreads very fast within the crop.

HARMFUL EFFECTS:

Causes wilting of the crop due to excessive water absorption by the fungi (ii) Leaves and parts of stem turn brown and become spotted thus reduce the photosynthetic surface area (PSA). This reduces crop yields.

Figure 72: The mildews of kales (Bad Maro…literally meaning..the arm of a mother in law)

PREVENTION & CONTROL:

(i) Use seed dressing (ii) Crop rotation (iii) Use certified seeds of crop varieties resistant to the disease

VIRAL DISEASES

Examples: Maize dwarf mosaic virus on sorghum; maize streak virus on maize, groundnut rosette virus, cowpea aphid-borne mosaic virus, Tobacco mosaic virus on tomatoes.

GROUNDNUT ROSETTE VIRUS

Transmitted by an aphid, Aphis crassivora as it sucks fluids from the crop.

Effects on crop: (i) Distorts plant leaves; (ii) causes leaf patches thus reducing PSA and yield;

Prevention & control: Apply any appropriate insecticide (e.g. Malathion) to kill the aphids; Use clean, certified seeds ; and Use resistant crop varieties

MAIZE STREAK VIRUS DISEASE

This is a viral disease transmitted by leaf hopper, Cicadulina mbila, as it sucks fluids from the crop.

Effects on crop: (i) Stripping of leaves. Parts of the leaves are stripped yellow, thus reducing the photosynthetic surface area and yield of crop;

Prevention & control: Apply any appropriate insecticide (e.g. Malathion) to kill the aphids; and Use resistant crop varieties

MODULE 7: CROP PROTECTION: PESTICIDES

MODULE 7: CROP PROTECTION: PESTICIDES

Pesticides are chemicals meant to control the farm enemies such as weeds, insect pests, parasites and pathogens.

HISTORY OF PESTICIDES (BIOCIDES)

A pesticide is a chemical, organic or inorganic, used to control population of major farm enemies / pests (disease pathogens, insects, nematodes, weeds and related organisms) that cause economic loses in crops and livestock. They act by maiming, killing repelling or rendering impotent the victims. At best, they can be called 'Biocides' as they work against life; they are all capable of terminating life. They are meant to kill, except the actual effect they cause is dependent on the dose and site of application (the route).

CLASSIFY PESTICIDES BASED ON WHAT THEY CONTROL

Pesticides can be classified based on what they control. These include:
Insecticide: kill insect pests;
Acaricides: Kill ticks (parasites and pests in the acarine family
e.g. ticks)
Herbicides: Kill weeds;
Fungicides: Kill fungal pathogens;

CLASSES OF PESTICIDES

Insecticides may kill by a number of methods. Contact ones kill by causing itching / inflammations, wounds, dehydration or similar effects of the parts of the pest it comes into contact with. The insect becomes restless, lose appetite and dies. This targets all categories of pests.

Systemic insecticides are those which are first absorbed by the host organism in the form of a fluid, then it is spread (translocated) within the victim's system, and it intoxicates most internal systems. The victim ends up dying. It is also called translocated insecticide. It targets mostly those pests with a piercing and sucking mouthpart such as aphids, bugs (bagrada, stink), etc.

Stomach insecticides are those normally first eaten up and ingested in the solid form by the pest, then it causes stomach problems, leading to the death of the victim. It targets those pests with a biting and chewing mouthparts such as grasshoppers, corn cricket, etc.

WORLD USE OF PESTICIDES

In 1989, the per capita pesticides consumption in Kenya was 0.5 kg. A total of 12,500 tons was used then. By 1996, this is estimated to about 25,000 tons, giving a growth rate of 14%. At global level, the 1981 use was 1.8 billion kg, with the United States alone using 20% of this (About 0.36 billion kg). By 1994/95, however, USA alone was using 544 million kg of active ingredient of pesticide, having shot more than ten-fold from the 1993/94 and 1994/95 periods which were 100 pounds and 1.2 billion pounds respectively. By inference, the global pesticide expenditure in 1993/95 was 400 million pounds and 4.8 billion pounds respectively.

Herbicides use increased most rapidly compared with other pesticides during the first decade of the green revolution. However, in the 1990's farmers have tended to use more fungicides and insecticides, seriously reducing the prominence of herbicides characterizing the onset of green revolution. Due to the increased pesticide use at global level from 135 million kg of active ingredient to over 2 billion kg of active ingredient in 1993/94 and 1994/95 respectively, a lot of concern is already being raised. WHO estimates that up to 375,000 persons are poisoned annually by pesticides.

CLASSIFICATION OF PESTICIDES

In this context, they are broadly grouped as:-

Herbicides

Others (fungicides, insecticides, nematicides)

GENERAL CHARACTERISTICS OF PESTICIDES

(a)Kill or suppress one or more groups of living organisms / farm pests (i.e. are biocides).

(b)Many are broad spectrum, i.e. work against a wide range of organisms, including non – targeted ones.

FATE OF PESTICIDES AFTER APPLICATION

Highly soluble ones may tend to be washed by rains into the soil and thereafter leached to deeper layers beyond reach of plants. This is wasted.

Some could be washed into the soil and get adsorbed into soil colloids' exchange sites. If it is relatively inert and does not meet an active reactant, it can be slowly released and absorbed by plants or washed into soil solution.

Insoluble ones could be carried away by erosion, and carried to the nearby water bodies such as dams, rivers etc. This becomes toxic to aquatic organisms as well as water consumers/users such as livestock and man.

Some easily get volatilized, i.e., turning into vapor thereby affecting plant shoots. Animals also get exposed via oral, topical, and respiratory routes. It becomes an atmospheric pollutant.

Some of the chemicals, especially dust, and get carried away from target sites by drift, thereby posing dangers to any organisms that directly depend on the atmospheric purity for survival and livelihood.

Some are so toxic that they kill even the organisms they were meant to protect. This is particularly when the dose higher than necessary. This is most common when the pests have developed some resistance, and higher more lethal doses are applied.

In relation to these, some additives are normally applied to impart the desired characteristics on pesticides. This is meant to make them more effective. These additives include:

Stickers: Are added to dust or spray to improve the adherence capacity (tenacity) of a pesticide. These include starch, emulsions oils, etc.

Safeners: Reduce toxicity of the chemical. This includes lime added to copper to reduce copper toxicity to the organism it is meant to protect.

Spreaders: Are added to spray mixtures to establish an even spread between the pesticides (chemical) and host surface. They also reduce sedimentation in the spray tank. Examples include minerals oils, sop amines, etc.

Carriers or Diluents: are added especially to dust. It necessitates the dilution of technical material with any fine powder called a carrier or diluents. These include kaolin (clay.

CROP PROTECTION: HERBICIDES

Herbicides are chemicals meant to control the growth of plants (mostly unwanted plants called weeds). They can be selective or non-selective; pre-emergence or post-emergence; or contact or systemic

TYPES OF HERBICIDES

Selective herbicides will kill one category of plants (crops and weeds alike); it is upon the parson applying to know whether it will affect the crop or not. The chemical cannot decide to spare your crop- it can be killed just like the weed. Selective herbicides include 2,4-D, 2,4,5-T, MCPA , Dalapon . Non- selective herbicides include: paraquat, sulphuric acid etc.

CLASSIFICATION OF HERBICIDES

This can be on basis of either mode of action chemical structure or effect on weeds.

(I) EFFECTS ON WEEDS

Herbicides kill weeds either by contact or translocation. They are either classified as contact or translocated herbicides respectively.

CONTACT HERBICIDES

Kill weeds upon contact, points, i.e. if the herbicide comes into contact with leaves it is the leaves that are killed; if it gets in contact with the stem, the stem is the victim. It can therefore be a good means of selecting plants or parts of plants to kill. If the whole field is to be freed from weeds, complete herbicide coverage becomes essential. This type of herbicides is more effective against annual that is most cases do not have perennating organs. Most are non-selective systemic herbicides - another name for translocated herbicide. Upon getting in touch with a weed, it is absorbed and translocated within the phloem and xylem up and down the plant. Thus the entire plant at least has a taste of the chemical. Upon reaching delicate parts such as flower buds young pods, meristematic tissues, their effects are immediately noticed. They kill this tissue slowly; and gradually kill the rest of the plant as its functional parts degenerate. This, however, is only possible if correct concentration is applied.

SYSTEMIC HERBICIDES (SH)

SH in lower doses act as growth modifiers (e.g. as hormones). They largely affect the plants' physiological system such as photosynthesis, respiration, photo – respiration, nutrients metabolism, protein synthesis, etc. Some kill or cause abnormal tissue development (carcinogens and mutagens); or inhibit nitrogen metabolism e.g. DNA or RNA formation such as the triazine. Due to their translocated nature they are more effective against perennial weeds than their contract counterparts. They are also cheaper to use since one does not need full or complete coverage for them to be effective. When they are applied to the soil as soil sterilants, they prevent weed growth when they are still on the soil pre-emergence stage.

PRE-EMERGENCE AND POST-EMERGENCE HERBICIDES

In this regard, herbicides can be further classified as either pre-emergence or post – emergence. Pre-emergence herbicides are those that are popularly applied to a soil surface just before crops emerge, even though they may have germinated. They are believed to kill any weeds that may have emerged. Emergence means the actual appearance of the young plant above the soil level.

PRE-EMERGENCE HERBICIDES

These are largely non-selective because they are meant to kill any plant that has emerged on the soil surface. The crops then are safe, as they may not have even germinated.

POST-EMERGENCE HERBICIDES

These are those herbicides applied to a field after crops have appeared above the soil surface. Since the crops are presumably mixed with weeds, the herbicides applied at this stage have to be selective. They should kill the weeds without having a significant effect, if any, on crops.

CHEMICAL STRUCTURE:

This is based on whether an herbicide is organic (of carbon origin) or inorganic (non carbon origin).

(A) INORGANIC HERBICIDES:

The earliest herbicides used included inorganic compounds such as brine (concentrated solution of sodium chloride); salt / ash mixture; Copper Sulphate; sodium arsinite (a sterilant); salts of Iron, copper, sodium, ammonium or sulphuric acid. The latter class/group was introduced as foliar sprays. Inorganic herbicides however prove highly phytotoxic and are currently being replaced.

(B) ORGANIC HERBICIDES:

These largely include phenoxyls: aliphatic (straight chain); aromatic acids (with a benzene ring); aniclides, amides; nitrophenyls; nitroaniline; areas and other few unclassified herbicides.

MODE OF HERBICIDE ACTION

Herbicides can also be classified on basis of their mode of action – either as selective or broad spectrum (non-selective). Many factors contribute to the selectivity of herbicides, among them being dose and time of application. A detailed account for herbicides selectivity considers the following as important components.

SELECTIVITY OF HERBICIDES

SECRETS BEHIND THIS PHENOMENON

Many factors are responsible for selectivity of herbicides. These include physiological factors, morphological factors, dose etc. These are discussed below.

(A) PHYSIOLOGICAL OR METABOLIC FACTORS

Some plants are capable of breaking down an herbicide, or suing it as metabolic. An example is maize, which is capable of detoxifying 2, 4-D into 2, 4-B.

(B) MORPHOLOGICAL FACTORS:

This refers to such aspects as location of growing points (meristems); root system differences (some deep, some shallow); differences in height between crop and weed; leaf angle; and leaf composition (some highly cuticularised others softer).

© Stage of plant growth; the smaller/younger plants are more prone to herbicides than older ones.

(d) The herbicides: these include characteristics such as concentration, application method, and mode of action (contact or otherwise).

(d) Environment – edaphic and climatic factors determine the state of the crop and the fate of herbicides.

(C) CHEMICAL STRUCTURE:

This is based on whether an herbicide is organic (of carbon origin) or inorganic (non-carbon origin)

The earliest herbicides used include inorganics such as brine (concentrated solution of sodium chloride): salt/ash mixture; copper sulphate; sodium arsinite (a sterilant); salts of Iron, copper, sodium, ammonium or sulphuric acid. The letter class/group was introduced as foliar sprays.

CLASSES OF HERBICIDES

1 Organics

PETROLEUM OILS

These are contact, non – selective (broad-spectrum) herbicides that work by penetrating and disrupting plasma membranes. Are very effective at seedling stage.

(B). PHENOXY HERBICIDES/PHENOXYLS: These are largely hormonal in actions, being phenoxy acetic acid derivatives. The common growth hormone, thus this class has an auxin –typed or growth regulating hormones, and very effective even at low doses. Examples are :Dichlorprop; Silvex; Mecoprop; 2,4-D; 2,4,5-T; 2,4-DB; MCPA; 2,4-DES; MCPB; MCPP; Di and tri methyl amines; and Di and tri ethanol amine

MODE OF ACTION:

Phenoxyls affect cellular division, activate phosphate metabolism, and modify nucleic acid synthesis. 2, 4, 5 – T is phenoxy compound of 2,4-D. When it is being purified, it produces a dioxin. Dioxins are known to be serious carcinogens and mutagens.

PROPERTIES:

Phenoxyls are non-staining, non-poisonous when in pure form, and not flammable. They dissociate in water and react with divalent cations such as Calcium, Magnesium etc. This makes the ions not available to plants, i.e. they are fixed. Thus they are not good for soil application.

The general structure of a phenoxyl is O-(CnH)-COOH

(C) ALIPHATIC HERBICIDES (straight – Chain). These are largely acids and used against broad-leaved weeds, such as seed grasses. They are most effective at a dose of 2.5 –4.5 kg Ha-1

EXAMPLES; Trichloro acetic acid (TCA); 2,2 – dichoro propionic Acid (2,2 –DPA); and Dalapon.

GENERAL STRUCTURE: CnHmHLX COOH, Where n, m and x are whole numbers while HL is a Halogen.

AROMATIC ACIDS:

These have a kekule (Benzene) ring and a carboxylic acid group, -COOH. They are therefore non-biodegradable due to the presence of the Benzene ring. Besides, they have long-term soil persistence due to their non- biodegradability. This makes them highly toxic to both plants and animals and therefore, suitable as soil sterilants. However this is not recommended from environmental health and biodiversity point of view. Examples are 2, 3, 6, -TBA; Dicamba; Chloramba; Dacthal, among others.

(e) ANILIDES: Includes Alachlor etc. This can poison fish if it gets into aquatic systems.

(f) NITROPHENYLS: These are nitrogen compounds with 2 benzene rings. The latter makes them have high environmental persistence, high toxicity and non-biodegradability characteristics. Their names have 'Fen' as prefix or Suffix.

(g) NITROANILINE: this is an organic nitrogen compound. It is highly volatile, rendering it as a fast atmospheric pollutant. The herbicides have "lin' as part of their names.

(h) UREAS: Are of Urea origin. The specific examples have 'UR' on their names such as diuron, monuron

(i) URACILS: Have CIL as part of their name e.g. Uracil.

(j) PYRIDINES: Have 'uat' as part of their names e.g. paraquat, they are heterocyclic Nitrogen compounds. They inhibit photosynthesis, and affect synthesis of enzymes and proteins; they are contact herbicides more effective in the presence of light. Are largely used against broad – leaved weeds and woody plants. Simazine is a popular pre plant or pre-emergence herbicides in maize fields. Other examples include triazines, piclorum, Atrazin etc.

(k) POLYAMINES: These have a mysterious role in growth and development of plants. They interact with nucleic acids and membranes. They are also associated with plant responses to a wide range of stresses, including pH; osmotic shock; cold; Sulphur-based pollution; cadmium toxicity and magnesium deficiency. It facilitates accumulation of putrescine in response to these stresses.

INORGANIC HERBICIDES

Brine (concentrated NaCl), Salt/ash mixture

Copper compounds e.g. Bordeaux mixture, CUSO4

Sodium Compounds other than NaCl e.g. Sodium arsinite, a soil sterilant.

Arsenical compounds, which check Sulphydryl group-containing enzymes, as well as uncoupled oxidative phosphorylation.

Salts of Copper and Iron

Salts of ammonia (NH_4^+) e. g. NH_4NO_3 which cause osmotic imbalances in contact with plants;

Sulphuric acid – Which is a desiccant and oxidizing agent; dehydrates the plants. They are applies as foliar sprays and largely act against weeds by plasmolysis.

CROP PROTECTION: USE OF FARM CHEMICALS

THE EARLIEST CONCERNS ABOUT CHEMICAL USE

DDT's (Dichloro – diphenyl – Trichloroethane) insecticidal properties were not discovered until 1939 even though it had been in the market serving armament roles for some time. It became the typical chemical of the green revolution. This was by virtue of its availability, affordability, cost effectiveness, broad coverage (non-selective nature) and durability / persistence. When its non - biodegradable qualities were discovered, it was soon realized that the resulting environmental persistence was a serious hazard to reckon with. It was therefore banned in many western countries.

For Africa, many tropical and developing world regions, the DDT banning was never to be effected until after many years. This was because the Chemical's immediate benefits far outweighed immediate loses. When clear signs of pest resistance development and the subsequent need for use of larger, more lethal and expensive doses arose, even the laggard Countries had no otherwise. This was despite the fact that there were no appropriate substitutes for DDT. However, a foundation for Chemical agriculture had been deeply laid despite concerns of side effects of these chemicals. In the Kenyan Highlands or enclaves, use of chemicals picked up so fast in production of tea, coffee, pyrethrum, wheat, maize among other cool – season crops. This high chemical input made East African Countries to acquire a dubious record of using almost 25% of all pesticides used in Africa South of the Sahara by 1962.

ERA OF ORGANIC PESTICIDES

The year 1942 marked the beginning of the present era of organic pesticides causing a diversion from the hitherto popular organo-metallics. The first chemicals discovered were 2,4-D and MCPA, both largely used as herbicides. To date, and more pesticides have been discovered and manufactured for very specific uses. The include insecticides, herbicides, nematicides, viricides, rodenticides, fungicides, bactericides and acaricides. The general name for all these chemicals is pesticides.

MAGIC AND PHILOSOPHY OF PESTICIDES

Biocides have been used for centuries to kill crop pests. The Greek poet and philosopher, Homer (1000 BC) mentioned pest – averting sulfur as a fumigant. Sulfur used to be burnt before being used in fumigation. Cato (200 BC) used smoke from a mixture of bitumen and sulfur to kill grape fruit insects. Demo Critus (470 BC) recommended sprinkling plants with a solution of pure amuria Olives to reduce blight attack and kill plant – root adhering worms.

ROBIGALIA

In 79 AD, Pliny reported the insecticidal value of arsenic. The Chinese were using this same chemical to control garden pests by 900 AD. The Romans have been long aware of wheat rust, and used to have a festival called Robigalia. This was held on 25th April annually to appease Robigus, one of the Roman Gods believed to cause the disease epidermics. They believed that this gave them good wheat- rust free harvest. Robigalia involved pouring red wine and sacrificing a red dog on the sacred grove of the god to get of the rust disease. These were some of the early efforts to understand and control pests. Pesticides were just one class of class of farm chemicals. Put together all Chemicals used in the farm for the purpose of boosting farm production are agrochemicals.

EQUIPMENT FOR APPLYING PESTICIDES

Pesticides can be applied by use of tractor- drawn equipment e.g. boom sprayer, or a knapsack sprayer shown below.

PARTS OF A KNAPSACK SPRAYER

Nozzle ; Cap; Tank; Handle; Delivery hose; Lance handle; and Trigger

Activity: Dismantle and re-assemble a knapsack sprayer

Figure 73: The boom sprayer

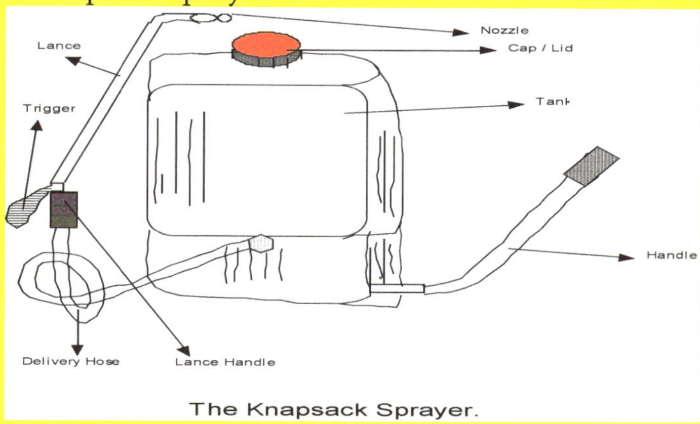

Figure 74: THE KNAPSACK SPRAYER

SOME DEFINITIONS

Active ingredient: This is the poisonous portion of a substance, pesticide or chemical. Example caffeine is the active ingredient of coffee.

Inert material: Is the non- poisonous part of a substance or formulation. It may be added just to dilute, spread or improve the physical properties of the active ingredient and the pesticide as a whole. Clay is commonly used. It is sometimes called filler material.

PESTICIDE FORMULATIONS

This means the physical form in which pesticide exists or packaged to operate. These include dusts, granules, fumigants, sprays and aerosols.

Dusts: Are formulations in which the physical state is in fine dry powder form. May be applied directly on plant or on seeds for preservation e.g. malathion dust.

Granules: Is a formulation whereby the state of the pesticide is in small solid particles such as pellets, and may take different shapes and sizes. May be applied directly in the soil or used as baits. Kill pest when they are eaten up (i.e. biting and chewing pests.

Fumigants: The active ingredient may be initially in liquid or solid form, but quickly changes to vapour (fumes) upon application. Often applied against soil nematodes, farm stores etc. The area of application may have to be covered for some time for the chemical to take effect, as otherwise the vapour would soon disappear. And only inert material remains. Example are furadan (soil applied), ethyl bromide etc. They kill the pest by inhalation.

Sprays: Are formulations in liquid forms which are then released through fine nozzles to give small particles of liquid. All chemicals applied using knapsack sprayer are usually sprays. Are often diluted with and applied

Aerosols: Are formulations stored as liquids under pressure. Toxic gas released when pressure released. Sometimes, the oil is the medium of pesticide preparation

HANDLING AND STORING FARM CHEMICALS

It is important to take precautions when handling and applying chemicals to the farm. The following are general guidelines:

Read all instructions before handling;

Use only for the correct purpose;

Use the correct dose;

Use protective clothing e.g. hand gloves, respirators, goggles, helmet, gumboots and overall; (Plastics are preferred because they are water repellant and drain easily without absorbing the chemical). This is illustrated in the picture below.

Do not drink, eat or smoke when handling farm chemicals;

Wash hands thoroughly with soap after handling farm chemicals;

Keep chemicals out of reach of unauthorized persons e.g. children;

Label the chemicals properly;

Dispose of containers safely after using the chemicals.

Figure 75: The Personal protective clothing used during pesticide handling sessons

ENVIRONMENTAL HAZARDS OF FARM CHEMICALS

May pollute water sources such as dams, lakes, etc;
Some are poisonous to man and livestock;
Some may reduce fertility of organisms;
Some may cause birth defects e.g. may lead to lame or mal-formed offsprings;
May accumulate in the bodies of animals (bioaccumulation) and along food chains..
May kill useful insects and birds;

BIOLOGICAL PEST CONTROL

This means the action of predators, parasites or pathogens in maintaining another organism's population density at a lower average than would otherwise occur OR the study and utilization of parasites, predators, and /or pathogens for the regulation of most population densities Or any condition under which the activity of the pathogen, pest or weed is reduced through the agency of any other living organism or life – forms, other than man himself.

WHAT BIOLOGICAL PEST CONTROL INVOLVES

Research on pest and predator habits.
Ecological field tests; such survival ability in as many environments or habitats as possible.
Introduction / importation and establishment of natural enemies, especially if within the locality those existing cannot be used for one reason or another.
Argumentation of natural enemies, i.e. ways of helping the already existing natural enemy (predator) to be more effective by manipulating the environment. This can be done by:
Changing cultural practiced by either ensuring main or intermediate host to exist throughout.
Releasing more predators (e.g. those that have been cultured in a laboratory).
Use genetic means, such as sterilized individuals.

KEY QUESTIONS IN BPC

How many attacks can a predator make for any given density of the host?
Does the attack rate depend on the predator's searching behavior, and is this searching behavior influence by prey density?

Does the predators' appetite for the parasite egg change with supply?
Does the proportion of prey killed depend on the prey density?
However, this trend almost always maintains an economic pest / insect level. Therefore BC strives to change this balance and maintains the set / achieved equilibrium by releasing predators when the prey population tends to increase.

ASSUMPTIONS IN THE BPC.
The predators do not attach a prey for fun. (2) As soon as the predator encounters a prey it attacks; and every attack results into death.

CHARACTERISTICS OF A SUITABLE PREDATOR
Must show high searching ability.
Must be highly specialized (most – Specific) to the host, i.e. monophagous.
Have good reproductive capacity relative to the host.
Show good adaptability to a wide range of environmental

FACTORS THAT HINDER SUCCESS OF BPC
Adverse environmental conditions
Inadequate supply of supplementary food, (i.e. when the prey population is low, there should be an appropriate
Lack of alternate host
Bad cultural practices (e.g. burning)

ADVANTAGES OF BPC
It is cheap to run where successful
Usually persistent without the need for recurrent expenditures
Do not entail any significant genetic counter – attack in the pest / host
No resistance develops.
Leads to no incidence of secondary pests.

USE OF ATTRACTANTS / HORMONES AND STERILE MALE TECHNIQUE (SMT).
(a) STERILE MALE TECHNIQUE (SMT).
This involves rendering insects, especially males, unable to produce by sterilization. It is very effective among organisms that only give birth or lay once in their lifetime. This is because, as they become ready to copulate, they are not effective. A whole generation can

thus be eliminated, with improved pest control rates. However, this may kill a species, causing serious ecological gaps that then may need to be filled by some alternative means e.g. organisms.

(b) USE OF HORMONES / ATTRACTANTS

The basis of this method is: insect behavior of looking for food, place of laying eggs or searching for mates, etc are each controlled or influenced by chemical stimuli in the environment. These operate through the olfactory sense. The natural attractants are called phenomes, and are of different kinds. Air-borne pheromones are generally volatile, are of low molecular weight, and many are derivatives of terpenes and fatty acids. Pheromone receptors are located mainly in insect antennae, but the mechanism whereby small molecules interact with some component in a receptor cell in such a way as to cause a cell to respond are uncertain.

Pheromones fall in two general groups:

Releasers or

Releaser pheromones initiate immediate behavior responses in insects upon reception, whereas primer pheromones produce physiological changes in an animal, which ultimately result in a behavioral response. Releasers act mainly through the nervous system, while primer action usually involves the endocrine system as well. Release pheromones may cause sexual attraction, alarm behavior, recruitment and aggression.

(I)ALARM PHEROMONE:

These are applied primarily by social insects to warn other members in the colony of impending danger. In combat among social insects, the contents of the mandibular glands are generally discharged onto the enemy, which then becomes labeled as an aggressor. . By diffusion of the pheromone vapors through the air, the alarm is communicated to other members of the society. The alarm substances of the ants are primarily terpenes and non-terpenoid hydrocarbons. Insects show far less sensitivity and chemo-specificity for alarm pheromones than for sex pheromones. Thus for alarm pheromones, selectivity is based more on volatility than on unique structural characteristics. Advise members of impending danger. This makes them prepared for either flight or a fight. If this is sustained for a long time they become tired, being on the offensive throughout. They, therefore, forget about sex and food.

(II) TRAIL-MARKING / RECRUITMENT PHEROMONES

This is common among insects that travel long distances to look for food, and use the marked trails as a guide to their residence, e.g. bees. The pheromone functions as a navigational aid in directing other members of the colony to a distant location. Recruitment by chemical communication is widespread in social insects such as bees, termites and ants. Most frequently, the pheromone trails are laid by foraging workers as they return from a food source e.g. in honeybee (Apis mellifera). Glands produce the chemicals for marking rich food sources, and the nest. Such pheromones can be mimicked (i.e. produced synthetically or artificially), placed in space to attract or aggregate the pests either to a bait, away from the hosts, or a place from where they can be killed or incapacitated. Alternatively, the artificial pheromone can be spread to disorient the pests.

(III) AGGREGATE PHEROMONES

These are used to aggregate (assemble) the insects to a feeding or mating site, and occur mainly among the coleoptera (the beetle, grubs, slug and weevil family). In such cases, feeding on the host triggers the production and release of pheromone. This frequently causes a mass infestation at the particular site. In pest control, this characteristic of the pheromone can be manipulated to attract the individual pests together to a bait from where they can be killed using any means such as chemicals, etc.

(IV)SEX PHEROMONES

The sex pheromone is usually released by the female to lure the male from a distance, and to further excite it sexually. Sex pheromones released by males primarily act in sexually exciting the female. Insects are highly sensitive and specific for the sex pheromones of their species, and require specific geometry and optimal isomers. When using sex pheromones to control pests, the idea is to release a chemical that causes excitement, but has no orientation. The males look for females in vain.

(V)INSECT GROWTH REGULATORS

This involves the release of a hormone- juvenile hormone released by corpus allatum. As long as this is kept in good supply and concentration within the insect, no Ecdysis (molting) can occur. Normally, insects undergo a number of molting stages before acquiring an adult status. Any mechanism of trying to suppress molting can successfully and permanently withhold development, so that the insect remains permanently in pre-adult stages, and therefore unable to reproduce. If this is successfully done for a long time, the lifespan of the insect can end without it reproducing.

Juvenile hormone analogues are insect-specific. Therefore it is not possible to have wide or broad-spectrum coverage of many insects using them. Ecdysis is another hormone released by the insects' prothoracic gland. It induces pupal and adult molting. As long as juvenile hormone is in good supply, its production is suppressed. This means the two act antagonistically- one can only be there if the other reduces or is absent. An example of juvenile hormone analogue is methoprene. Their field success is still being tested, but they are very effective in the laboratory.

CONCLUSION ON HORMONES

Pheromones, however, break down very fast and cannot be effective for long. Also, mechanisms to synthesize artificial pheromones need complex steps that are a preserve of only few specialized scientists.

ALGAE AS PATHOGENS

(i) Cephaleuros virenscens causes red rust (in India); (ii) Black spot of peppers (iii) Algal spot in citrus

(B) PLANT PATHOGENIC SEED PLANTS

Mistle toes: Is a parasitic green plant which gets water and minerals from the host plant

Dodders: has no chlorophyll, causes physiological imbalances in the hoist plant so that it may cause equivalent of mutations. It manipulates the plant to serve its own needs (like a plane hijack).

Striga spp (Witch weed).: Infects parasitically the roots of cereal crops (4) Orobache spp: Witches broom / broom rapes) affects maize and tobacco.

MODULE 8: ORNAMENTALS, LAWN MANAGEMENT & LANDSCAPING

8.1 ORNAMENTALS
8.2 LAWN MANAGEMENT
8.3 LANDSCAPE DESIGN

MODULE 8: ORNAMENTALS, LAWN MANAGEMENT & LANDSCAPING

8.1 ORNAMENTALS

Ornamentals are plants with beautiful outlook, and is a term often used interchangeably with flowers. This topic is about production and management of flowers. Ornamentals are useful in (i) Decoration (ii) Environmental conservation- i.e. some of the flowers in your garden could face the risk of extinction if they were elsewhere (iii) Improves the value of land- i.e., you earn more from your land if you landscape it well with flowers (iv) Provide shade especially if they are tree or shrub-like, e.t.c. At the end of it, you should know the meaning, types / classes, examples, and production requirements of at least one ornamental plant. You should also be able to relate ornamental plants to landscaping. Three kinds of ornamentals exist: Pot flowers, bedding and flower cutting.

POTTING (POT PLANTS):

Are plants raised or grown in pots for display purposes because of their attractive foliage, growth habit or their flowers. Examples are chrysanthemums and oxalis

(i) Pink star / oxalis plant (Oxalis semiloba): a groundcover plant that is evergreen. Grows to 20 cm tall, and likes sun. It has clover-shaped leaves, and produces pink flowers in spring and summer

(ii) Chrysanthemums: A popular houseplant with colourful, tightly bunched petals radiating from a dense central eye (like the sunflower). They are available throughout the year in all types of colours. Flower in 2 months.

BEDDING PLANTS:

Are flowering or foliar plants planted in flowerbeds to provide a temporary colourful display. Examples are: (i) Marigolds (ii) Dianthus

GROUND COVER:

Ground cover plant;

Bedding ornamental plants

FLOWER CUTTING:

Roses e.g. white banksia rose (Rosa banksia). This is a climber and semi-deciduous. It grows to 3m tall, and produces showy double white flowers in spring. Roses are cut flowers for decoration due to their wide variety of colour, scent and hardiness.

Figure 76: The cut flowers in a table pot

(ii) Bougainvillea (orange-red)/ Bougainvillea spp. is an evergreen climber which grows to 3-5 m tall. It is sun loving, and flowers most of the year. Can also be grown as a flowering hedge. It is a good security plant. Other varieties produce dark purple, and others white flowers.

NOTE: Can use pots, seedbeds and seed boxes in planting ornamentals (flowers). Pots should be perforated to improve drainage. Seed boxes are mostly wooden with spaces between the planks of wood used. These improve drainage. Seedbeds can also be aligned at the edges using bricks or even long planks of wood to effectively contain any additives e.g. water, fertilizer etc. The equipment has to be kept clean as a precaution against disease.

MARIGOLD:

Like periwinkle, Marigolds provide yellow and orange colours in a bedding. They are easily raised from seed and are long flowering. They vary in size from French marigolds to the taller African marigolds. They are planted out into the flowerbeds before the flowers open.

The periwinkle flower (pink and white flowers)

PREPARATION OF STANDARD SOIL MIXES:

Done when suitable soil is not available. Soil is improved by mixing with other types of or the addition of compost and sand. If soil too sandy, can be mixed with clay or compost at 5 sand: 1 clay. If it is too clayey, it can be mixed with sand at 5 sand: 1 clay.

PRICKING OUT:

This means removing the seedlings of 2-3 leaves from the nursery / beds into pots awaiting transplanting into the field. Not common in marigolds.

HARDENING OFF OF SEEDLINGS:

The exposure of seedlings within the nursery to harsher conditions so that they become hardened and acclimatized to the outside conditions. This is done prior to planting to help them reduce transplanting shock in the seedbed (field). May be done by reducing amount and frequency of watering, removing shade, or partially uprooting to reduce moisture absorption by the roots.

Plants need to be watered immediately they are transplanted to reduce transplanting shock

TIMING OF PLANTING:

Should be done at the beginning of rainy season. However, all season production can be guaranteed with irrigation. This gives the advantage that some market can be tapped in time, e.g. by timing the production so that the flowers are ready just when the season is on, e.g. Christmas being a festive season. In such cases, plating time depends on the expected crop maturity in synchronization with the market.

TRANSPLANTING:

Marigolds are moved from nursery to the seedbed (field), i.e., transplanted when 5cm tall before their flowers open, but after hardening off. The seedlings should be watered well before uprooting, which should be done with a hand trowel leaving enough soil, and the entire root intact. The planting holes should have 2-5 g of super phosphate as basal dressing well mixed with soil. The transplants should be watered immediately.

SUITABLE SPACING:

15 cm x 45 cm (This may vary depending on variety).

FERTILIZER APPLICATION:

Apply 10 g of super phosphate as basal dressing at transplanting. When there are good signs of the plants becoming well established (2-3 weeks later), apply 30 g of 2:3:2 mixed with 5 litres of water and spread at 125-200 ml / plant (i.e., 0.75-1.2 g of 2:3:2 / plant OR averagely 1 g of 2:3:2 per plant).

SUITABLE IRRIGATION METHOD:
Overhead mist system; provides a uniform and regular water supply, automated.

TIMING AND REQUIRED FREQUENCY OF IRRIGATION:
A good method is desired since marigolds derive their beauty from the firm, colourful flowers. They readily wither with water stress. Overhead mist system is the best in a shade house as it gives a uniform cover and reduces labour. Can add some nutrients to the water to keep the plants healthy. Water mature plants after 3-4 days, but young seedlings need more regular watering. Can use electronic timers and moisture meters. Evening or Night watering encourages fungal diseases.

COMMON PEST:
The black bean aphids strike especially in warm moist weather. Plant leaves and stems become distorted and weak, and leaves covered with sticky honeydew, which attracts ants, and later become a sooty mould. Controlled by Spraying with dimethoate.

COMMON DISEASE:
Damping off is a fungal infection of the neck of the plant at the soil level, which makes the seedlings to topple over. Mostly common in damp places; therefore over-watering should be avoided. Any copper-based fungicide e.g. copper sulphate can be used at planting or seedling emergence stage to control it.

PRUNING AND TOPIARY OF ORNAMENTALS:
Topiary is the judicious pruning of ornamental shrubs into geometric shapes (e.g. sphere, cube etc) or shapes of animals such as birds. Plants are pruned methodically and intentionally to form the desired shape.
Pruning is the removal of unwanted parts of plant e.g. diseased and dead parts, and is meant to maintain size, shape and beauty of the plant. In ornamentals, the objective is to regulate the plant to its most effective production of flowering, fruit and foliage- and maintain the right balance among these 3. Need a sharp pruning equipment (pruning saw, secateurs, bow-saw, pruning shears) to reduce size of wounds. Pruning improves aeration and lighting of the flower, thus giving it the best appearance.

PREPARATION AND MARKETING OF MARIGOLD
This can be done at 2 levels: seedling or flower stage. First placed in disposable cheap seed trays (each normally contains 12). This allows easy pricing and counting to the customer. Should not take long between cutting and selling, and are at times harvested on order.

The plants must look healthy, and disease-free with the flowers not yet quite open. The seed trays should be moist to maintain turgid and sturdy plants. Can be sold to individuals, through plant nurseries, at garden, super-markets, (open-air markets and at times in the streets), or flower shops

8.2 LAWN MANAGEMENT

INTRODUCTION:

Lawn is a decorative turf, or simply, the thick mat of grass maintained in a short, presentable form for sports, beauty, etc. It can be established using sexual methods (seeds or seeding) or vegetative material such as turf (sod, plug, stolon etc) by asexual methods. A sod measures 1m by 50 cm, and is a thick mattress-like stuff comprising live lawn plants uprooted with roots intimately held up together with the soil. It thus gives an instant utility product- i.e. one which can be used immediately it is laid. But sod is difficult to handle- e.g. carrying / transporting. It can therefore be cut into pieces of no standard size and laid on the ready ground. The small pieces are laid (plugged) into the prepared surface, and is why the stuff are called 'plugs'. You can be a landscape contractor if you practice the lawn establishment procedures well and internalize them. What do you think are some of the disadvantages of seeding? Give examples of plant species used for making a lawn. Also suggest in what ways the lawn and ground cover are related, i.e., common features.

SOME DEFINITIONS

LAWN:

Lawns are decorative turfs. Turf is a covering of mowed vegetation growing intimately with an upper soil layer intermingled with roots and stems. Lawn comprises low-lying and spreading plant species (mostly grass), which cover the soil surface and are planted to give an attractive and evergreen landscape. Alternatively, ornamental grasses can be used.

ORNAMENTAL GRASSES

These are grasses needing very little maintenance, are drought resistant, and grow well in most soils.

A well groomed landscape with lawn, well laid and decorated stones and flowers

TURFS

Turf is a covering of mowed vegetation growing intimately with an upper soil stratum of intermingled roots and stems. They are of different kinds

Lawn turfs: are for decoration, and are sometimes called ornamental turfs. Are usually cut slightly taller than the sports turfs

Sports turfs: For use in football, golf course etc. Are usually the shortest cut. Golf courses may need even more fine quality lawn species.

Utility turf: Exist primarily for soil stabilization; along road sides, they absorb toxic vehicle emissions; along airport runways, they reduce dust and thus help prolong lifespan of the engines. Are usually the tallest cut among the different turfs.

IMPORTANCE OF LAWNS.

Makes an attractive site / beautifier e.g. decorative lawns as in parks, homes and gardens.

Control soil erosion (i.e., soil stabilization) as in hillside where a thick mat has formed.

May reduce dust and increase lifespan of engines e.g. the utility lawns used along airport runways.

May make a ground more suitable for a sporting activity e.g. sporting turf used in golf courses and football pitches.

Increase the value of estates / property.

PREPARATION OF LAND INTO A TILTH SUITABLE FOR LAWN:

The first thing is land preparation i.e. slashing to create a proper environment for cultivation. Then soil preparation follows. The steps are:

(i) Till the soil to 30 cm deep to loosen it, uproot weeds, and expose pests

(ii) Amend the soil by adding appropriate chemicals to correct pH, e.g. lime

(iii) Grade the soil, i.e. make the site level, or give it the desired slope / terrain

(iv) Control weeds by watering the soil deeply, leave weeds to emerge, and then kill them using any appropriate method. Repeat this a number of times until there is no more active emergence of weeds.

(v) Floating: filling in low spots and deep marks that may have resulted from watering during weed control. Can use clean, weed-free material e.g. sand.

(vi) Rolling: Running a cylindrical heavy material e.g. a tank filled with sand or water over the soil to make it firm and bring soil particles together, thus having a mature surface / landscape.

(vii) Water the soil: to ensure there is no small or large hole prior to planting, i.e. a mature planting surface.

(viii) Establish/ install the lawn

If using seeds, rake the soil lightly; if using sod, place them as they are then water thoroughly so that the soil in the roots tie up with the soil below; if using plugs or stolons, dig up small holes using fingers and refill accordingly.

USE OF FERTILIZERS AND MANURES:

The big 3 nutrients- N, P and K are most important and are needed in large quantities (macro). When establishing a new lawn, apply enough phosphorous-based fertilizer to help in root establishment and growth, e.g. TSP. DSP, rock phosphate etc. Then to maintain the green beauty associated with lawn, regular top-dressing using nitrogenous fertilizers is vital. This promotes rapid shoot growth. Deficiency leads to pale and yellow lawns. Potassium determines the lifespan of the lawn. Other micronutrients are also vital e.g. magnesium, Sulphur, calcium etc. Kraal manure usually contains some weed seeds. The better alternative is poultry manure, and should be used in reasonable amounts to avoid scotching the plants. Compost manure is even better.

PLANT SPECIES USED FOR LAWN

Bermunda grass (star grass) - Cynodon dactylon
Kikuyu grass (Pennisetum clandestinum)
LM grass (Swazi)(Lawrenzo Marques)
Golf-course green / Canada green
Couch grass (Digitaria scalarum)
Broad leaf
Wonder leaves
Florida

PLANTING MATERIALS FOR LAWN:

Selection is based on (i) how the lawn is going to be used, (ii) where it is to be installed, (e.g. under shade or sunshine or partial shade) (iii) the desired maintenance e.g. intensive or moderate or minimum care). Four main types of planting material can be used in lawn establishment- the seel, sod, plug or stolon.

Seeding- means the use of seeds. Buy from a reliable source to avoid weeds. Canada green-use 500g seed / 46m2; Kikuyu grass-use 1kg of seed mixture in 50m2. Then use the rake head to gently press the soil down to improve seed-soil contact.

Sod- the shoot plus the root covered with soil; exist in size of 1m x 50 cm sections. Can be rolled as a mat for easy transportation. This gives an instant beauty backyard, which is mature and mostly disease-free. But is expensive.

Plug- These are trays with individual cells, each cell comprising a seedling and soil. These can be made from a sod. The cells may die, creating patches in a lawn.

Stolonising: this is the use of stolons (above ground stems running parallel to and mostly on top of soil surface). These are cheap and easily obtainable, and take less time to establish. However, they may die.

LAWN MANAGEMENT PRACTICES:

Large lawns are beautiful, but costs are high. They often need to be mowed, trimmed and raked free of tree leaves and mow clippings. Reducing lawn size while maintaining the beauty of the landscape can be ensured by introducing ground covers.

LANDSCAPE MAINTENANCE HINTS

Simple landscape designs are easy to maintain. Smooth curves, for instance would be better than sharp curves in lawns.

TIME OF PLANTING:

Can establish at any time as long as there is enough water source. However, speed of establishment depends on season- as some lawn materials are winter-based (i.e. cool-season) while others are warm season (summer-based). In Botswana, it should be at onset of rains (October) when soil is moist. Mix small seeds with sand (20g seed: 1 kg sand), then spread evenly at 1kg/50 m2.

WEED CONTROL:

Removing unwanted plants from the lawn; mostly broad-leaved weeds such, as Oxalis spp is common. Weed control is done to maintain the beauty of the lawn. Can do regular handpicking / uprooting, can use a selective herbicide etc.

MOWING:

Means cutting the lawn material to a short, attractive height and at the right time. Mowing can be done to 3 cm to 8 cm. It is vital to maintain the mowing height.

EDGING:

Cutting the overgrowths at the side of the lawn; to remove grass that has over-grown the main plot. Can use nylon cord trimmer, long handed lawn shears or long-handed edging shears.

SPIKING:

Use spike-toothed equipment to make small holes (up to 7.5 cm or 3 inches deep) in the lawn so as to improve water penetration through the thick mat of lawn material, especially the mature lawn. It is also meant to improve soil aeration.

SCARIFICATION/ DETHATCHING:

Refers to the vigorous use of a rake or rake-like tool to remove thatch (usually done in autumn). The thatch is a layer of fibrous (sometimes dead) material on top of the soil surface. It facilitates further and more vigorous growth of the lawn.

IRRIGATION:

Supplementary watering of the lawn to keep the plants free from water stress and therefore keeping them green, attractive and healthy. Time and how much water depends on experience.

PEST CONTROL:

The pests include ants, crickets, termites etc

(i): ANTS- these hide under the lawn, and also on trees. They cause discomfort to (bite or climb onto) those resting on the lawn.

Control: Use Ant-Kill or Ridant (400 ml / 10 litres of water). In trees, apply up to 60 cm tree trunk.

(ii) CRICKETS: affect ornamentals and lawns. Eat the foliage and make it unpreventable.

Control: Use carbaryl 85% wettable powder at 10 grams / hole.

DISEASE CONTROL:

Examples of diseases of lawn include: downy mildew, rust.

Downy mildew:

Affects lawns and ornamental plants, and make the colour of leaves to turn powdery white.

Control: Use dithane-M 45 or Pedza Nhamo dust weekly.

(ii)Rust: affect ornamentals, and change their colours to brownish- orange.

Control: Use Dithane -M-45 or copper oxychloride weekly.

EQUIPMENT USED IN MANAGEMENT OF LAWN:

The following is a list of equipment required for lawn management, as well as their individual uses.

i) Reel or rotary mower: for cutting the lawn to the desired height.

ii) Detaching machine- for removing dead and / or fibrous material from the lawn.

iii) Mechanical rake: used to pick up clipping sand other surface debris after mowing.

iv) Edging equipment: a metal or plastic, which aids to stop grass spreading into flowerbeds and to keep the edges firm. Can also use a mechanical edger, or a long-handed edging shears. The latter is used to trim the edges after mowing. It is much easier to use than the ordinary hand shears.

v) Watering can: can be used in small scale, and useful in application of herbicides, liquid feeds and other pesticides.

vi) Small hand trowel: used for removing / digging out isolated weeds- as pulling them by the hands at times leads to roots being left behind intact.

vii) Hollow-tine fork: useful for deep aeration of the lawn to improve the drainage in heavy soil.

viii) Garden fork: a simple equipment for aerating the turf / lawn deeply enough to relieve compaction

vix) Mechanical trimmer: A power-driven trimmer which cuts horizontally. Useful for cutting grass around trees or walls- the nylon cord trimmer is light and easy to use.

GROUND COVERS

These are plants that spread across the ground but do not grow tall, so no cutting is required. There is therefore little maintenance. They act as mulch, and some are nitrogen fixing. Some have beautiful flowers as well. Groundcovers need an edge barrier to contain them. A low brick or wood edging is good enough.

EXAMPLES OF GROUNDCOVER SPECIES:

Tapien (Verbena), Bishops weed, Juniper, clover, and Allysum. These are all perennials. Clover has very little water needs, is nitrogen fixing, and is very fast growing. Varieties of clover available are Red clover, Dutch white and yellow blossoms. Dutch type almost resembles lawn; the other 2 varieties grow taller and are more recommended for pasture application.

8.3 LANDSCAPE DESIGN

Figure 77: left; rural landscaping using hard paving (rocks / stones)

DEFINITION:

Landscaping is the laying out of ornamental features in the outdoor face or space (i.e., outside the house) so that it looks attractive to the eye. It is meant to improve the look and appearance of the outdoor space. Landscape design is the actual planning of how to lay the outdoor features in place, while landscaping covers both the design and the practice (the art).

OR improving the appearance of a piece of ground with decorative planting (limited because non- plant material can also be used in landscaping);

OR laying out of grounds attractively or in imitation of a natural scenery

OR gardening

Beautiful school compound landscapes, as a result of ESD intevention

Figure 78: A formal landscape in Nairobi, Kenya Figure 79 A garden landscape in Botswana.

KINDS OF LANDSCAPES

These include: formal landscape, garden, rockery, cemetery, Park, playground, waterfront.

Natural landscape in form of 'snow' enjoyed by the winter countries

Formal landscape: may comprise one or more of the following: lawn, flowerbed, trees etc. Often found in government or institutional quarters, and state house etc.

Garden: may have one or more of the following: lawns, crops, ponds, rockery; and is often found in homes, hotels and schools.

Rockery comprises well-arranged rocks of different sizes with plants between them. Often found in roadsides, focal feature in private and public quarters, and in gardens
Parks may comprise lawns, trees, shrubs, ponds, and are found in recreational places.
Playground comprises lawns, flowerbeds and shade trees. Commonly found in schools, clubs etc.
Waterfronts are made up of ponds, fountains and / or fishponds. Often found in schools and hotel areas.

MATERIALS USED IN LANDSCAPING
These may be plant or non-plant material.

PLANT MATERIAL
Include trees, groundcovers, grass (lawn), flowers and shrubs.
(i)Tree (Setlhare): Used for cover (shade), fragrance (beauty), smell, and to imitate nature. Choice depends on mature size, shape, danger to human health, how low from ground it branches,
deciduous or evergreen, and whether it bears fruits or not

Figure 80: Forest plantation used in landscaping the naro Moru forest office compound.

(ii) Shrubs: used in traffic control, wind breaking and fragrance. Selection criteria are as in trees.
(iii) Flowers: May be annuals, biennials or perennials. Selection is based on life cycle and the desired colour.
(iv) Ground covers: are low-lying plants that cover the ground. Are sometimes used as lawn material in areas that are hard to mow e.g. hillsides, where they also help prevent soil erosion.
(v) Grass (Bojang)/ Lawn. This is already described in good details in the previous subsection of this chapter.

NON- PLANT OR SURFACING MATERIAL

These are meant to cover the outdoor flow. Mainly the type of use the flow is going to be subjected to determine the selection of all these materials. In heavy traffic, we can use soft or hard paving.
(i) Soft paving: includes asphalt and other loose aggregate material like sand and gravel
(ii) Hard paving: Include bricks, stone and concrete. Selection of paving / surfacing material depends on cost, availability, amount of use the surface will get, aesthetics (beauty), maintenance requirements etc. (NB Maintenance requirements often again depend on landscape design) (see bricks below).

USE OF PLANT LIFE IN LANDSCAPE DESIGN

1) Plants give life, natural beauty (e.g. as a focal feature), fragrance, and even plant products such as fruits, etc to the landscape.
ii) Trees, shrubs, ground covers, or flowers can be used depending on the design.
iii) They also provide privacy e.g. in cases where hedge is used as live fence.
iv) Acts as boundary and windbreak in many cases.
v) Should choose plants adapted to the climate (temperature, rainfall, light/ day length, wind, sunshine hours, relative humidity, aspect (direction of slope with respect to the sun; angle at which sun strikes the plant- may give different faces of the same plant), topography (nature of terrain-whether flat or sloppy or rugged) and soils of the site.
vi) One should also look at maintenance needs of the plants, appearance and other growth habits (e.g. deciduous/ evergreen, lifespan, etc).

ENVIRONMENTAL IMPACTS OF LANDSCAPE DESIGN

Often where there are some unwanted natural features such as trees, grass, rocks etc, we remove to suit our own personal interests. This may reduce biodiversity, and cause soil erosion. The design is at times too artificial, and some vital plant or animals may become extinct due to the interference and the changed landscape. Death of some plants may result e.g. when hard paving material is used which reduces or stops infiltration of water into the soil. See picture below taken from Botswana.

Figure 81: A well groomed and well maintained combined landscape. Source: Jackson and Hutchinson pp 12

Figure 82: Non-plant material used in landscaping in Maun, Botswana at Motsamai Mpho Hall, Maun Senior.

Figure 83: A tree killed by using hard paving material (bricks) to landscape its surroundings.

In some cases, even barren, wastelands have been completely rejuvenated by landscaping

In some cases, new microclimate is created conducive for some species of plants and / or animals e.g. a pond for aquatic organisms and some benefit wildlife.

It creates an attractive site (i.e. aesthetically pleasing; beautiful)

Some landscapes may be too monotonous e.g. football pitches monocrop etc.

DETAILED DESCRIPTION OF SELECTED LANDSCAPES

The following is a detailed description of two examples of landscapes. They re selected because are likely to cause confusion, unlike the rest which look straightforward.

THE ROCKERY

A rockery is an attractive arrangement of rocks with plants growing on them. It thus creates a contrast with the surroundings by adding colour, form and depth. Suitable plants grown between the rocks add beauty from their natural look, colour and smell of flowers. This mix makes the features blend very well with the surroundings. Rocks that occur naturally in the area should be used. The rock arrangement is such as to ensure there are sufficient soil pockets where the plant roots would grow. The layout should follow simple principles of design such as simplicity, massing, repetition, balance, proportion, and dominance. This makes it possible to maintain the landscape. Some plants or rocks can be used to give a focal point.

Figure 84: A beautiful environment created by landscaping in CAVS, University of Nairobi

Figure 85: A rockery combined with water front (pond and fountain all in one), at this popular University of Nairobi fountain near the great court.

WATER FRONT

This involves using water feature in the landscape e.g. pond or fountain. An example is a pond. It is a fairly easy task- taking into account the nature of soil, site (must look natural), nearby features such as trees, water availability, and presence of building materials. Need to lay a strong foundation, then use concrete to form the bulk of the pond. Big trees can make the pond start leaking too soon, may shade the pond, leaves may fall in water and make it less attractive It is recommended to use a waterproof cement to reduce chances of pond leakages. Can also use a plastic sheet at the bottom to make the pond leak-proof. Can establish simple and easy-to-maintain water plants such as water lilies, papyrus and water hyacinth. One can have a mixture of submerged, floating and half-submerged plants to add beauty. Place these water plants into the pond with their roots still in their containers to avoid them being too stressed; the roots soon break the containers and spread. Should avoid shade since water plants are sun-loving, and also since a lot of sunshine is desirable to cause reflection which adds beauty. Need to do regular pruning to control the growth of the plants. Can also introduce fish. Also can lay some rocks at the periphery to make the scene look natural.

A water front for decoration in KEMU, Meru. kenya

Figure 86: Pond with aquatic plants representing a water front. Figure 87: A pond without plants in it.

EXAMPLES OF EXISTING LANDSCAPES

Golf-courses in any city, The many fountains scattered in various cities in the USA (the one in Chicago city near the millennium park is especially outstanding; SANITAS (in Gaborone, near Gaborone Dam), private compounds e.g. schools and teachers' compounds in your own country; Fountain in front of Delta medical (Maun, Botswana), at the government enclave and near parliament, Gaborone; Rockery between lion and Springbok houses in Maun senior secondary school; and also in front of the administration block (Maun SSS), Uhuru park in Nairobi, Kenya; Megacity in Kisumu, Mega plaza central view a the base, Kisumu city, Kenya; the barracks in any country, including national youth service compound, Kenya etc. etc.

Natural ‘snow’ landscape in the USA during winter

LIVESTOCK HUSBANDRY: INTRODUCTION

MODULE 9: LIVESTOCK HUSBANDRY

9.1 LIVESTOCK MANAGEMENT

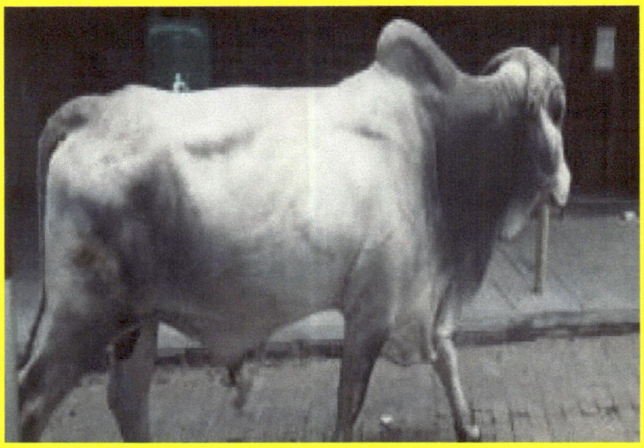
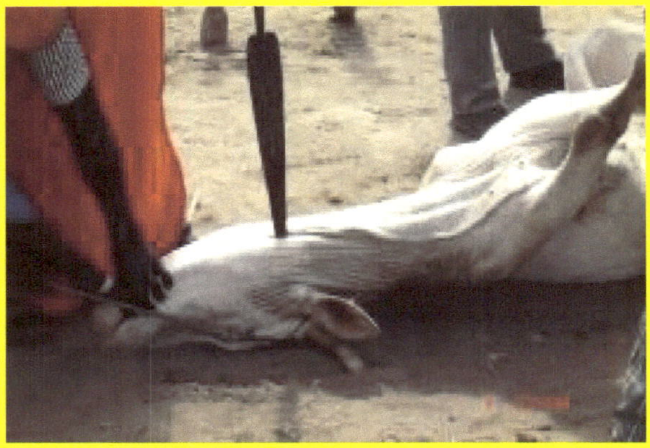

LIVESTOCK HUSBANDRY: INTRODUCTION

Livestock are domestic animals mostly kept for food and / or industrial products. Examples are Goats and sheep (sometimes-called (shoats), cattle, pigs, poultry, fish etc. Livestock are a source of (i) Food such as meat, eggs, milk etc; (ii) Products such as wool, hair, skin, etc for industrial use; (iii) By-products such as manure, bones (for making bone meal), blood (making blood-meal) etc. Production of both crops and livestock in the same farm is called mixed farming. A Person owning livestock is generally respected - i.e., livestock give us social prestige. This is why in Botswana, everyone tries to at least have a cattle post (cattle post complex). The food products from livestock are very rich in proteins, sometimes called first class proteins because they are made up of essential amino acids. Proteins are digested to give rise to amino acids. There are over 22 amino acids, some of which are essential and others are not. Essential amino acids are those which the body cannot manufacture, and therefore has to be from one's diet. Non- essential amino acids are those that can be synthesized by the body in case of need.

TASK: List any 6 amino acids you can get from your reading. Why do you think feeding concentrates such as blood and bone meal to cattle is not popular right now? Conduct a survey to establish which part of your country produces what livestock.

9.1 LIVESTOCK MANAGEMENT

The aim of this topic is to enable students and practitioners understand the essential aspects of livestock housing and general well being. Livestock management includes all daily requirements such as housing, feeding and nutrition, rearing of young stock, breeding, general handling on daily basis (stockmanship), and record keeping. All these are meant to ensure the general well being of the livestock expose them to minimum stress, if any, and ensuring a fast and un-interrupted growth and production. Specific aspects of management will be discussed under specific classes of livestock commonly reared in Eastern and southern Africa. These include poultry (chicken) and pigs. Management is by age of livestock.

LIVESTOCK HOUSING

Reasons for housing livestock:

1. To give shelter against extreme conditions of cold, heat, wind, rain etc."
2. To offer protection against thieves and predators.
3. To control breeding.
4. To enhance growth rate by controlling movement (reducing or minimising wasteful movements).
5. To ease other management aspects such as feeding, health etc.

QUALITIES OF A GOOD LIVESTOCK HOUSE

Durable
Well ventilated and well lit
Spacious enough
Constructed easily and cheaply
Safe, i.e. not slippery
Easy to clean

TYPE OF BROILER HOUSING

Housing of broilers is often discussed as management systems. These include:
(i) Free-range (ii) Semi-intensive or Fold system, and (iii) Intensive systems. This is because the nature of management system dictates the type of house for use by the broilers.

Figure 88: Chicken under deep liter system of management

A: Free-range:

Birds allowed roaming freely over a large, wide-open area which may or may not have a perimeter fence. Only a simple shelter is provided. 200-300 birds can be kept per hectare of land. They get food for themselves.

Free range poultry management (Left..chicken; right....ducks)

B: Semi-intensive:

This involves the use of the fold system. Birds are confined in small movable houses called folds or arks (name acquired from Biblical Noah's ark). A fold may be 3.3m long, 1.5 m wide and 1.5 m high, and takes the shape of a gable roof of a house. This is enough for 10-15 birds. One third of the fold is roofed to provide shelter, while the other area is left open and only covered with some wire netting. The folds are moved daily to a fresh area. Birds are thus controlled to some extent; are kept in a smaller area than free-range but larger than intensive system Food and water are provided.

A small movable fold (Arch) for keeping / managing chicken

C: INTENSIVE SYSTEMS:

A large number of birds are kept in a very small area and are controlled all the time in all respects. Are 2 types- deep litter and battery cage systems. Each is described below.

DEEP LITTER SYSTEM:

Birds kept in a house with a deep litter for layers. Roosting perches, feed and water troughs are provided. At times, a welded wire or slated floor made of wooden slats is used instead of deep litter. A solid floor 60-90 cm above the ground is used. The droppings fall below the floor in the pit and can be removed at times. It is possible to keep 4-5 birds / m2 in this system

Figure 89: Poultry under the deep liter system- layers Figure 90: broilers

BATTERY CAGE SYSTEM:

Birds are kept in wire cages for the rest of their lives. The cages are hung from the roof of the house (about 60-90cm, or maximum 1 metre high) and thus do not touch the ground or floor. This facilitates cleaning. 1-4 birds can be kept in one cage. The cages are made of wire and placed in tiers, generally one above the other.. These run from one end of the house to the other. The cages have a sloping arrangement so that the eggs (in case of layers) roll gently to the edge where they can be easily collected. Watering and feeding systems run continuously from one end of the cage to the other.

Figure 91: Chicken being fed under the battery cage system of intensive management

TYPE OF HOUSE (REARING PEN) NEEDED FOR PIGS (A LACTATING SOW AND ITS PIGLETS).

The house should be in a well drained area, well ventilated, free from draughts, dry and warm, easy to clean, and be spacious. It should ideally have farrowing, rearing and fattening pens. Farrowing pen is for giving birth. One compartment can serve both as farrowing and rearing. There should be guardrails about 1 foot from the wall, and above the floor. There should be a creep area with special feed and heat for the young ones.

CHARACTERISTICS

Spacious:

Should have enough space for movement (6 m2 for adult and 1.3m2 for each grower / fattener). Can make a twin unit of 2 separate pens suitable for sow and weaners. In a farrowing pen you can increase the area a little bit due to the presence of piglets. Can have a pen of 3m by 2.5m plus a weaners' pen of 3 m by 3 m. This is shown in a diagram below. If a boar is kept, a separate pen is needed.

2. Durable with strong floor, walls and roof.

Roof should be 1.7m high, and made of corrugated iron sheets, asbestos or thatch. But asbestos is a health hazard, while thatch is not durable. It should slope to one side to let rainwater run off easily.

Floor: made of concrete with good foundation 40 cm deep (shallow floor is easily dug by the pigs). It should slope slightly towards the door to facilitate cleaning and drainage. Surface should not be slippery.

Walls: made of cement/ sand mixture (concrete). Should be strong so that they do not wear out as pigs scratch themselves against the walls. Should have no cracks or lumps. Diamond mesh fence and pole delicate and easily weakened because pigs put their snout between the wire.

Door: 120 cm high and 80 cm wide and easy to remove / open and effectively keeps pigs in. Pigs should not be able to open doors.

Feed and water troughs

Should have outlet pipes fitted to ease cleaning with hosepipe. Is made of concrete fixed to the floor, and allowing 30 cm of trough space per pig. Should be 15 cm deep and rounded at bottom with no corners to collect dirt

Guard rails

Placed inside about 30 cm from the wall and above the floor. This protects piglets from being crushed by the sow. It helps form a creep are where piglets can rest; it also may have a source of heat controlled at 320C to keep piglets warm in winter; and it has creep feed for young ones.

HANDLING AND STOCKMANSHIP IN LIVESTOCK

Stockmanship is the art of handling and caring of livestock. A stockman is the person doing the art / practice of stockmanship.

Figure 92: Diagram of pigsty showing ground plan of a rearing pen for a lactating sow and its piglets

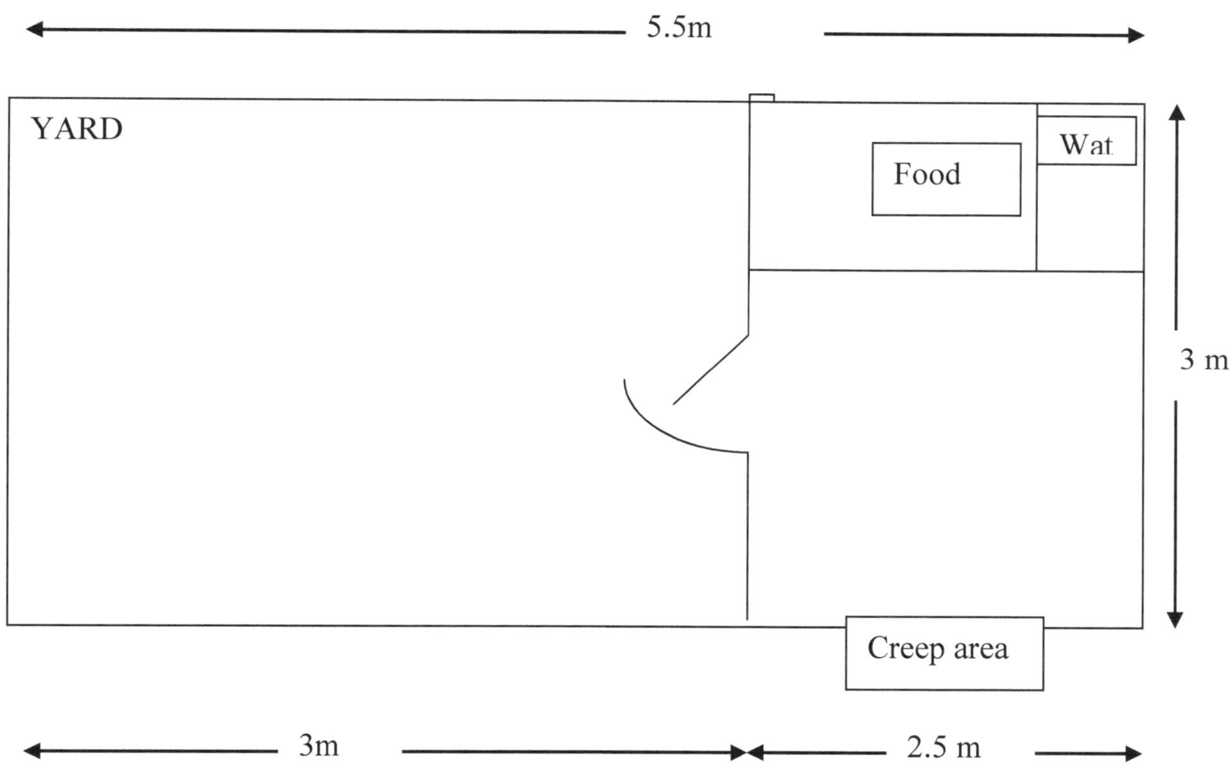

Figure 94: cattle being trained for an annual trade fair (international show)

A good stockman should build a good, stable relationship of confidence, trust and reliability with his/ her livestock. This facilitates their growth and production, as they are not scared. This art should be consistent in all aspects of daily management, including feeding, rearing young stock, record keeping and general handling.

LIVESTOCK HANDLING

Handling should be friendly and consistent. A timetable for routine management activities should be made and followed. The stockman should approach livestock in a consistent manner, e.g. when singing a soothing song, daily; proper timing of each activity is vital to avoid stressing the animals. There should be no unnecessary noise around, as this may make the animals nervous. The same type and colour of clothing should be used daily e.g. when feeding the animals, etc.

Other types of livestock:

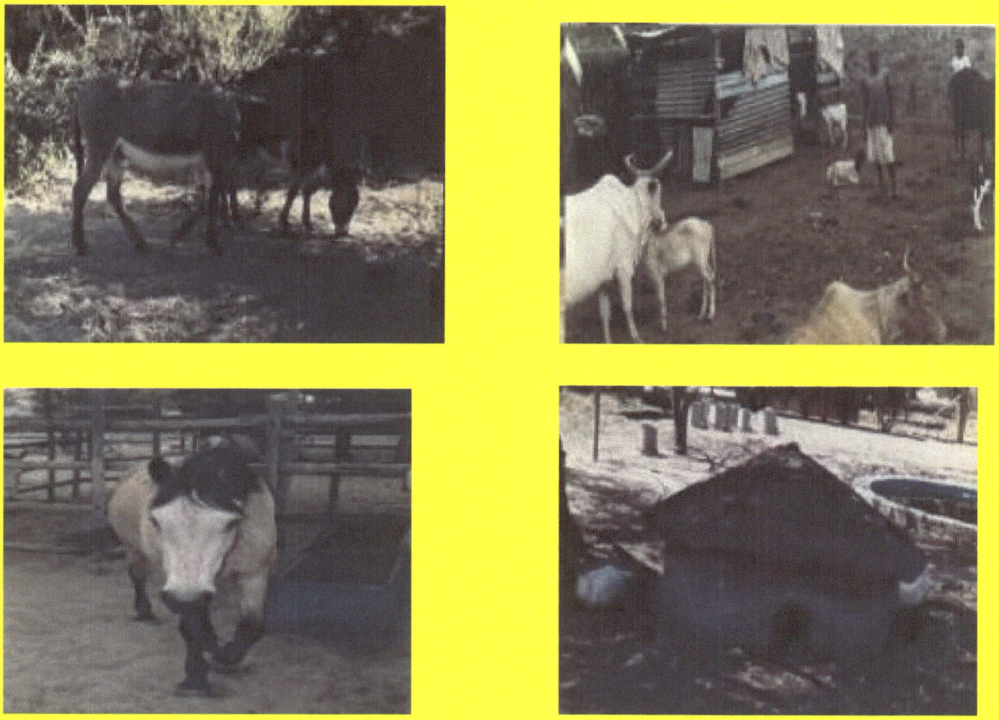

Figure 95: other livestock

MANAGEMENT OF PIGS

REARING OF YOUNG STOCK:

Days 0-3: Young piglets are delicate and can often die within the first 3 days of their birth. Care should thus be taken by doing the routine activities properly and timely. To reduce early deaths, (i) they should be warm enough but not too hot (320C) (ii) sow has enough milk; hungry piglets are noisy and do not settle on the teats. Therefore feed the mother well.

Day 2-3: IRON INJECTION: Piglets kept in pens can become anemic (scouring, slow growth and death) since the mother's milk never has enough IRON. At day 2 or 3, give them IRON in form of a paste put in mouth, or an injection of Iron.

Day 3: TEETH CLIPPING: Piglets have sharp wolf teeth, which can easily injure the mother's teats during suckling, as well as other piglets. These are clipped using teeth clipper or pliers.

IDENTIFICATION:

Each pig is given a number for identification using an ear notch. These are small pieces cut out of the ear using an ear notching equipment.

CASTRATION:

Male piglets that are not wanted to form part of the breeding herd are culled by castration at 3-4 weeks old. The testicle is removed with a knife or scalpel. This serves to (i) increase growth rate (ii) remove smell (iii) make them docile (iv) control breeding

WEANING:

This is done at 8 weeks old. From day 10, the piglets are given pelleted creep feed in the creep area. Between this and weaning time, the average feed taken by each piglet per day is 0.25 kg. The mother sow is removed from the piglets to reduce stress on the young ones. The sow is given reduced amount of water to help it stop producing milk quickly.

PEST CONTROL:

After weaning, the piglets are dosed with piperazine to kill internal pests and worms.

NUTRITION AND FEEDING

Adult sows consume 3 kg feed per day, while piglets feed on averagely 0.25 kg/day- with a smaller amount initially, and up to 1 kg /day just at weaning age. The table below shows food requirements for different age groups and weight of pigs:

Table 12: Feed requirement by pigs by age.

Age (WKS)	8	10	12	16	20	25	30
Feed requirements (kg/day)	1	1.25	1.5	1.8	2.25	2.5	2.8
Weight of pig (kg)	15	19	25	38	56	74	90

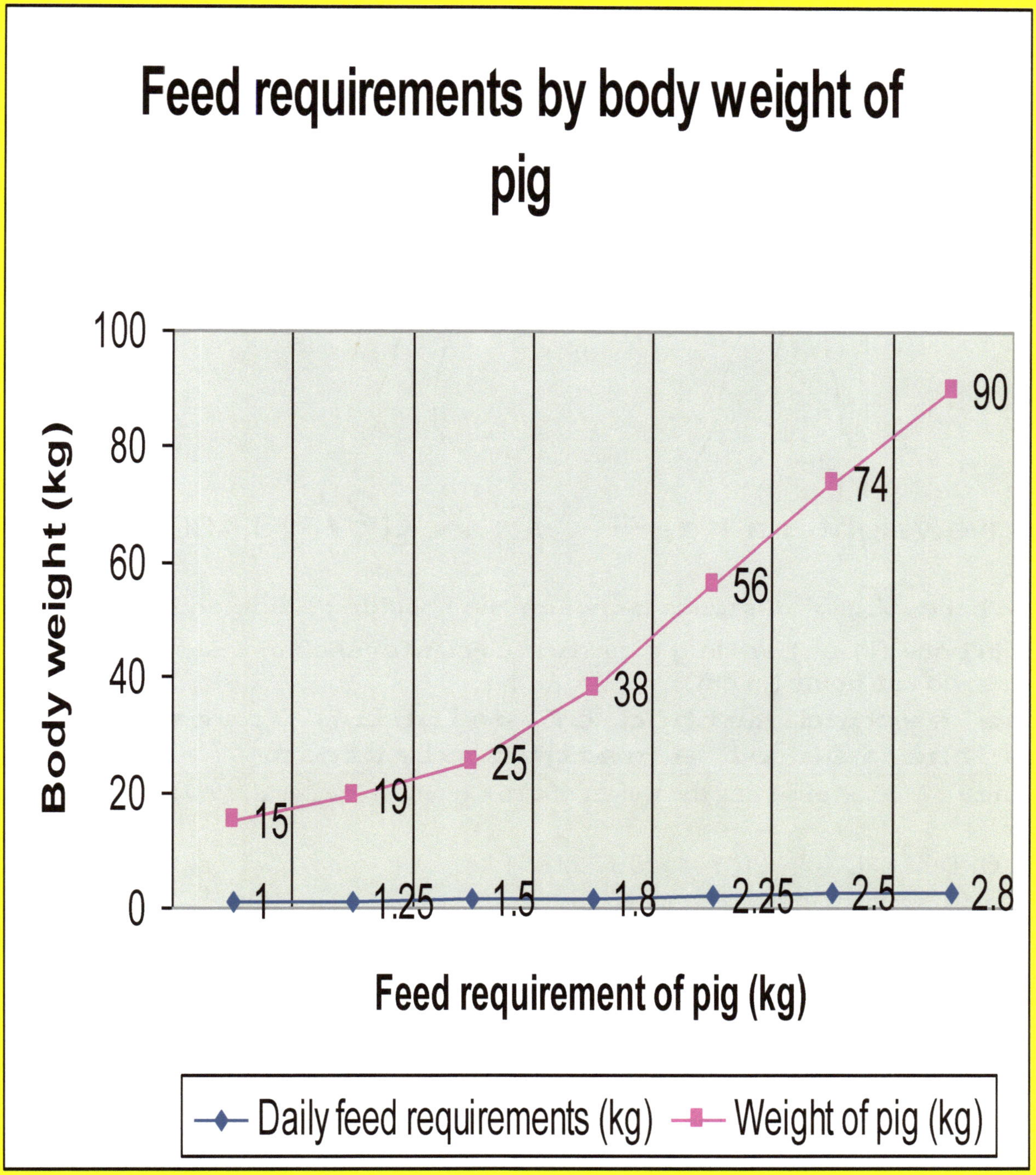

Figure 96: Daily feed requirements for pigs

Alternative method of estimating the weight of cattle

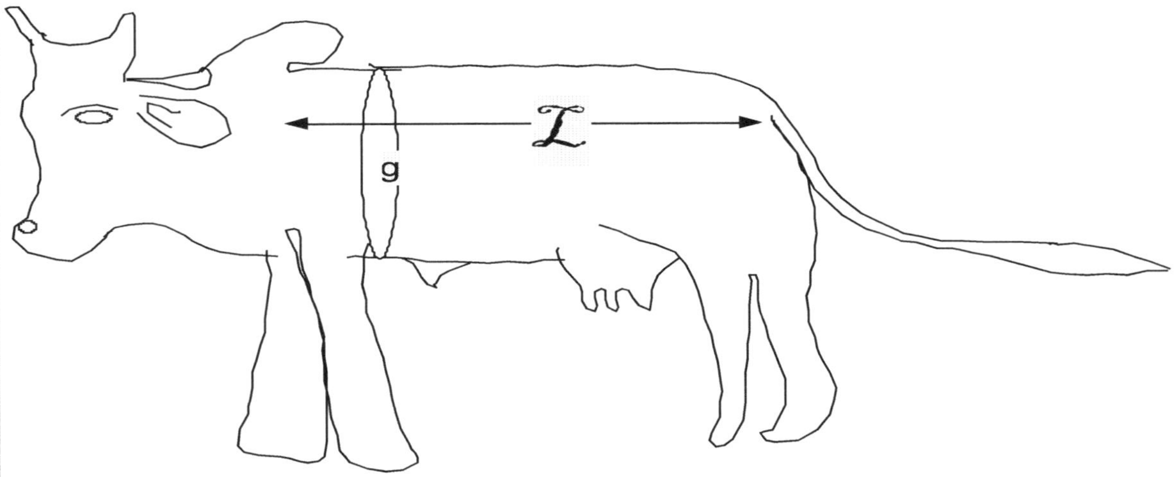

Liveweight (in Kg) = $\mathcal{L} \times g^2 / 10{,}400$

Where $\mathcal{L}$ = distance between the shoulder blade and the pin bone (in cm), while g = girth or circumference of the body at its widest point (in cm)

The second method of weighing livestock is the use of a weight platform

Figure 97: Method of getting the weight of a cow (Source: Ngui et al, 1991).

RECORD KEEPING IN PIGS MANAGEMENT

In pig management, the following records are important.

1. Breeding records: show the sows identification number, her parents, date of birth, details of each litter produced e.g. date of service, boar used, farrowing date, Number of piglets born alive, Number of piglets born dead, average weight at 3 weeks and at weaning, and date of weaning. The performance of each sow can then be compared with the rest, and used to decide on which ones to cull.
2. Feed record: How much feed (and what type of) used in each pen daily.
3. Purchases and sales records: Gives the pig number, price for each sale and/ or purchase, If slaughtered it should give live weight and killing out weight, as well as killing out percentage.
4. Health record This shows date and disease that affected the pig, treatment, and results. A pig that regularly falls sick should be culled.

5. Financial record: To show all the returns and expenses incurred in the pig enterprise, thus helping decide whether a profit or loss has been made.

STOCKMANSHIP IN BROILERS

Broilers are chicken kept for meat production only, and are slaughtered at 60-70 days old when weighing 2-2.5 /3 kg.

Key management practices:

Selection and culling:

The following are characteristics of good broilers:

(a) Quick growth:-The bird must put on a lot of weight in the first 8 weeks. Producing a meat carcass from a fast growing broiler takes less time, less labour and less food

(b) A good food convrsion ratio: This should be about 2:1, meaning that if the broiler eats 2 kg of food, it puts on 1 kg of body weight.

© Goos skin and flesh colour: Some customers like chicken with a yellow skin, others prefer white skin. Skin colour depends partly on breed, but feed containing yellow maize gives the skin a yellow colour.

(d) Good size and plenty of flesh: Customers want meat, not just skin and bone.

(e) A high viability: This means that not many birds die when they are small. If birds die, there is less profit for the farmer.

For good Layers, the following are the indiactors:

Space between the pelvic bones : these are the bones beside the vent where the eggs come out. In a laying hen the pelvic bones are more widely separated than in a hen that is not laying. . Hold the bird and gently feel for the pelvic bones under the skin. If they are close together (the width of one finger), she is not laying. If you can feel a wide space (2 or 3 fingers widths), she is probably laying.

A large, moist vent: This should also be light in colour and should be a sign of laying.

A small, dry vent shows that the hen is not laying. Use thumb and forefinger to assess this.

Large brigt eyes: These are sign of good health, but also show that the hen is in lay.

A bright red comb is also a sign of good health.

There should be no yellow colour: A yellow colour around the skin around the vent, the eyes, the openings of the ears and around the beak indicates a bird not laying.

trap nesting: To be sure that a hen is not laying, keep her in trap nest by herself for a few days. Give her food and water, and watch to see if she lays an egg. This method is also effective in broody hens; just let cold draught pass through the nest; the bird will come to lay soon.

ANIMAL HANDLING

Animal handling should be gentle, friendly and consistent. A timetable for routine management activities should be made and followed. The stockman should approach livestock in a consistent manner, e.g. when singing a soothing song. daily; proper timing of each activity is vital to avoid stressing the animals. There should be no unnecessary noise around, as this may make the animals nervous. The same type and colour of clothing should be used daily e.g. when feeding the animals, etc.

ROUTINES FOR LOOKING AFTER CHICKEN

Chicken are nervous birds, so they should be kept calm and quiet. To avoid upsetting them, do the following:

(a) Let them know you are coming: Never go into the chicken house suddenly. Make some noise or say something as you come near the house. The birds will then know you are there and will not be starlled.

(b) Move slowly and quietly: When you are in the chicken house, sudden movements will frighten the birds. Do not throw paper sacks about when you are filling the feeders or throw grass around when you are putting down more litter on the floor.

© Follow a definite route each time: This is s that the birds learn what to expect. Let them think you are the frind who brings them food, not he eenemy who is going to catch them and eat them.

RECORD KEEPING:

Records are needed in the following areas:

Cost of heating the brooder;

Daily feed consumption and its cost;

Labour and medicine costs;

Number of dead chicken (i) up to end of brooding and

Ii thereafter up to slaughterDates for major management

iii.activities such as arrival of chicks, vaccination etc;

Number, breed, and value of birds, as well as source

(i.e., name of breeder or hatchery);

Health records to enable the manager conduct culling and selection more

This information helps the farmer to calculate the rearing cost per bird. He can calculate percentage mortality for each period; average feed consumption; feed consumption per dozen of eggs; etc.

REARING OF YOUNG STOCK:

One has to prepare to receive young chicks by doing the following:

Have a well-built and well-ventilated house;

Clean and disinfect the house and equipment;

Install feeders and drinkers;

Purchase enough of the right feed and drugs/ vaccines;

Put in place brooder lamps for chicks (NB brooding unit is a unit provided with heat where chicks are kept from when received at one day old to about 5 weeks when they have grown feathers. This protects them from extreme weather conditions, as well as from predators and thieves). It should be well ventilated, with clean water and feed. Initially, the temperature is maintained at 25-30oC, and thereafter, heat can be gradually withdrawn.

(IV) NUTRITION AND FEEDING PRACTICES:

The day-old-chicks (DOCs) are given broiler starter mash until 3 weeks old. This is often in powder form. Thereafter, they are given Broiler finisher mash in form of powder or pellets. The starter mash is finer; richer in proteins, vitamins and minerals; and contains additives like coccidiostats. The finisher, on the other hand, has bigger particles; richer in carbohydrates but has less of the other nutrients; and has more fiber that help digestion.

(V) FEEDING:

Broilers are often given food ad-lib, i.e., they feed on as much as they can (unrestricted). However, food wastage should be avoided by:

Giving pellets instead of mash to older chicks;

Half-fill feeders; and

Adjust the height of feeders to be equal to the height of the back of chicken.

Figure 98: The feeding systems for poultry: left-Deep litter system; Figure 99: The Battery cage system

A farmer can avoid relying fully on commercial feeds by going for local home mixing. The fishing remnants small fish omena (refered to as Ochonga in LUO) is a rich source of proteins, and can be mixed with a good energy source such as grains (sorghum, millet etc) in a ratio of 1:5 respectively, thoroughly moixed and ground. This gives good feed for most of the chicken regardless of their age and status.

STANDARD USED IN THE FORMULATION

Chick starter marsh: Given 0-5 weeks. Finest quality; richer in proteins, vitamins and minerals.

Growing / grower mash: From 5 weeks to laying time (18-22 weeks); need less proteins but more carbohydrates to the active chick;

Laying marsh: Protein and mineral rich; birds are not so active, and should not get too fat.

Broiler marsh: Rich in Carbohydrates and proteins for fast growth and energy, including fattening.

Generally 10 g of protein is contained in 105 g of sorghum, which contains 10 kcal of energy. Similarly, 298 g of sorghum floor contains 1000 kcal.

The table below further gives the nutritional requirement of chicken by age.

Table 13: Nutritional requirement of poultry by age.

CLASS AGE (weeks)	Chicks 0-8 weeks	Growers 9-18/ 22 weeks	Layers 22 + weeks	Broiler Starter Finisher
Energy (Mcal / kg	2.97	2.70	3.13	
Protein (DCP %)	20	14/16-17	15-20	24
Calcium %	1.1-1.2	1.1-1.2	2.4-2.9	6
Phosphorous (%)	0.6-0.9	0.6	0.6-0.8	08-1.3
Vitamins (μg/kg)	650	1650	4730	

(Source: Ngugi et al, 1990)

Table 14: Sources Of Other Nutrients Other Than Energy And Proteins

Vitamin A	Thiamine	Riboflavin	Ascorbic acid	Nicotinic acid	Calcium	Iron
Ododo	Sorghum	Fish	S.P. Leaves	Sea Fish	Ododo	Millet / sorghum
Sweet potato leaves	Maize /	S P Leaves	Ododo	Sorghum	S.P. leaves	Maize / Wheat
Dry fish	Millet / wheat	Ododo			Dry fish	Ododo
	Ododo	Maize				S.P. Leaves

Some alternative home mixing materials are shown in the table below.

Table 15: Nutrient Sources from different feeds:

	Sorghum	Ochong'a	Ododo	Sweet.Potato Leaves	Sand / oyster shell	Maize
CHO	√					
Proteins		√				
Ca		√				
Fe	√		√	√		
Nicotinic acid	√	√				
Ascorbic acid			√	√		
Riboflavin		√				
Thiamine	√		√			
Vitamin A		√				
Grit					√	
P		√				

OTHER CROPS OF HIGH NUTRITIONAL POTENTIAL

Sunflower, soybeans and cotton can have oil extracted from their seeds and the remains thereafter used as livestock feed (cake).

The cotton crop

COMMENTS ON FOOD MIXING:

Sorghum is generally the poorest in proteins among the major cereals. However, it is more superior to most in terms of digestible proteins.The order of decreasing superiority in protein among cereals is: rice, millet, maize, wheat and sorghum.

Crude protein levels are only significant in cereals where they are not to be provided by another source such as the richer proteinous foods such as fish. In our case, omenalets / ochong'a is the key source of proteins. Dry fish provides other nutrients such as Vitamin

A, Riboflavin, Nicotinic acid, Calcium and proteins. Thiamine is also provided by sorghum, maize and Ododo (Amaranthus spp). Ascorbic acid is provided by sweet potato leaves and Ododo. Iron is provided by millet / sorghum, Ododo and sweet potato leaves. Sorghum provides CHO / energy SE at the rate of 70 units of nutrient density per 100 g of dried edible feed. It is also a source of Iron, Nicotinic acid and thiamine.

MAJOR LIMITATION OF HOME MIXING:

Mixing may not be thorough

Grains may not be available on the farm.

NB: For pigs, production target is 10 piglets per 6 months, or 20 piglets per year.

(vi) Watering:

Clean water must be given daily - unrestricted.

(vii) Housing:

Table 16: Housing Space requirements of chicken:

Animal	Floor area per bird	Reason
Small chicks in brooder	50 cm2 each	Need to be close together to conserve heat, to avoid effects of extreme cold on the delicate chicks.
Laying hens, heavy breeds	3 birds / m2	Live under stress from 18/22- 72/84 weeks old (i.e. for 50- 66 weeks laying / reproductive life)
Laying hens, light birds	4 birds / m2	
Broilers	6 birds / m2	Normally slaughtered at 6-10 weeks old; only a short duration of stress

Chicken need density of 2-5 birds / m2 (1 stride X 1 stride) of floor space. Larger birds (e.g. adult layers) need lowers spacing, while younger ones can be more densely housed, partly to save them from extreme cold. Overcrowding leads to stress, easy transmission of diseases (e.g. infectious and contagious ones) and vices (e.g. fighting, egg pecking and cannibalism).

CHICKEN REARING SYSTEMS:

Extensive system: Free range;

Chicken left to roam over a wide area which may or may not have a perimeter fence;

Simple shelter is provided;

Requires enough land;

200-300 birds can be kept per hectare;

2. Semi-Intensive System: The Fold Or Ark System:

The fold system is a simple method for keeping a limited umber of birds.

The birds can be confined in small movable houses called folds or arks.

Most convenient size is 3.5mX1.5m X 1.5m for 10-15 birds

1/3 of the fold is covered/ roofed to provide shelter and laying place; the rest is open and covered with wire net.

The folds are movable daily to fresh areas using the handles at the top or the ones at the base if the width of the fold is reasonable (shorter than 1.5 m).

Disease control:

Table 17: Vaccination program for poultry (Chicken)

Age	Vaccination	Vaccine
DAY 1	Marex Disease	Marex vaccination
Day 1-14	New castle disease (NCD)	NDV intraoccular
14th day	Gumboro disease	Gumboro vaccination
3rd-4th weeks	Fowl pox	Fowl pox vaccine
5th week	Gumboro disease	Gumboro vaccination
6th week	NCD	NDV Komorov or lasota vaccination
16th week	NCD	NDV Komorov for point of lay
18th week	Gumboro disease	Gumboro disease for parent stock
Every 6 months	NCD for adults	NCD for breeding stock after 4th month

KEY MANAGEMENT PRACTICES FOR DAIRY CATTLE

The management practices aim at achieving the following targets as shown in the table below:

Table 18: Target weights at critical stages in the life of cattle:

Weight at birth	Average live weight	Weaning weight	Weekly weight gain	Breeding weight of heifers at 3 years	Slaughter weight at 4 years
22 kg	450 kg	120 kg	2-3 kg	320 kg	400 kg

ARTIFICIAL INSEMINATION OR SERVICE:

Done to a heifer at either 15 months or 270 kg whichever comes first. However, the first service should have been done by 18 months of age. Those heifers that constantly require repeat services should be culled.

ANIMAL HANDLING:

Animal handling should be gentle and friendly and consistent. A timetable for routine management activities should be made and followed. The stockman should approach livestock in a consistent manner, e.g. when singing a soothing song, daily; proper timing of each activity is vital to avoid stressing the animals. There should be no unnecessary noise around, as this may make the animals nervous and may inhibit milk let down process. The same type and colour of clothing should be used daily e.g. when feeding, milking, vaccinating the animals, etc

MILKING:

A cow is milked continuously from when it gives birth, to 3 months later when it is served again, to about 7 months into pregnancy. This gives about 9-10 full milking months, with a break of only 2 months awaiting its giving birth again (also called dry period). It is milked twice a day.

PREGNANCY CHECK:

If after insemination, the cow comes to heat again, it may be a sign it did not conceive. One should inseminate again, or call a veterinary doctor. Animals with many repeat services should be culled.

DRYING OFF A COW:

As 7 months of pregnancy approach, one should prepare the cow for drying off. This is done by: Milking less; (ii) milking only once a day. Because of a possibility of mastitis infection during drying off, antibiotics should be used on each quarter of the udder.

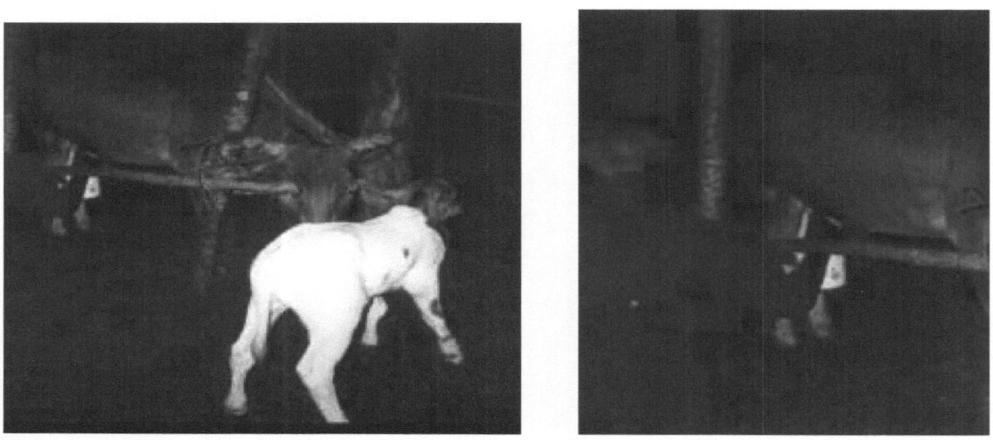

Figure 100: Milking of cattle: Left-Indigenous cattle; Right- Exotic dairy cattle

Figure 101: The milking Machine

STEAMING UP:

The cow should be given feed commensurate with its production level (challenge feeding). However, during the dry period, the animal should get an increasing quality of feed (rising plane of nutrition) rich in proteins until parturition. This is called steaming up.

FEEDING YOUNG ONES:

The young one is not yet ready for rough fibres. Within 487 hours of its birth, it should be fed on colostrums (The first mother's milk). +This is believed to be rich in nutrients such as proteins, as well as containing antibodies and vitamins. Also the digestive system of the calf is such that absorption of macronutrients large particles is possible without digestion, i.e., the pores are still large enough to allow these particles to pass through directly. Thereafter, the following feeding program should be followed:

Table 19: Feeding schedule for a calf:

Age (weeks)	1	2	3	4	5	6	7	8	9	10	11	12	13	14	15	16
Amount of feed (kg)	-	3	4	5	6	7	7	7	6	5	4	3	2	2	1	1

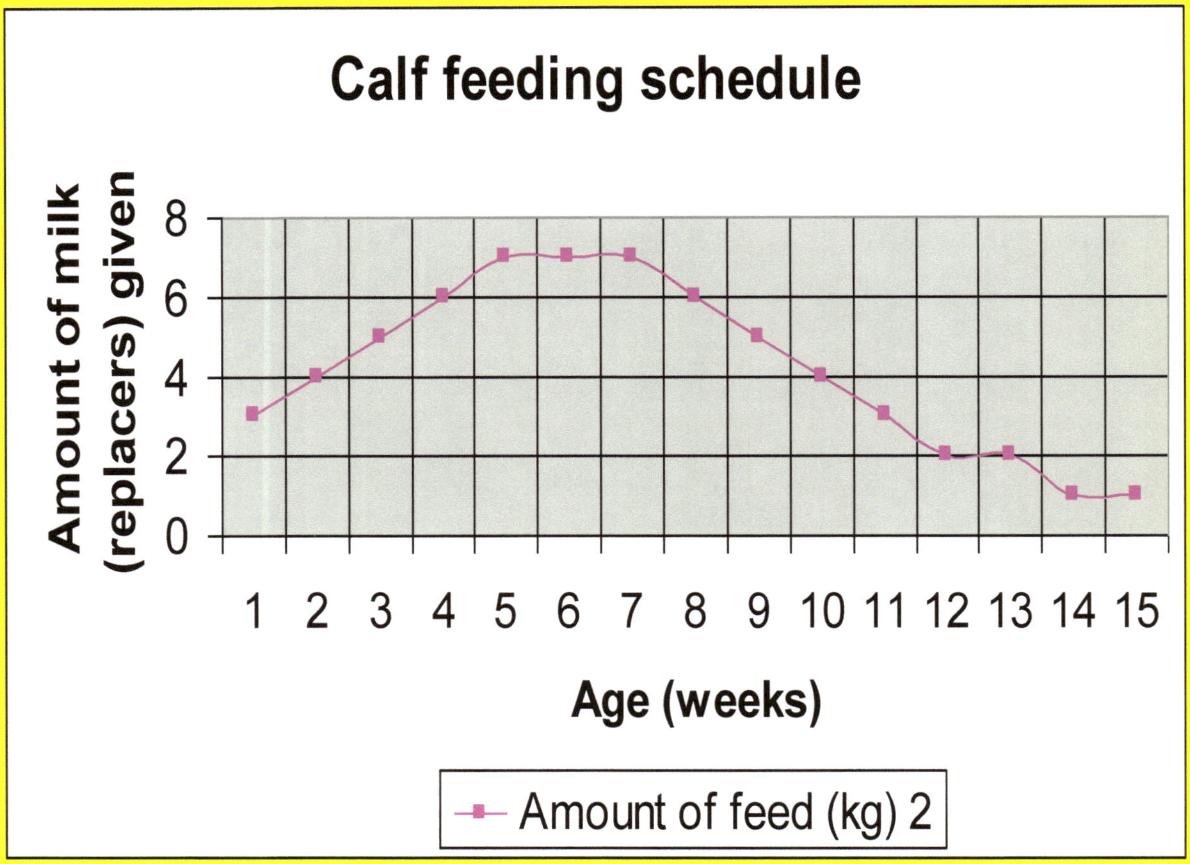

Figure 102: calf feeding schedule (in kg)

NB: The above can be whole milk or milk replacers, milk equivalents, or early weaning concentrates. From the third week, solids may be introduced into the feeding program. Hay can be introduced at 5 days of age so that the calf gets used to it gradually. About 1-2 kg of concentrate can be given to calves 1-2 months old. Weaning should be gradual to avoid shock to the calves. The first month milk can be given thrice per day; in the second to fourth month, it is given milk twice per day. At 4 months, the calf should be fully weaned.

Lucerne (leguminous) hay | Non-leguminous hay

OTHER ACTIVITIES:

Debudding or dehorning is done at two weeks of age. However, this can be varied depending on method to be used- as follows: spoon dehorning at 2 months old; hot iron rod at 6-30 days old; and chemicals at 3-14 days old. Castration of bull calves is done at 3-14 days of age; deworming (also called drenching). Vaccination against a wide range of diseases, identification (e.g. ear tagging or tattooing), and removal of extra teats etc should be done at an early age as recommended by the veterinarian.

FEEDING ADULTS:

The adults should be fed maintenance and production rations. This should comprise roughage and concentrates at the following rates:

Roughage: About 3 kg per 100 kg live weight of animal;

Concentrate: 1 kg for every 2.5 kg of milk produced;

Water: Ad libitum

Mineral licks to provide calcium, among others.

Record keeping:

Records are needed in the following areas:

Expected parturition date;

Daily feed consumption and its cost;

Labour and medicine costs;

Dates for major management activities e.g. vaccination etc;

more easily;
Diary- to show major activities conducted;

Table 20: Some useful management data for different livestock.

Livestock	Cattle	Sheep/ Goats	Pigs	Poultry	Rabbit
Earliest age at puberty	9 months	7 months	6 months	4-5 months	5 months
Weight at puberty	200-250 kg	25-35 kg	70-80 kg	1.5-2 kg	2-2.5 kg
Preferred age at first service	15 month	10 month	9 month	5 month	6 month
Length of heat period	18 hours	Sheep – 36 hrs Goat – 39 hrs	2-3 days (60 hours)	-	
Length of oestrous cycle (days)	21	Ewe-17 Goat -20	21	-	
Gestation period (days)	270-280	Sheep-150 Goat- 151	114/5	21	28-30
Weaning age	4 months		6-8 weeks	4-6 weeks	

MODULE 10: LIVESTOCK ANATOMY AND PHYSIOLOGY

Animal Reproduction
Reproductive Organs Of A Bull
Functions Of Parts Of Reproductive System Of A Bull
Accessory Glands
Male Sex Hormone
Reproductive Parts Of A Cow
Ovaries
Fallopian Tube:
Uterus / Womb
Cervix:
Vagina
Vulva
Heat Period
Signs Of Heat
Importance Of Signs Of Heat
Preparation Of Uterus For Pregnancy
The Oestrous Cycle In Cows
Female Sex Hormones
Reproduction In Poultry
Ovary
Infundibulum (Funnel)
Magnum: Is The Site Where Albumen Is Added To The Ovum
Isthmus
Uterus
Vagina
Cloaca
Vent
Reproductive System Of The Male Bird
Digestion In Livestock
Ruminants And Non-Ruminants
Mouth
Oesophagus
Stomach
Digestion In Non- Ruminants
Duodenum
Ileum
Caecum:
Colon (Large Intestine)
Rectum
The Liver
Anus
Digestion In Ruminants
Mouth:
Oesophagus
Stomach
Rumen
Reticulum
Omasum
Abomasum:

MODULE 10: LIVESTOCK ANATOMY AND PHYSIOLOGY

ANIMAL REPRODUCTION

Reproduction is a biological process which involves the production of young ones. It keeps different species to continue existing, i.e. it is for procreation. Sexual reproduction involves the male and female gametes (sex cells) fusing to form a zygote, which develops into an independent being. In animals, the male sex cells are called sperms or spermatozoa, and the female ones are called eggs or ova (one ovum). Reproduction ensures that species continue to exist (procreation), as well as pass genes from one generation to another.

REPRODUCTIVE ORGANS OF A BULL

This comprises the scrotum, testis / testicles, epididymis, sperm duct, prostate gland, seminal vesicle, seminiferous tubules, urethra and penis. Each of these is shown in the diagram below.

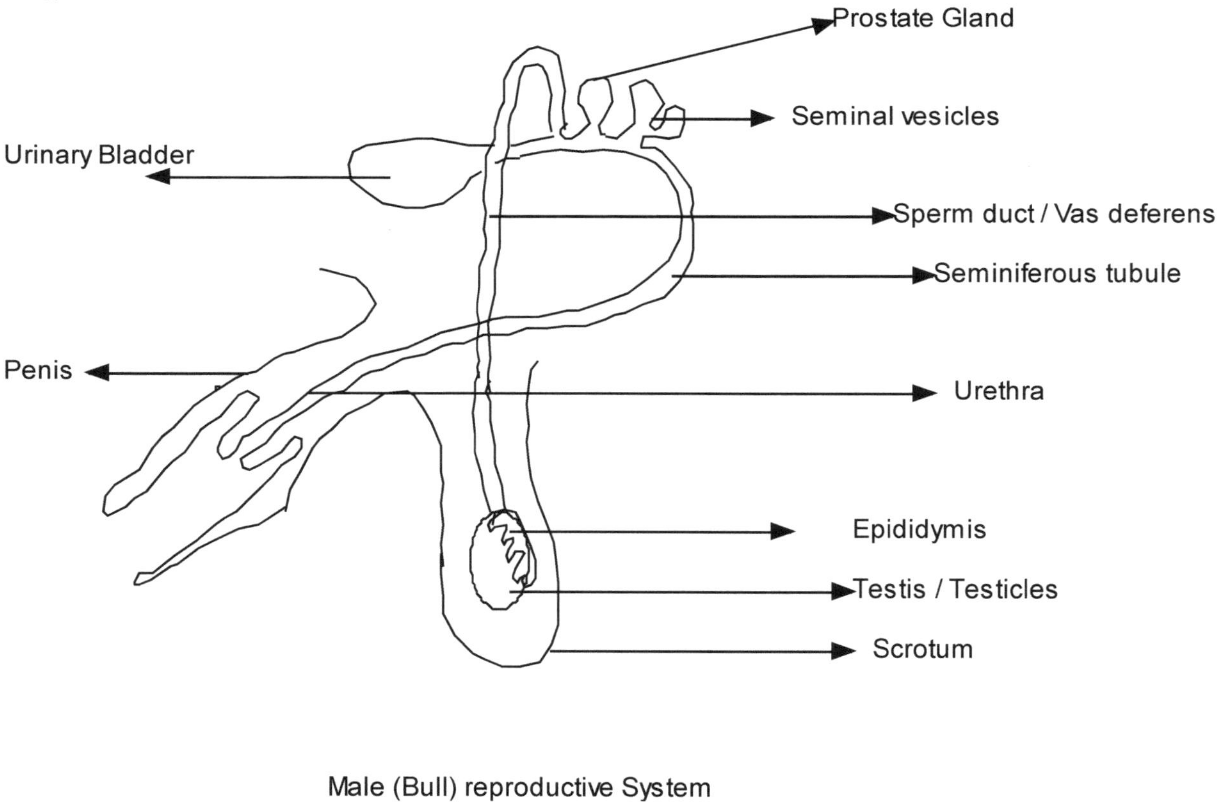

Figure 103: The male reproductive system

FUNCTIONS OF PARTS OF REPRODUCTIVE SYSTEM OF A BULL

Testes: Are two oval-shapes organs enclosed in a sac called the scrotum. They are where male gametes (spermatozoon) are produced. They also produce a male hormone, testosterone which is responsible for the male secondary sexual characters.

Epididymis: A large coiled tube in the testes which stores sperms

Vas deferens/ sperm duct: Receives the spermatozoa from the epididymis, and release it to the urethra.

ACCESSORY GLANDS

Prostate gland: Found at junction between reproductive and urinary systems. Provide fluid to neutralize the urine remaining in the urethra, nourishes sperms, and also acts as a medium for sperms swimming. The mixture is then called semen.

Seminal vesicles: Short coiled tubes where semen is kept ready for mating.

Seminiferous tubules: A passage for semen from the seminal vesicles to the urethra during mating.

Urethra: A tube which passes semen outside the pennies. Also serve in urine elimination.

Penis: A male organ used for mating. Has air-spaces (sponge) which get filled with blood at mating to make the organ firm and strong.

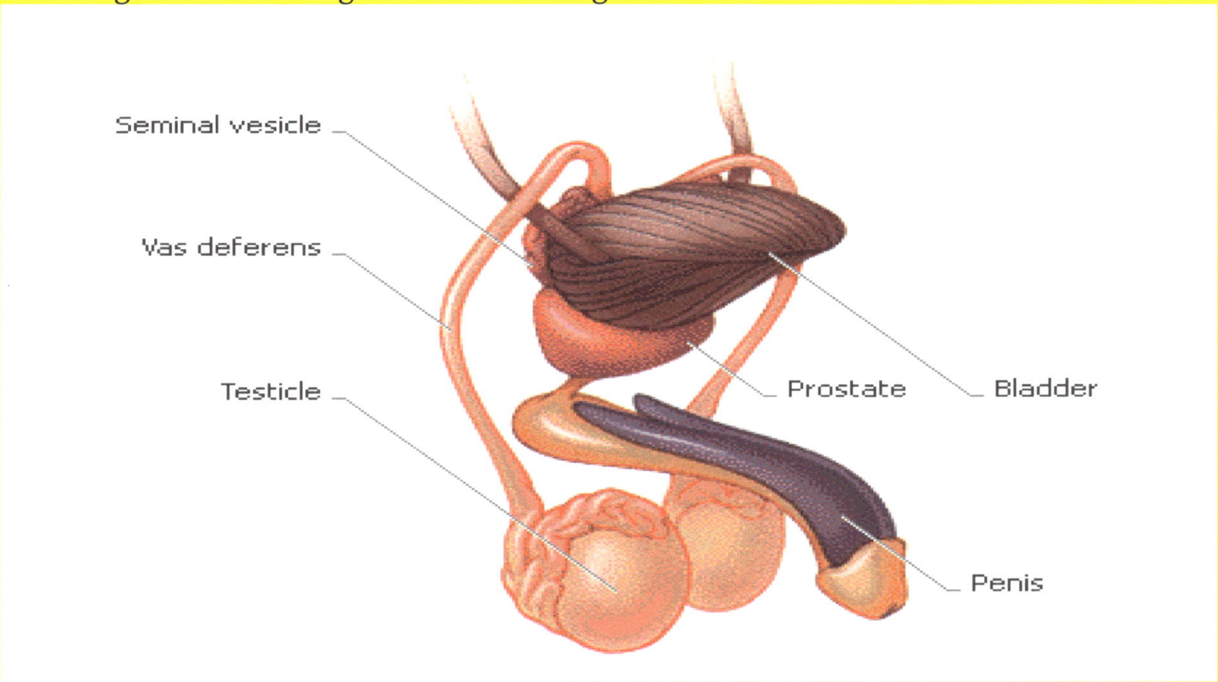

Figure 104: Reproductive system of a male animal.
SOURCE:http://encarta.msn.com/media_461516400_761560628_1_1/Male_Reproductive_System.html

MALE SEX HORMONE

Testosterone is produced in the testis and plays the following functions:
Responsible for male secondary sexual characters e.g. body shape, development of the reproductive system, deepening of voice, (in man- growth of hair in private parts)
Encourages male sex drive or urge (Libido), thus making the mating process to be possible.

REPRODUCTIVE PARTS OF A COW

This comprises the ovaries, the funnel of oviduct, the fallopian tube, the uterus, vulva, vagina and the cervix. These are described below.

OVARIES

Are 2 small rounded organs located in the abdominal cavity near the kidneys. The functions are:

1. Produce ova; and

FALLOPIAN TUBE:

Provides a site where fertilization occurs;
Carries egg from ovary to the uterus (When fertilization does not occur)

UTERUS / WOMB

Nourishes the foetus
Protects the foetus
Site for implantation of the zygote which develops into a foetus

CERVIX:

Closes the uterus during pregnancy and thus prevents pathogens from entering the uterus

VAGINA

Receives the pennies during mating
Site where semen is deposited during mating
Serves as birth canal through which the young one passes when it is being born

VULVA

The external opening of a cow's reproductive system
It facilitates the mating process by directing the penis into the vagina

Female mammals have 2 small round organs called ovaries. They are located in the abdominal cavity near the kidney. Ovaries are responsible for production of eggs. The egg cells begin to develop inside the ovaries and other cells surrounding the egg receive more blood supplies and hence promote the egg development. The region swells and is called the follicle. When the egg is mature, the follicle ruptures to release the ovum into the oviduct. The release of eggs from the ovary into the oviduct is called ovulation. It is at this stage that fertilization can occur if mating occurs. It is a period when a cow is receptive to a bull (cow is said to be on heat). The ovulation process repeats itself regularly if fertilization does not occur. In a cow, it repeats after every 21 days. This is called the oestrous cycle.

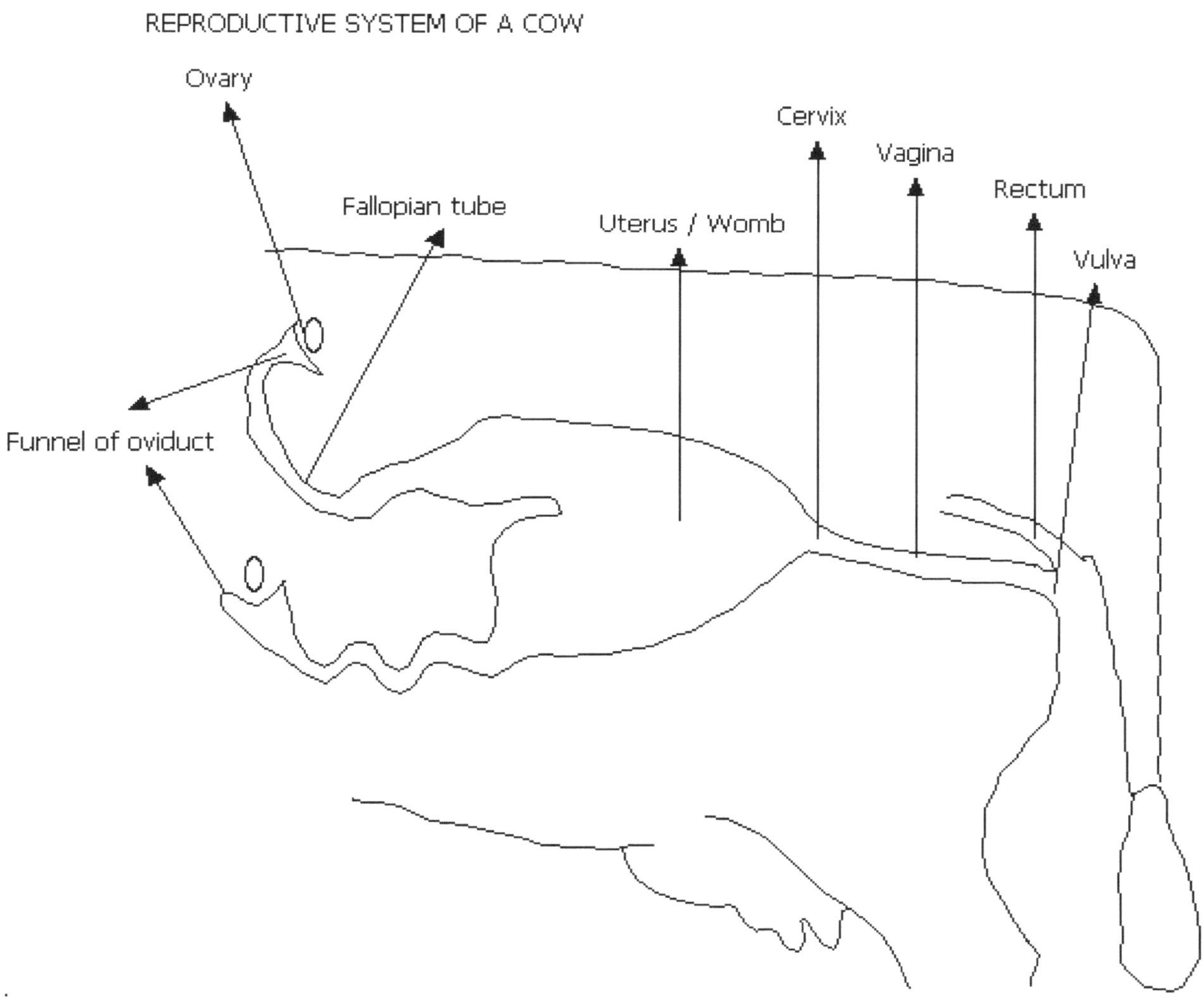

Figure 105: The female reproductive system

Oestrous cycle is a period between one heat period and the next. During this period, a series of changes occur under control of hormones. These include development of the eggs, ovulation, heat period, preparation of the uterus to receive the zygote, preparation of the vagina for mating etc. If fertilization takes place, the cow becomes pregnant and the cycle stops. The duration of pregnancy is called gestation period- if from the time fertilization occurs until the day the young one is born. Thereafter, the cycle resumes.

HEAT PERIOD

Is the time when the cow becomes ready to be served by a bull- i.e. it is receptive. It stands still and accepts the bull during mating. ALL the other times, a cow does not mate (because it is not receptive).

SIGNS OF HEAT

Cow may try to mount other cows herself
Vulva becomes red and swollen;
Slight rise in body temperature;
Allows a bull to mount her without running away;

IMPORTANCE OF SIGNS OF HEAT

It is important for a farmer to detect the signs of heat because that is the only time the cow can be served by the bull; when it has a ready mature egg, and the only time artificial insemination can be done. Losing the chance means waiting for another 21 days- i.e. feeding an unproductive cow for another 3 weeks- which is quite a loss and delay in farm production.

PREPARATION OF UTERUS FOR PREGNANCY

THE OESTROUS CYCLE IN COWS

During ovulation, the uterus of a female animal prepares to receive a fertilized egg from the oviduct by growing a thick spongy lining where the zygote attaches itself (implantation) in the uterus for protection and nourishment. If the egg is not fertilized, it dies, or is dissolved by the tissues around, while the uterus lining, mostly blood, comes off and is expelled through the vagina. In human females, this is the time they say they are in their periods.

FEMALE SEX HORMONES

Oestrogen

This is the chief female sex hormone. Is produced by the ovaries and has the following functions:

Responsible for development of secondary sexual characters e.g. female features (body shape, udder etc);

Controls the oestrous cycle, thus making it occur in the regular interval e.g. 21 days in a cow;

Develops the female sex drive urge which makes her more receptive to the bull.

Progesterone

Produced by the ruptured follicle in ovaries and also by the placenta during pregnancy.

Functions

Prepares the uterus for implantation of zygote;

Stops the release of more eggs from ovary during pregnancy;

Development of udder during pregnancy to be ready for milk production

Follicle Stimulating Hormone (Fsh)

Produced by the pituitary glands situated at the base of the brain. It is responsible for:

Development of follicles in ovaries;

Controls the production of oestrogen in ovaries

Luteinising Hormone (LH)

LH is also produced by the pituitary gland. It controls the rupturing of the follicle, thereby causing ovulation to occur. It leads to the production of corpus luteum (hence the name corpus luteum), a yellow body that develops to take place of the follicle immediately after ovulation and fertilization.

BOX 17: Female Reproductive System

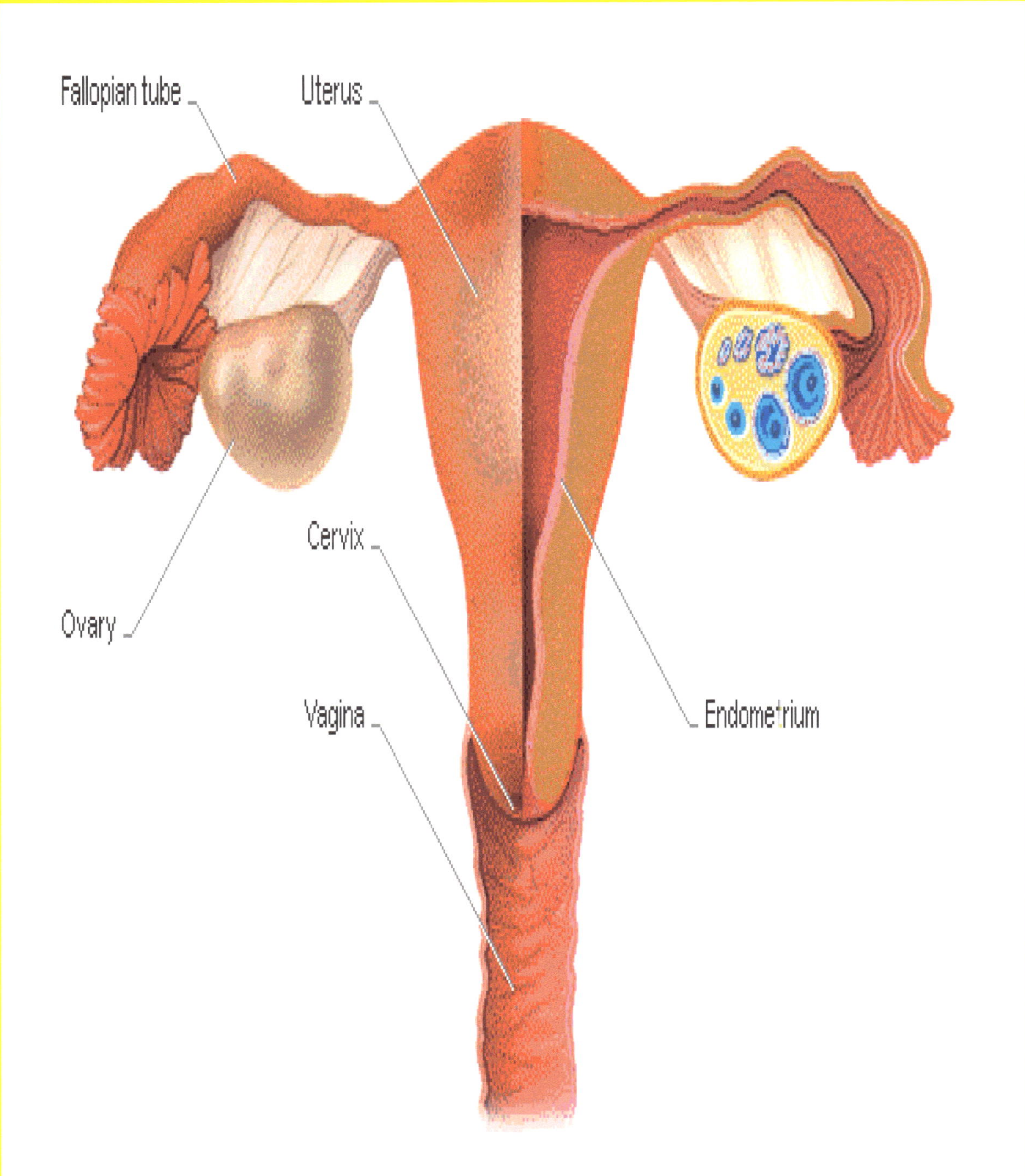

The bones of the human female pelvis form a bowl-shaped cavity that supports the weight of a developing fetus and encloses the organs of the female reproductive tract. Two ovaries, the female gonads, produce mature eggs. Leading away from the ovaries are the fallopian tubes, or oviducts, the site of fertilization. The uterus, a muscular organ with an expandable neck called the cervix, houses the developing fetus, which leaves the woman's body through the vagina, or birth canal. SOURCE:http://encarta.msn.com/media_461545224_761560628_1_1/Female_Reproductive_System.html

MENSTRUATION

An average menstrual cycle begins with three to five days of menstruation, the shedding of the uterine lining, during which hormone levels are low. At the end of menstruation, a pituitary hormone stimulates new follicles to develop in the ovary. These secrete estrogen as they mature, causing cells in the lining of the uterus to proliferate. Mid-cycle, one mature follicle releases an egg. The empty follicle forms the corpus luteum, an endocrine body that secretes progesterone. Under the added influence of progesterone, the uterine lining thickens further and swells in preparation for the implantation of a fertilized egg. If fertilization does not take place, the corpus luteum dies and hormone levels fall. Without hormonal support, the uterine lining disintegrates and discharges, beginning a new menstrual period and cycle. http://encarta.msn.com/media_461554919_761559881_-1_1/Menstruation.html

Reproduction in poultry

The mature female bird (hen) has a single large ovary which contains many eggs at different stages of development. At maturity (4-5 months old), eggs with yolk are already formed and are released into the infundibulum one at a time. A fully developed egg takes 26 days to be ready. The following are the different parts of the female reproductive system of chicken: The ovary, infundibulum, magnum, isthmus, uterus, vagina, cloaca, and the vent.

OVARY

Produces the ovum

Site where egg develops and matures before being released into the

INFUNDIBULUM (FUNNEL)

(i) Receives egg from the ovary

Stores sperms

Site for fertilization

MAGNUM: Is the site where albumen is added to the ovum

ISTHMUS

Water, minerals salts and vitamins are added to the ovum

UTERUS

Shell added to enclose the egg

More water and mineral salts also added

VAGINA

Adds cuticle to the egg

CLOACA

Site for sperm deposition

VENT

This is an opening which passes the egg out during the laying process. Once the hen begins to produce eggs, the process goes on throughout the rest of her life whether the cork is present and mates her or not. The cock serves only to fertilise the eggs so that chicks can come out. But in case no chicks are required (as in the case of layers), then one can continue producing without a cork. Female sex hormones are responsible for the development and release of eggs

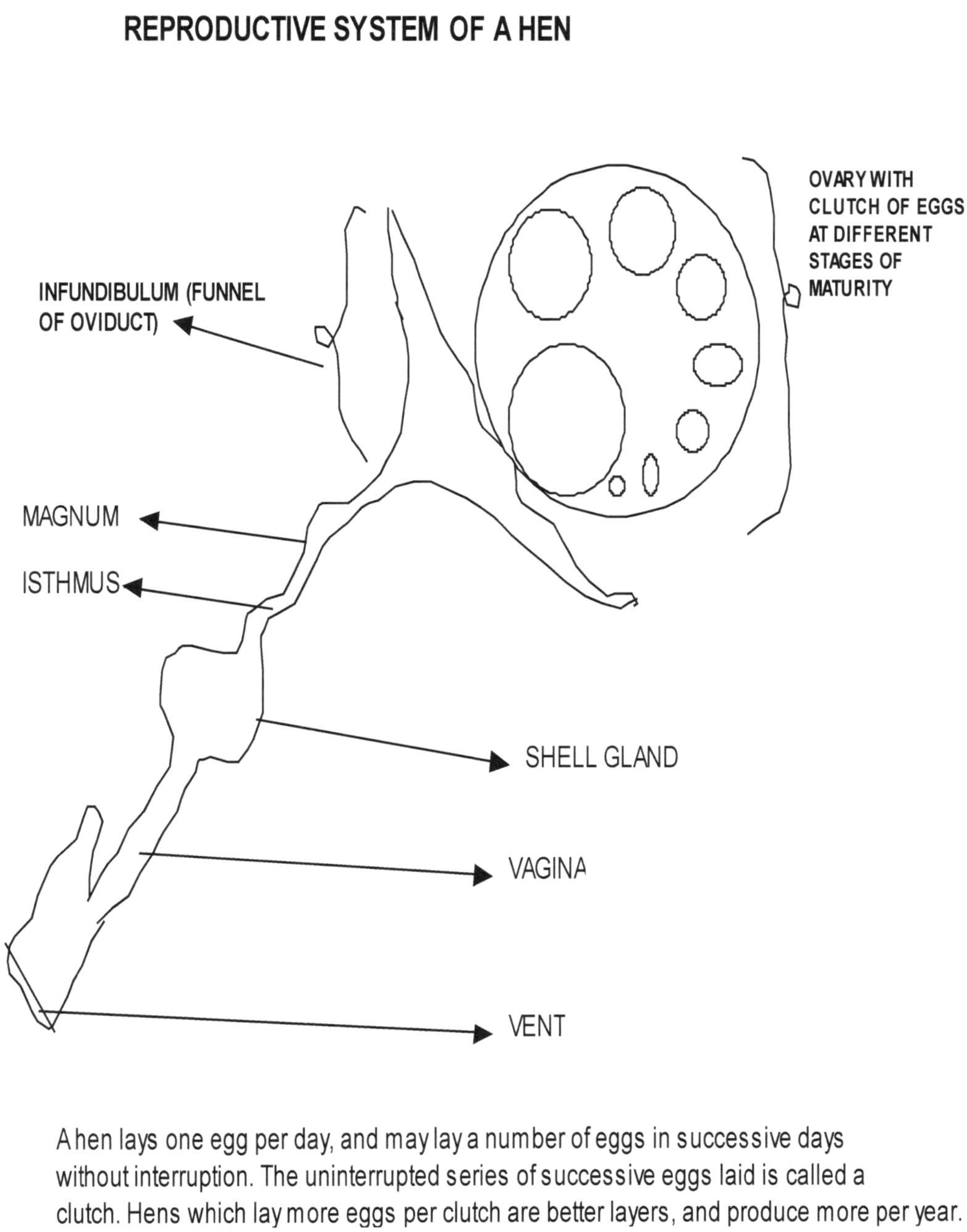

Figure 106: The reproductive system of a hen

REPRODUCTIVE SYSTEM OF THE MALE BIRD

The main reproductive structures in the male bird (Cock) are:

Two testes which produce sperms;

The vas deferens which is the sperm duct- a passage for release of sperms

The vent, which is the organ for mating.

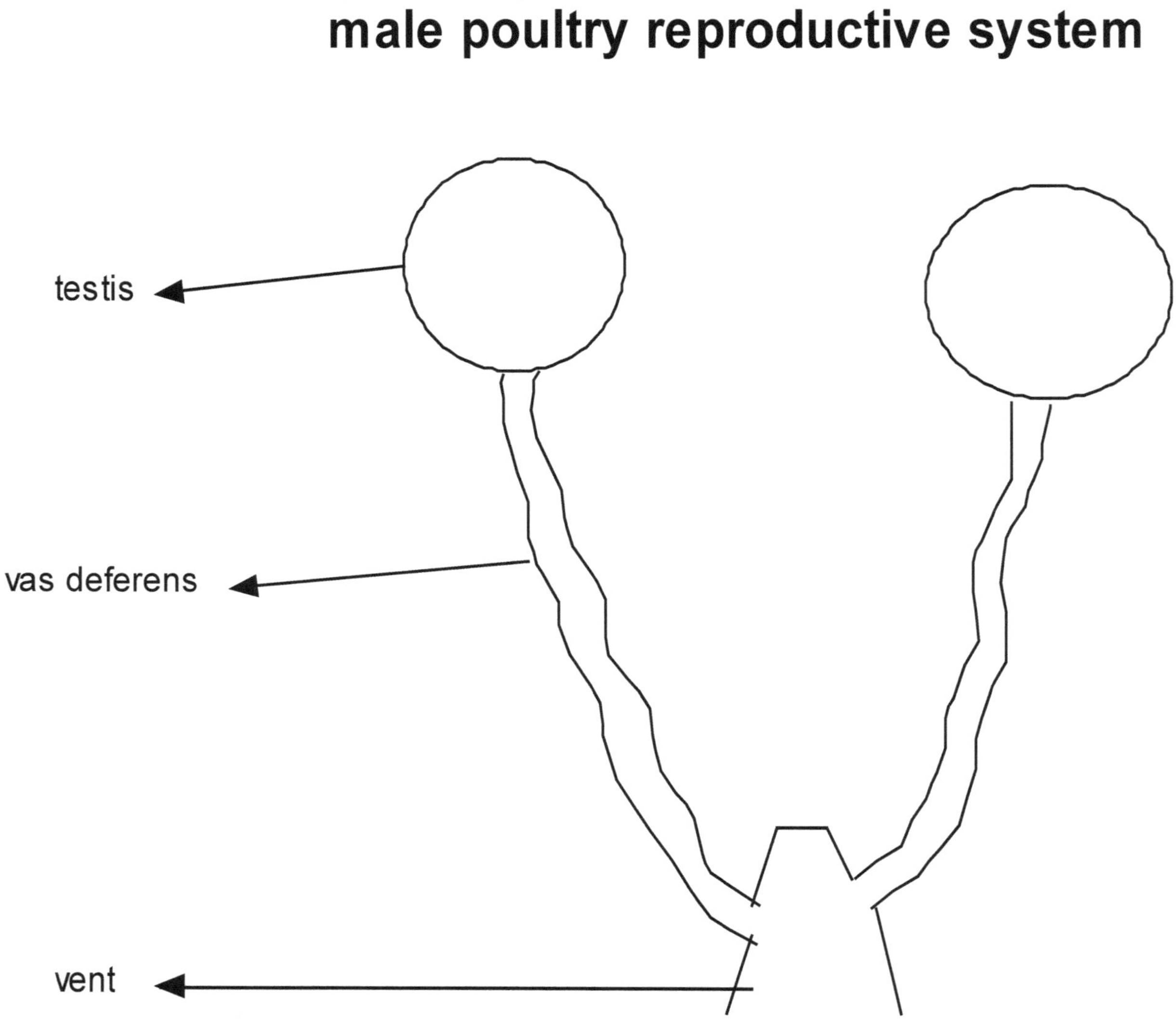

Figure 107: The male reproductive system of a cork

The male bird (cock) has no pennies. When stimulated to mate by the male sex hormone, the sperms travel down the vas deferens in a fluid form. The male bird presses his vent hard against that of the female and the sperms are released. When the sperm enters the female they begin to swim up the oviduct towards the ovary. If a ripe yolk is found, fertilization occurs. And this forms a fertile blastodisc attached to the yolk. In this case a fertile egg is laid. A hen can produce a series of fertilized eggs without mating the cork fro some time because the vent stores extra sperms that are then used to fertilise a number of eggs as they become ready.

DIGESTION IN LIVESTOCK

Digestion is a process by which complex food substances are broken down into smaller absorbable forms. It involves mechanical, chemical and enzymatic (biological) processes. It takes place in the alimentary canal

RUMINANTS AND NON-RUMINANTS

Farm animals can be classified as either ruminants or non-ruminants. This classification is based on the nature of stomach that an animal has. Ruminants are herbivores with four-chambered stomach comprising the rumen, reticulum, omasum and abomasums. Goats, sheep and cattle are all ruminants. Non-ruminants are animals with simple one-chambered stomach, e.g. in pigs (same as in man). The following is a description of functions of parts of the digestive system of livestock.

MOUTH

This is the food entrance. Receives food, breaks it down by chewing using teeth (mastication), and mixes it with saliva which softens and lubricates it. The tongue does the mixing and rolling of feed in the mouth. The chewing process increases the surface area of the food so that as it is mixed by the tongue, it gets in closer contact with the salivary amylase (an enzyme) from the saliva, which breaks down starch into maltose. The saliva is alkaline, thus helping neutralize mouth for proper functioning of the salivary amylase enzyme (ptyalin). The mucus in the saliva helps hold the food together to form a bolus that can be easily swallowed by peristalsis.

OESOPHAGUS

The food is then rolled into a ball called bolus, and forced into the gullet by the tongue. The food here moves by the systematic contraction and relaxation of the muscles- a process known as peristalsis. The bolus reaches the stomach at the cardiac sphincter, an entrance. The food goes through the following path as it undergoes different stages of digestion: mouth- gullet / oesophagus - stomach- duodenum- ileum-caecum- rectum and anus.

STOMACH

Here, the gut widens to form a large bag. It has 3 major functions:

Temporary food storage

Mixes the food with stomach fluids to form a watery fluid called chyme, and churns the food;

Stomach wall produces an acidic gastric juice which contains hydrochloric acid (HCl aq), rennin and pepsin.

DIGESTION IN NON- RUMINANTS

The HCl neutralizes the alkaline food from the mouth. Pepsin and rennin cannot act in an alkaline environment. Enzymatic digestion also continues. This is accomplished by the proteins being broken down into peptides by the pepsin; and the rennin coagulates milk so that its surface area for digestion increases, and also to reduce its mobility (i.e. retain it in the stomach for some time to have it digested). The food then moves through another gate, the pyloric sphincter, to the first part of the small intestine, the duodenum.

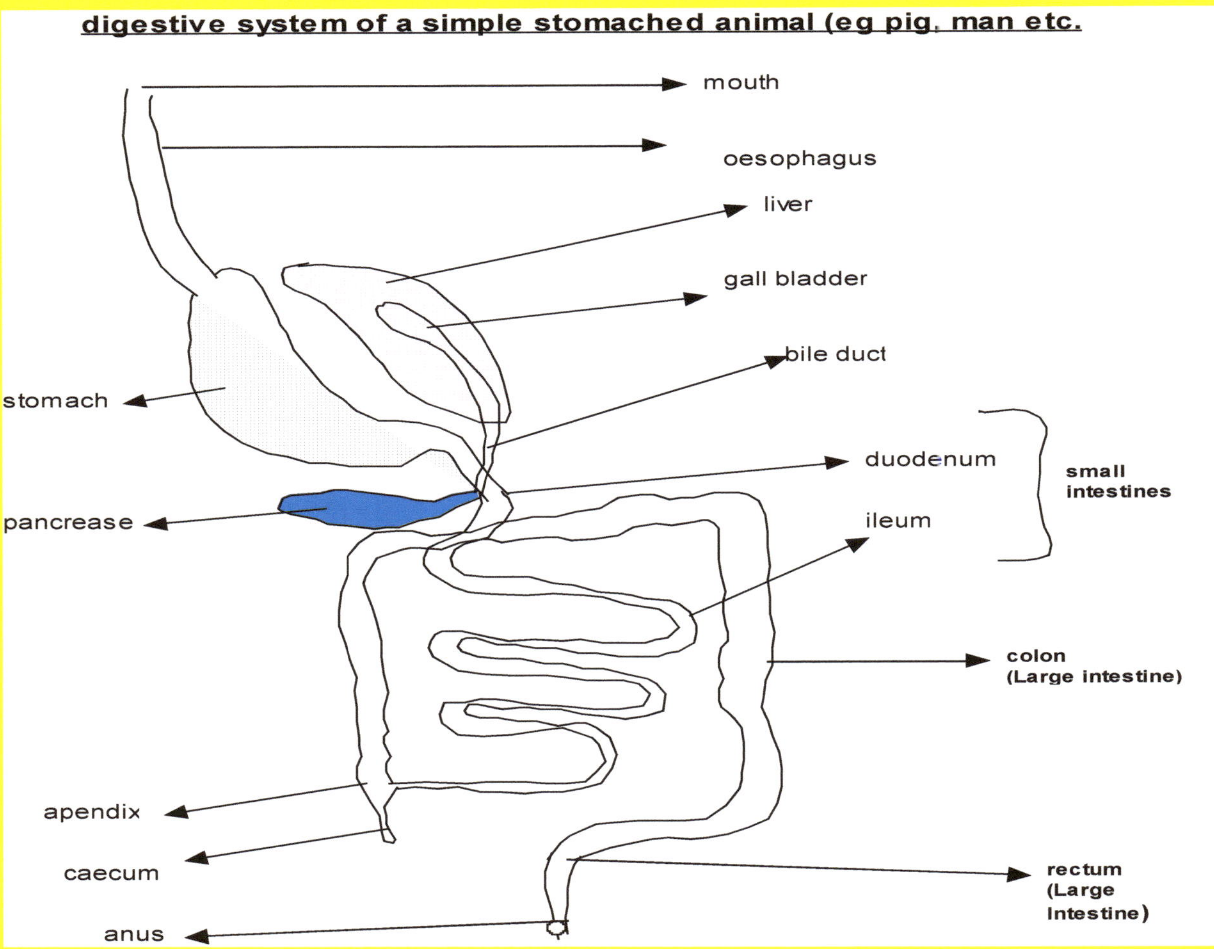

Figure 108: The digestive system of a simple stomached animal

DUODENUM

Food is mixed with a number of fluids:

Bile produced by the liver and stored at the gall bladder. This is an alkaline fluid which, besides neutralizing the acidic chyme from the stomach, also breaks down fats into smaller particles. This is called the emulsification process. These have a higher surface area, and are then ready for further enzymatic digestion.

Pancreatic juice is produced by the pancrease. This is an alkaline juice which neutralizes the acidic chyme from the stomach to make it possible for the enzymes to work. It contains lipase, trypsin and pancreatic amylase. The lipase acts on the emulsified fat particles to release fatty acids and glycerol. Trypsin acts on the peptides (proteins) to release dipeptides and amino acids. Pancreatic amylase acts on the maltose to release glucose. These final building blocks are then ready for absorption. Up to 90-95% of all digestion occurs in the duodenum. Material is then moved to the second part of the small intestine, Ileum, for absorption

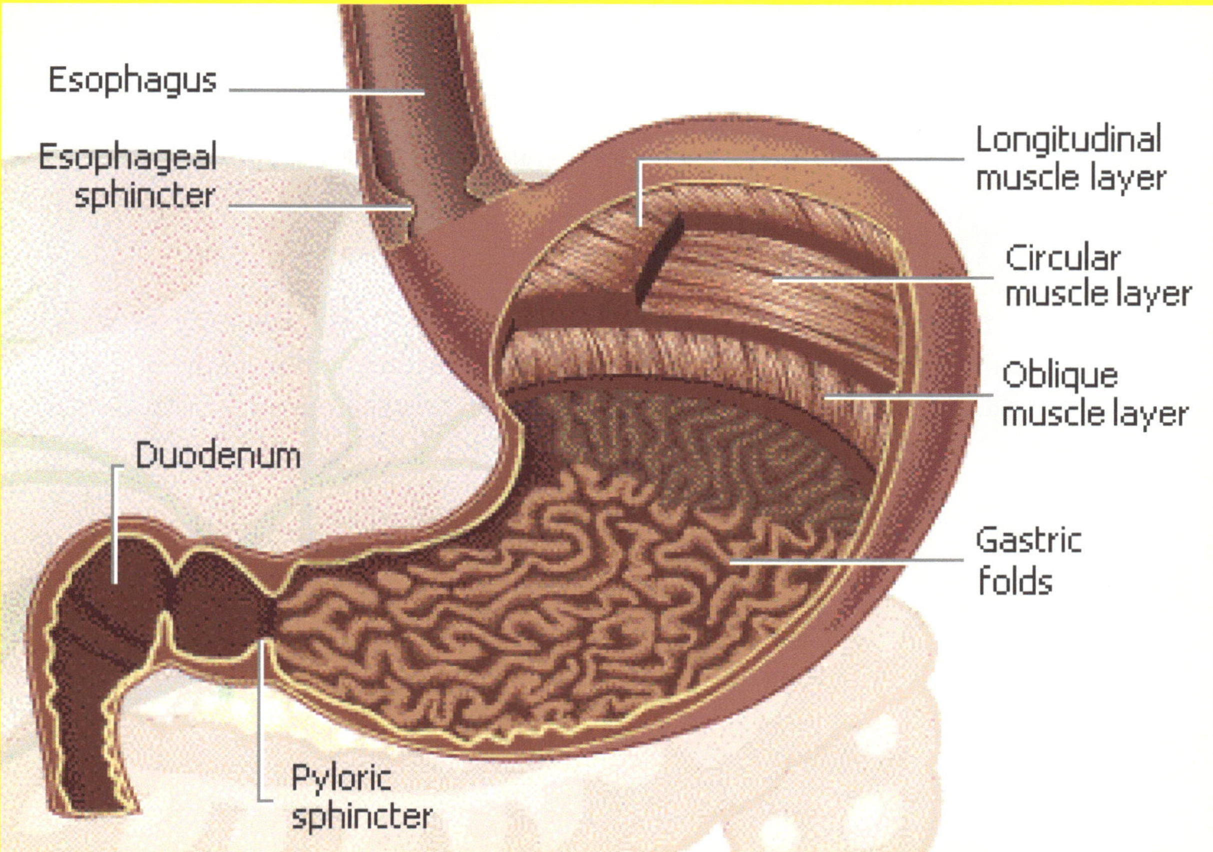

Figure 109: The animal stomach: Source:

ILEUM

This is where 90% of all absorption occurs. This is possible because of the following characteristics of the ileum:

It has a rich supply of blood (i.e., highly vascularised);

It is very long;

It has a very high surface area, increased by the presence of

the ileum wall also produces intestinal juice which contains almost all the enzymes needed to digest all types of food-stuffs. Thus all the proteins are broken down to amino acids; all carbohydrates are broken down to monosaccharides (glucose, fructose and galactose); and all lipids are broken down to fatty acids and glycerol.

CAECUM:

Lies between ileum and small intestines. It has microbes which break down cellulose. Thus caecum is very useful to herbivores which are not ruminants, such as horses and rabbits, as the bulk of their feed are converted to useful forms by the microbes here in a form of symbiotic relationship.

COLON (LARGE INTESTINE):

Water from the undigested food is absorbed into the bloodstream and passes the undigested food to the rectum.

RECTUM

Stores undigested food, dead cells, and mucus (together called faeces) before being expelled out of the body. Water absorption may still continue, and sometimes may lead to very hard faeces, and difficulties in their expulsion. It is the opening through which faeces are expelled out of the gut.

THE LIVER

Is the largest organ in the animal body. Although it is not directly part of the digestive system, it performs some functions which are related to digestion.

Functions

Produces bile which is stored in the gall bladder

Stores vitamins A and D

Detoxifies poisonous substances;

Converts excess glucose into glycogen and stores it

Converts excess proteins into carbohydrates by the deamination process- i.e. a process by which the amino group in the amino acid is removed and used to form urea, while the other product used for energy provision

Stores some minerals such as iron after the red blood cells are dead.

ANUS

DIGESTION IN RUMINANTS

A ruminant feeds on rough feeds, and has 4 chambers in the stomach to help it convert these into useful absorbable materials.

MOUTH:

For taking in the food (ingestion); the teeth chew the feed, and breaks it into smaller units. The ruminant saliva has no enzymes.

OESOPHAGUS

Is a passage between the mouth and the stomach. Food moves through it by peristalsis.

STOMACH

Divided into 4 chambers: Rumen, reticulum, omasum and abomasums. These are shown in the diagram below.

RUMEN

Is the first and the largest of the four stomach compartments in a ruminant.

Fuctions:

Food storage

Food churning thus mixing it with rumen fluid thus facilitating microflora action;

Produces microflora (bacteria, protozoa and some fungi) that ferment the feed;

Microbes have cellulace that starts breaking down the cellulose into smaller stuff.

The food then moves to the reticulum when the animal is resting.

RETICULUM

A stomach is temporary food storage. Food particles are separated so that the small particles are passed to the omasum, while the large ones are rolled into balls of cud and returned to the mouth (i.e. regurgitated). The teeth now grind the feed thoroughly, and when swallowed again, the fine stuff goes straight to the omasum, the third compartment. The wall of the reticulum looks like honey comb and therefore is able to trap foreign materials such as stones, plastics, nails, metal etc.

OMASUM

Made up of a number of layers or folds (like pages in a book). These strain the food and the juice go to the last compartment, the abomasums.

Figure 110: The digestive system of a cow

DIAGRAM OF THE FOUR CHAMBERS OF A RUMINANT STOMACH

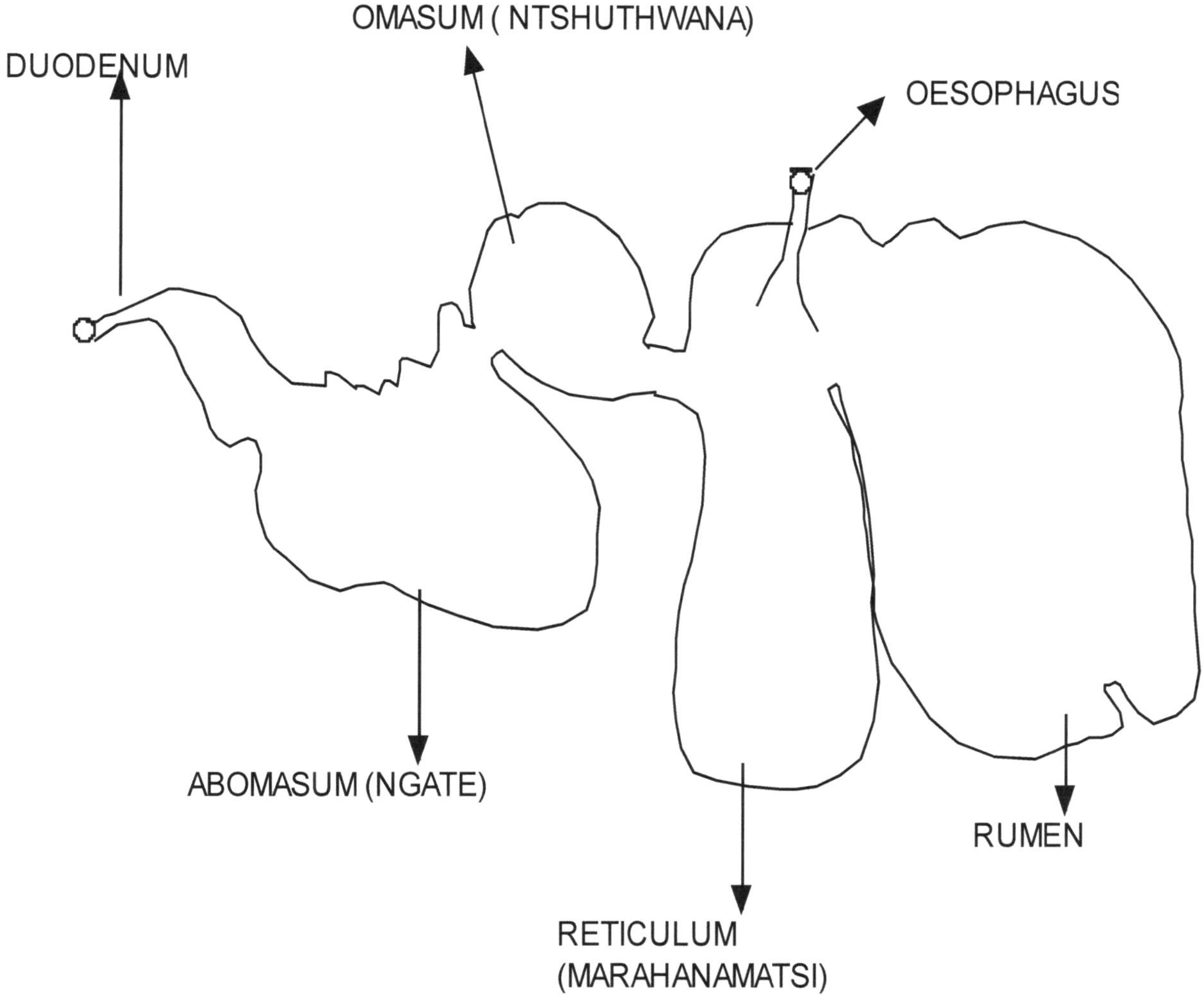

Figure 111: The four chambes of a ruminant stomach

ABOMASUM:

Is the true digestive stomach of a ruminant where normal enzymatic digestion starts.. It is the point where gastric juice is secreted by the gastric gland along the walls of the abomasums. The rest of the digestion is similar to that of the non-ruminant described earlier.

The young ruminant more or less behaves like a non-ruminant, when the other 3 chambers re not yet operational. However, final digestion of proteins and carbohydrates takes place in small intestines, while the large intestines largely absorb water.

Table 21: Differences between ruminant and non- ruminant

	Ruminant	Non-ruminant
Number of chambers in stomach	4 (Polygastric)	1 (Monogastric)
Saliva	Has no enzymes (ptyalin)	Has enzymes (ptyalin)
Feeding habit	Largely herbivore	Omnivore
Cud	Chews	Does not
Site for most digestion and absorption	In all 4 chambers	Small intestine
Cellulose digestion ability	Can digest	Cannot digest

MODULE 11: LIVESTOCK HEALTH AND DISEASE

INTRODUCTION
CHARACTERISTICS OF HEALTHY ANIMAL
SIGNS OF DISEASE IN LIVESTOCK
METHODS OF PREVENTION OF LIVESTOCK DISEASES
CLEAN HOUSING
SUITABLE DIET
INTERNAL AND EXTERNAL PARASITES
CONTROLLING LIVER FLUKES
AN EXTERNAL PARASITE (TICKS)
LIFE CYCLE OF A CATTLE TICK
CONTROLLING OF TICKS

MODULE 11: LIVESTOCK HEALTH AND DISEASE

INTRODUCTION

Health is the state of complete well being of an organism- in all dimensions such as physical, physiological, psychological etc. In such cases, all body organs are functioning properly, in a coordinated way. Disease is the opposite of health.

CHARACTERISTICS OF HEALTHY ANIMAL

The animal must have an upright stance and a steady walk
The animal must have a good appetite
The head held upright
The eyes clean and bright
Nose cold and wet
Ears warm and alert
Skin in good condition, without patches
Cattle 38.80C; Pigs 39.2 0C; Sheep 39.40C; Goats 39.40C; and Poultry 41.0 0C

Table 22:Cattle and chicken health Indicators

Livestock	Body temperature 0C	Pulse / minute	Respiration rate / minute
Cattle	38.5-39.5	40-70	10-30
Chicken	40.5-43	200-400	15-30

SIGNS OF DISEASE IN LIVESTOCK

The animal takes an abnormal posture e.g. when standing or lying,
Animal walks unsteadily.
The animal has poor appetite
The head not held upright
The eyes are dull and watery
Nose cold and wet (producing mucus)
Ears cold and don't seem alert
Breathing difficulties e.g. whistling sound;
Skin in poor condition, with patches

METHODS OF PREVENTION OF LIVESTOCK DISEASES

To keep animal in good condition attention should be paid to the following motto 'prevention is better than cure". Given below are some key strategies of ensuring livestock are kept healthy.

CLEAN HOUSING

Suitable living conditions for animals are important. There should be enough space, lights and ventilation in livestock houses, and no dampness, cold draughts and no overcrowding. Livestock such as pigs and poultry, which are kept in pens and houses, soon pick up disease if their quarters are not kept clean. Harmful organism can remain on dirty floor, droppings, congested air/atmosphere, and wall and pass from one animal to another. Livestock housing should be cleaned out regularly and disinfected from time to time to maintain animal health.

SUITABLE DIET

Livestock must be given the right amount of food and it must be of the correct nutritional value. To achieve this it is necessary to know the nutritional requirements of the stock at their various ages. Cattle on the open grazing may have all they require during the rainy season. Pigs need a careful balanced diet for efficient growth. Incorrect feeding can lead to ill health. Sheep and goats are capable of surviving even in times of drought. Poultry are much more healthy when fed correctly rather than being left to eat domestic scrap. Malnourished animals have a low resistance to diseases.

Sheep and Goats (Shoats) are some of the hardier livestock

INTERNAL AND EXTERNAL PARASITES

These are organisms that get its food from another, called a host, with the host losing and the parasite gaining. The parasite may be based inside the host (thus called internal parasite), or based outside (thus called internal parasite). Whichever the class of parasite, they have the following effects on the host;

Create wound as they pierce through the tissue of the host, thus causing pain and restlessness. This may even depress appetite.

The wounds thus created become points for secondary infection by opportunistic organisms e.g. bacteria, and other pathogens.

Once attached to host, they suck blood, thus causing anaemia to the host;

They take in feed from the host (e.g. tapeworms), thus causing malnutrition, emaciation, and stunted growth;

If they attach outside, they may cause the skin to loose shape and quality, thus fetching less in the market (e.g. ticks).

Due to the above effects, the parasites must be controlled if the animal is to grow well, and keep healthy.

Examples of internal parasites (endoparasites) are tapeworms, roundworms, liver flukes, etc

External parasites (Ectoparasites) are: Ticks, lice, flies, mites, tsetse fly etc. Given below are life cycles of some of the parasites.

An adult fluke is about 4cm long

The adult flukes live in the liver of the primary host. Their eggs are passed down the bile duct to the intestine. They pass out with the faeces.

The eggs may lie in the grass for any months without hatching if the weather is dry. As soon as water is present, they hatch and the larvae can swim about for up to two weeks.

If they meet a certain species of snail, the larvae bore through its skin. They live and feed on the snail swim up the stem of grass or the other plants in the thin film of water on the surface.

When they are on the grass, they form cysts, arresting stage of the life cycle. They are waiting for the cow to eat the grass.

When the cow eats the grass, the cysts hatch in to the flukes in the intestinal and reach the liver, where they stay.

Table 23: DESCRITION OF SOME LIVESTOCK DISEASES:

Disease	Etiology	Host	Symptoms	Prevention and control
Mastitis	Bacterial Streptococcus, staphylococcus, etc	Cattle, goats, sheep	Udder swollen; Milk stained with blood clots	Good milking practices e.g. hygiene, strip cup etc Injection with Antibiotics (IV or IM)
Foot and mouth disease (FMD)	Infectious virus (Myxovirus	Cattle, sheep, goats	Fever; Painful blisters around mouth, udder and between hooves Excessive salivation; Loss of appetite; Lameness	Vaccination; Quarantine
Anthrax	Bacteria (Bacillus anthracis)	All animals plus man	Fever, very quick death, Blood oozes from nose & anus, Blood does not clot,	Vaccinate using blanthrax ((Also used against black quarter).
Rinderpest	Virus Myxovirus	Cattle	Running nose, Frequent drinking Discharge from eyes Oral lesions after 3-4 days, resulting in excessive salivation Bloody diarrhea Difficulty in breathing	Vaccinate NO TREATMENT
New castle disease (NCD)	Viral	Chicken / poultry	Difficulty breathing Drinking too much but not eating; Watery mouth, eyes and nose;	Vaccinate with NCD vaccine at 2 weeks, 6 weeks, 16 weeks and after every 6 months thereafter
Black quarter/ Blackleg	Bacterial Clostridium chauvei	Cattle, sheep, goats	Fever, lameness, loss of appetite, shoulders and rear legs swell, (Gassy swelling under skin)	Vaccinate (with blanthrax vaccine)

Table 24: Some Disease of livestock:

Disease	Causative Agent	Host Animal
1. Coccidiosis	Protozoan	Chicken, rabbits, calves, kids & lambs
2. New castle	Virus	Chicken
3. Fowl Typhoid	Bacterium	Chicken
4. Heat water	Bacterium (carried by ticks)	Cattle
5. Trypanosomiasis	Protozoan (carries by flies)	Cattle
6. Tuberculosis (TB)	Bacterium	Cattle
7. Mastitis	Bacterium	Cattle, rabbit
8. Contagious abortion	Bacterium	Cattle, sheep, goats
9.Anthrax	Bacterium	Sheep, cattle, goats pigs and humans
10. Foot and mouth Disease (FMD)	Virus	Cattle, pigs, goats, sheep
11. Contagious bovine pleuro pneumonia	Virus	Cattle, sheep and goats
12. Rabies	Virus	Dogs and wild carnivores
13. Botulism	Bacteria	Cattle, sheep and goat
14. Piglet Anaemia	Lack of iron diet	Piglets
15. Aphosphorisis	Lack of phosphorus in diet	Cattle, goats, sheep
16. Rickets	Lack of Ca P and vitamin D	Piglets, calf lamb

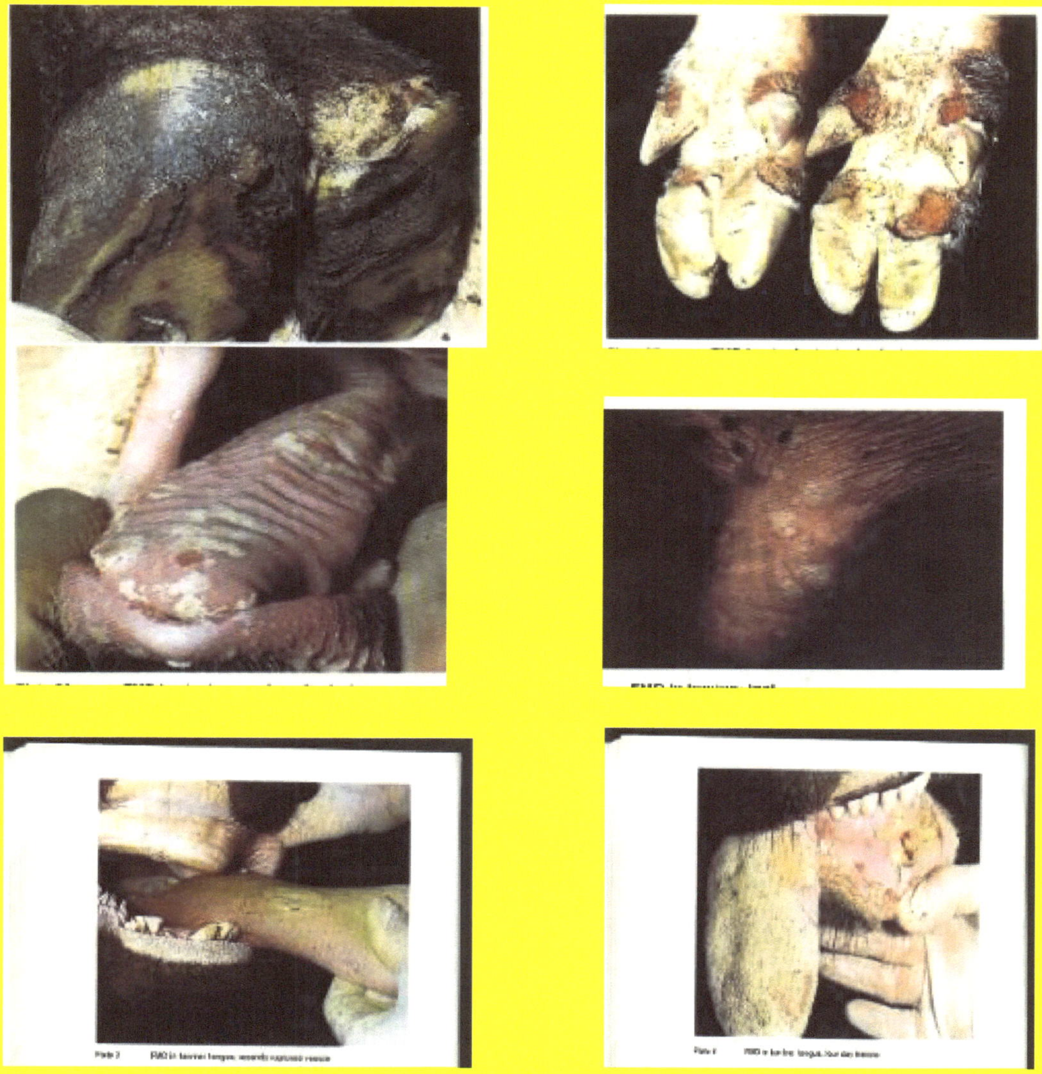

Figure 112: The foot and mouth disease of cattle, sheep, goat and pigs

CONTROLLING LIVER FLUKES

The best way to control flukes is to drain all wet and swampy land, or to keep grazing animals away from it. Then the flukes will die off. If necessary, fences should be erected. Snail- infected was may be treated with IPPM of copper sulphate (cuso4) to reduce the snail population. Drenching animals with dewormers such as NILVEN or sodium sulphate (Na2SO4) is an effective treatment of infected animals.

AN EXTERNAL PARASITE (TICKS)

A cattle tick is an example of an ecto- parasite. There are several species of ticks, all related to spiders, having eight legs- and belong to the acarine family.. Ticks are worst of all the external parasites. They cause a lot of harm to the cattle by sucking their blood and by carrying diseases like heat water, red water, east coast fever and anaplasmosis.

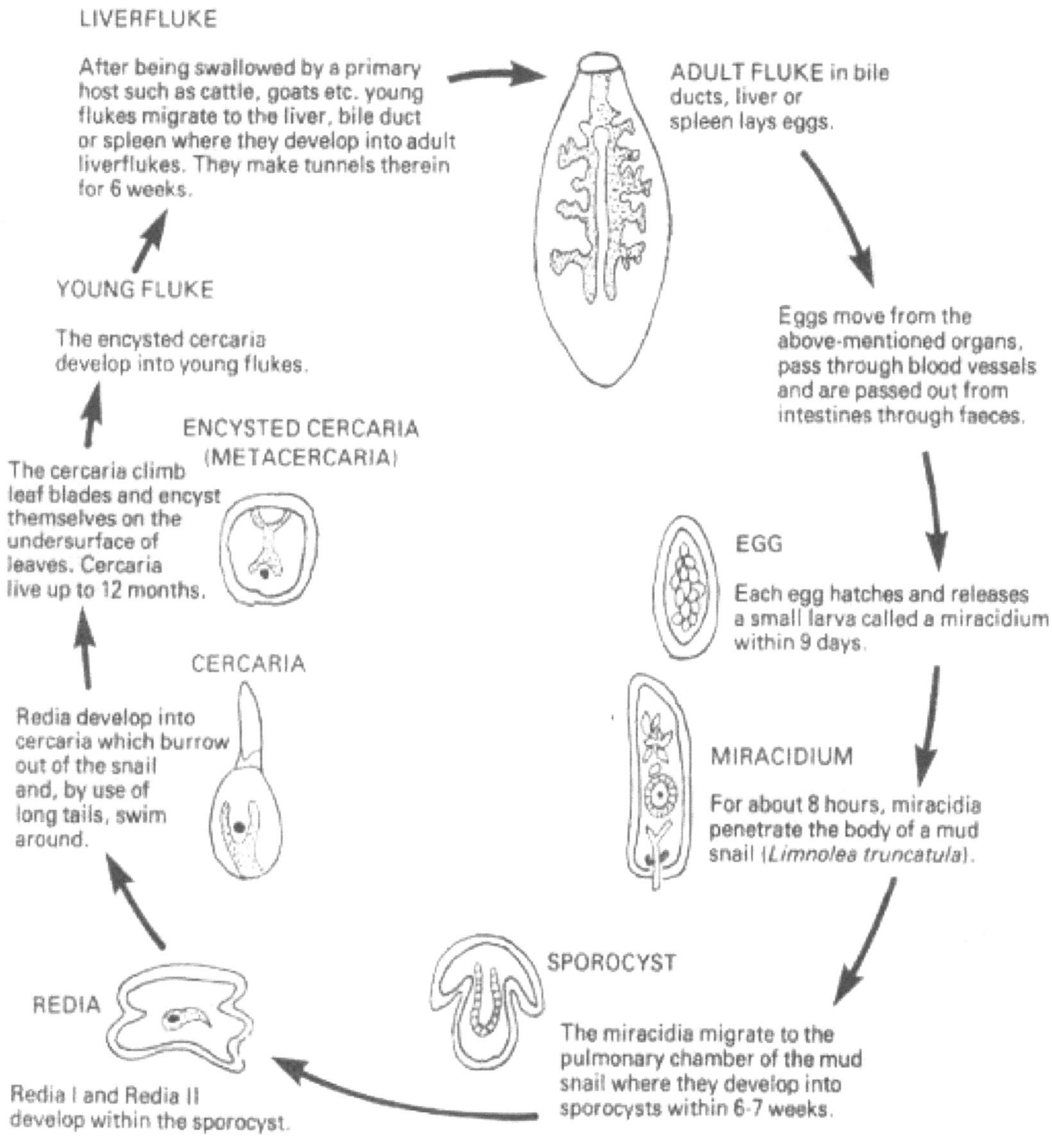

Figure 113: Life Cycle Of The Liver Fluke

LIFE CYCLE OF A CATTLE TICK

The female sucks blood from a cow. The cow is called the host of the parasite.

When the tick is full of blood (engaged) she drops to the ground She lays thousands of small eggs on the grass

These eggs hatch into larvae which have six (6) legs

The larvae climb up the grass and wait for another cow to brush pass them or lie down on the grass. They get onto the cow.

The larvae suck the blood of the host cow. They grow and change into a tick nymph, with eight (8) legs.

They grow still more, and eventually become adult ticks.

As it goes through its cycle, a tick may drop off one cow and get onto another. It may suck blood from two or three different cows. If one of these cows has a disease, the next one will catch it, because the tick may carry the pathogenic organism from one cow to another.

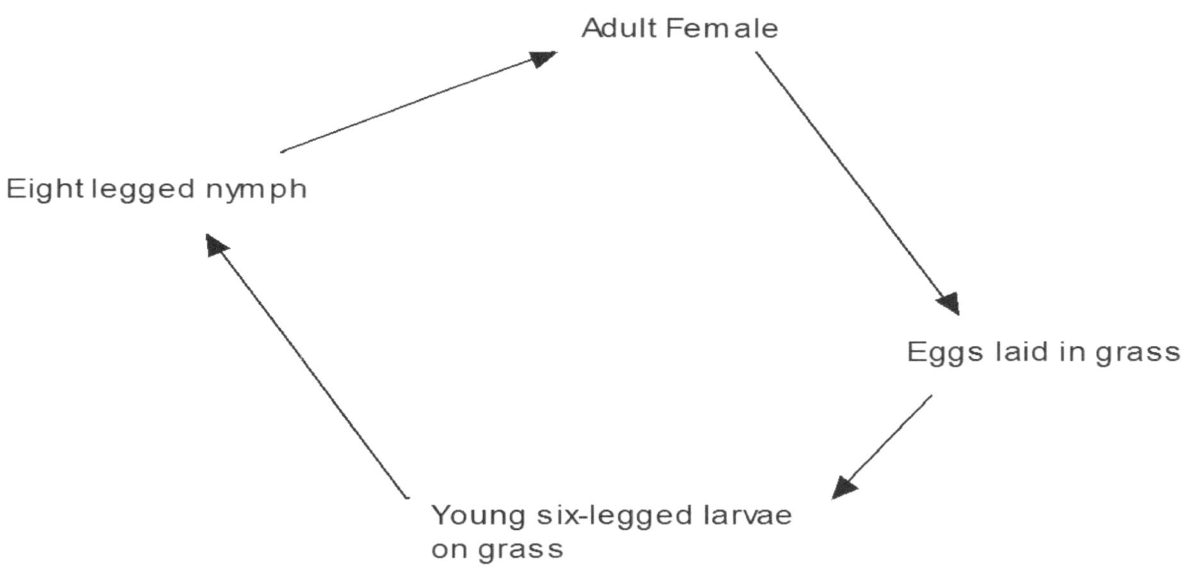

Simple life cycle of a tick.

Figur 114: Simple life cycle of a tick

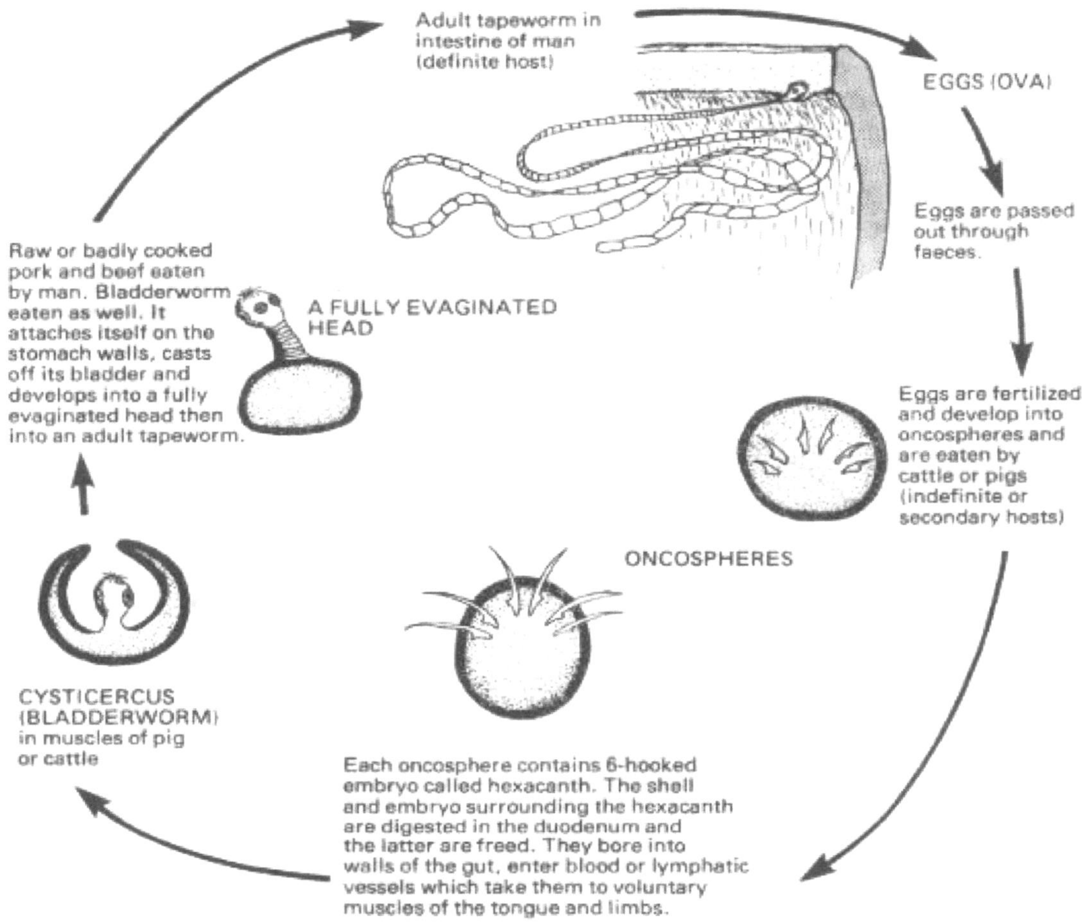

Figure 115: Life cycle of tape worms:

CONTROLLING OF TICKS

By total immersion of the livestock in a drip, or total body cover with an acaricide and spray.

By carrying out such a procedure regularly. It may be necessary, twice a week in some areas.

By maintaining the correct chemicals concentration in the tank. Different chemicals should not be mixed in the dip.

By using a head dressing

The best time to kill ticks is when they are on the cattle. Tick grease is applied on the cattle manually in places where the ticks like to attach and feed – in the ears, under the tails. Tick grease is an ointment that contains an acaricide. Ticks are also killed on the ground by ploughing and by grass burning.

MODULE 12: GAME FARMING AND GAME RANCHING

GAME FARMING
IMPORTANCE OF GAME FARMING
ECOLOGICAL PRINCIPLES USED IN SELECTING GAME FARM
THE ENERGY FLOW IN AN ECOSYSTEM
HABITAT / GEOGRAPHICAL PREFERENCES OF SOME GAME ANIMALS
CAPTURE OF WILD ANIMALS (GAME)
METHODS OF CAPTURING GAME ANIMALS
CHEMICALS
USE OF NET.
MIST NET
LASSOING
OSTRICH FARMING
EXTENSIVE SYSTEM
SEMI-EXTENSIVE
SEMI-INTENSIVE
INTENSIVE
EGG COLLECTION
EGG HANDLING AND INCUBATION
MARKING OF BIRDS (OSTRICH) IN CAPTIVITY
ADVANTAGES OF OSTRICH MEAT OVER BEEF
GOVERNMENT EFFORTS TOWARDS GAME FARMING DEVELOPMENT IN BOTSWANA
LEGISLATIVE MEASURES ENSURING SUSTAINABLE USE OF GAME

MODULE 12: GAME FARMING AND GAME RANCHING

GAME FARMING

Introduction

Game farming is the intensive and small-scale management of game animals. It is the managed production of semi-domesticated animals for production of market products such as skin, meat, and for tourist attraction, and educational purposes. Game farms may be in small compounds, or even inside houses, while a ranch is an open environment, with restricted boundaries e.g. fence. Game animals are undomesticated fauna often termed as wildlife, and may be useful for recreational purposes e.g. tourism, hunting etc. Examples of game farms in Botswana are Khama Rhino sanctuary, Nata Sanctuary, crocodile farm, hippo pool etc. In Eastern Africa, they are numerous, including Elephant, Kudu, Springbok, Eland, Buffalo, birds, ostrich, snakes, porcupine, lions, and leopards. In Kenya, birds such as the quale, guinea fowl, antelope, etc have been popular.

Game ranching is the extensive, larger scale management of game animals. It is a form of game farm, except on a much larger scale in an expansive area but with controlled boundaries that restrict the movement of the wildlife. The game are often offered veterinary attention, as well as controlling their population when need arises to match the carrying capacity of the ranch. These are the two key aspects that make it different from a normal wildlife or game park. It is the 'managed' production of non--domesticated animals for production of market products such as skin, meat, and for tourist attraction, and educational purposes. It is the same thing as livestock ranching, except for the type of animals in the ranch; livestock such as cattle, goats, sheep etc are kept in a livestock ranch, while wildlife such as kudu, elephant, lions, zebras etc are kept in a game ranch.

Figure 116 : The big cats

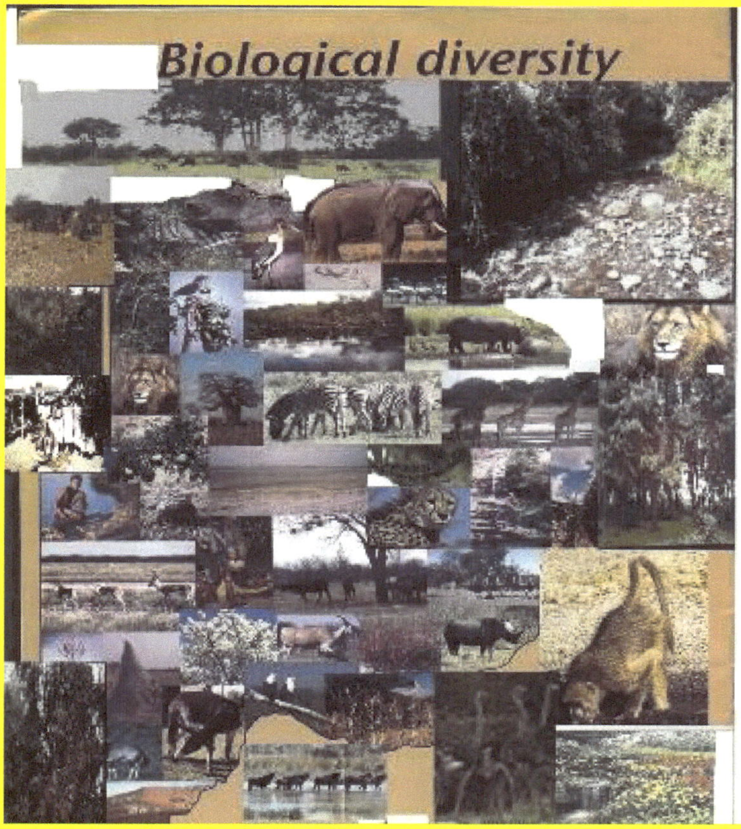

Examples of game animals

Figure 117: Left-The big CATS

Figure 118: Giraffe

IMPORTANCE OF GAME FARMING:

Source of income;

Source of game meat;

Acts as a means of conserving endangered wildlife species;

A recreational facility;

Used for research promotion;

ECOLOGICAL PRINCIPLES USED IN SELECTING GAME FARM

The suitability of an area for game farming will depend on factors such as:

Type of game to be kept (defines livestock unit);

Number of game to be kept (dictate carrying capacity of ranch);

Type of vegetation / food available- grass/ shrubs or bush (dictated by feeding habit of animal e.g. grazers, browsers, carnivores, omnivores etc);

Geographical preference of the intended game (climate, disease prevalence, vegetation etc)

Large animals such as the buffalo and elephants will prefer a place rich in vegetation to provide for their energy needs; hunters/ carnivores such as leopards may prefer an open area to enable them see their prey from a distance, and chase them undeterred.

THE ENERGY FLOW IN AN ECOSYSTEM

Producers are fed on by herbivores (primary consumers), which in turn are fed on by secondary consumers (carnivores), which are in turn fed on by others higher in the food chain hierarchy. Energy flows upwards along the food chain, and all energy originate from the sun (solar energy), and is trapped by plants during photosynthesis. As this energy is passed on to the next feed level, there is at least 90% loss at every stage. Thus the most efficient system is whereby the herbivores feed on the vegetation (i.e., only keep herbivores). Then most energy is conserved and recirculated within the ecosystem. However, for ecological balancing, the herbivores have to be preyed on to avoid overpopulation, overgrazing, and soil erosion. This is why a mixture of game with different niche is required. This also helps reduce risks, as game farming is an enterprise. The diagram below shows feeding relationship among organisms at different feeding levels.

Green plants (primary producers)

Herbivores (primary consumers)

Carnivores (Secondary consumers

Carnivores (Tertiary Consumers

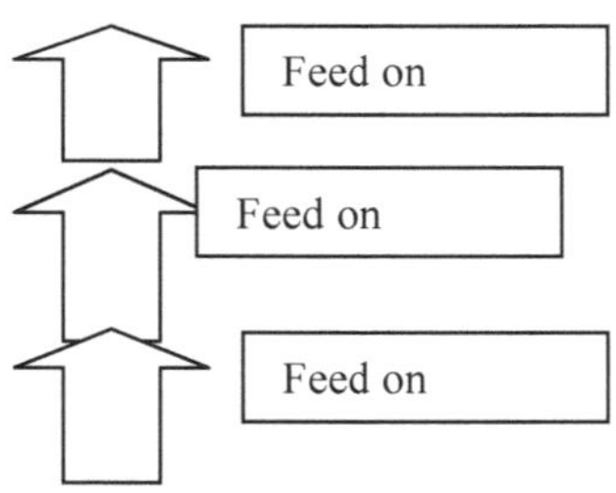

Figure 119: The food web

Generally, the denser the vegetation, the higher the food (energy) supplies. Such an area has a high carrying capacity (number of livestock / game units that an area can support sustainably without degrading it). The longer food chain ensures that most animals have a place in the niche, regardless of their position in the food chain. A game farmer should therefore cater for a wide range of feeding interests commensurate with the diversity of game to be kept.

HABITAT / GEOGRAPHICAL PREFERENCES OF SOME GAME ANIMALS

Generally, larger animals prefer heavy vegetation and vice versa- commensurate with their body size which is proportional with their food needs. Elephants, Buffaloes and Roan Antelopes are therefore found in the Chobe area comprising tropical woodlands and forests. The Impalas and Kudus are found in the Tuli block that has got moderate levels of vegetation. These are medium size game, which need both the grassland for their feeding, as well some taller browsing vegetation for shade and cover against their predators. Gemsboks, Eland and Springboks are found in the Kalahari area- which is almost the centre of Kalahari desert- thus has very scanty vegetation. It is a grassland savanna with no or very few trees. This open land also makes it possible for the springboks to see their predators from far, and the open vegetation facilitates their fast running to escape from the killers.

Figure 120: The antelopes like grass hahitats. The giraffes like habitats with tall shrubs they can reach.

Figure 121: The mophane woodland of southern Africa (often inhabited by larger wildlife) The aquatic environment (Hippos' habitat)

CAPTURE OF WILD ANIMALS (GAME)

Game ranching and farming is an extension of man's long time efforts to continue domestication. Capturing and caring of wild animals for domestication is one such a process

Game may be captured for a number of reasons:

Research (investigate diseases, medicinal or behavioural research);

Translocation especially the endangered species being moved to either a better, safer place e.g. the Rhino etc, or being transferred to a place with less population once an area has been overpopulated. This is currently tasking place in Kenya- transfer of about 400 elephants from Shimba hills to Tsavo game park, a program of 2 years.- effective August 2005.

Trade: for either zoos exchange, auctioning or pet;

Game ranching-for business, breeding endangered species, and for protection.

Treatment against an injury or disease.

METHODS OF CAPTURING GAME ANIMALS

Method depends on the size, speed of animal, terrain, budget and intention. If it is straight hunting, any crude method can be used to incapacitate or maim it maximally at the earliest opportunity. Some methods include:

Chemical immobilization using tranquilisers

From a moving vehicle or horse;

Rope snare;

Use of net (i.e. net capturing methods such as firing using a rocket or lining a net across a path)

CHEMICALS

Generally, a wide range of methods exist. These include the following: use of a darting gun that contains an immobilizing chemical or drug. This is introduced into the body of the animal either in the muscles or immediately under the skin. Cap-chur gun uses blank catridges to propel a projectile containing the drug. It can be fired from a vehicle or helicopter. Crossbow and special arrowheads can also be used to deliver the drug into the animal. Care needs to be taken on the dose of drug used. Chemicals can be used on any type of game. Tranquilizing chemicals / drugs include Ketamine, Combelene, Immobillion, Hyoscine and Etorphine. For most of these, there is no antidote; only natural recovery occurs. Therefore one should not overdose as some can kill.

USE OF NET.

This can be used both at day and at night. At night, small and medium game can be captured at night using 500,000- 1,000,000 watts spotlight or candlepower with a 4-wheel drive vehicle, and a net. The net is attached loosely on a long line at a point where animals are driven into at full running speed. This entangles the animal immediately it hits the net wall. At daytime, a net can be used to make a boma type of enclosure (kraal or corral) with a neck and a very wide mouth.

MIST NET

Birds can be trapped using this method. Mist nets are very delicate and fine that the birds cannot see them at daytime. Birds immediately get entangled, but should not be entangled for too long as they easily die. Water birds can be caught using stronger nets suspended in water and camouflaged, then birds driven into it. Similar method can be used to trap ground birds such as guinea fowls by laying the net around their feeding grounds.

The guinea fowls (Awendo)

LASSOING

This method uses the rope and targets large game such as baby elephants. Only natural fibre rope from sisal, hemp etc should be used, as nylon ones can strangle the game. You hold a rope on a light long piece of straight pole. The rope should have a slipknot at the other end. You run a horse or drive a vehicle near running animals and place the loop with a slipknot over the head or leg of the animal.

Products of fibre crops

OSTRICH FARMING

This is the Rearing or keeping of ostrich for their feathers, skin, eggs, meat and egg shells. Feathers are used for making feather dusters and decoration; skin used for making expensive leather products such as shoes, belts, bags and jackets; egg shells are used for painting etc. ostrich can be raised by extensive, intensive and semi-intensive systems.

The ostrich

EXTENSIVE SYSTEM

Males and females are put in large camps with minimum supplementary feeding and attention. Therefore birds have to get their other requirements from the wild. They find nests for themselves, lay eggs and young ones brought up by the natural parents. Feeding costs are low, needs less skill and labour, and is cheap etc. But productivity is low, as the number of eggs laid depends on the state of the environment, and some are predated on by enemies.

SEMI-EXTENSIVE

After chicks are hatched and counted, they are taken from some nests and placed in other nests with different parents to make up 20-30 chicks per breeding bird. Each foster pair rears the chicks as their own, and chicks get starter ration in enclosures that keep out older birds.

SEMI-INTENSIVE

Birds are given supplementary feed in addition to their natural feeds. Breeding birds are given breeding ration during breeding. Chicks are placed in runs or Lucerne fields during the day and at night in movable / potable houses. In this system, the chicks normally keep each other warm at night, but heaters may still be provided. Initially they are given 1kg of layers pellets, increased to 2 or 3 kg / bird/ day. Fresh feed always chopped and included with pellets to improve its palatability.

INTENSIVE

Chicks artificially reared in runs and houses with heat supplied at night up to 3 months, after which they are transferred to larger pens where they are grown under close supervision and management for slaughter or breeding stock. All feeds are provided.

EGG COLLECTION

In systems where eggs are not left to be incubated and hatched by the birds naturally, eggs must be collected at least once a day, preferably twice a day at 7-8 am and 5-6 pm. This is because most egg-laying occurs either early in the morning or late in the afternoon. Leaving eggs the whole day exposed to the direct sun can cause overheating and damage; exposed eggs also get bacterial attacks. Eggs should be collected with washed hands or gloves, and are placed with their blunted ends up in baskets with sponges that have been disinfected, fumigated or washed. A sticker should be placed on each egg with information such as name of collector, date and time collected, and camp / paddock from which the egg was collected.

EGG HANDLING AND INCUBATION

Before putting eggs in an incubator, the following handling and pre-treatment procedure are necessary;

All cracked and chalky eggs removed;

Dirty eggs washed in VIRKON disinfectant and cleaned with soft brush immediately after collection (brush use leads to higher incidences of late shell mortalities);

Use of sponge or cloth damages the mucilage cover and force pathogens in through the pores.

Eggs are then placed in fumigation rooms where they are exposed to formaldehyde gas for 20-30 minutes to kill pathogens on the egg surface. (Formaldehyde gas obtained by mixing 80 g of KMnO4 and 130 ml of formalin (40%) per m3 of incubation room). After 20-30 minutes, the incubator and ventilator doors are opened to release surplus gas. Fumigation is most efficient at RH of 60-75% and temperatures of 21-250C. Virkon solution is used at a concentration of 5mg/L for washing or 10 g/L for spraying the eggs and incubators. The washing should be lukewarm; if it is cold, shrinking causes harmful pathogens to enter the egg.

After washing, the eggs are taken to the receiving room, then to storage room (temperature of 20-250C, RH 75-80%). The eggs are stored with the air space facing up or on their sides, and should not be stored for more than 1 week. When in store, the eggs should be turned at least once a day, but if they have been transported a long distance, they should be left undisturbed for at least 12 hours to stabilize. This also enables candling, recording and incubator management so that eggs can hatch in batches at intervals instead of small numbers daily.

After 7 days, the eggs are taken to the heater for preheating. This is dome at 25 0C for 12 hours. Preheating is done to prevent eggs from sweating- caused by condensation of moisture on eggs, which enhance the growth, and spread of bacteria and viruses. In the incubator, eggs are kept at 360C at 26-34% RH and turned continuously to prevent the developing embryo from being attached to the inner shell membrane and ensures uniform temperatures for embryo development. On the 14th day in the incubator, eggs are candled to identify the position of the air sac and fertility. The air sac are is marked X with a pencil; eggs which do not show the air sac are usually infertile and thus should be removed. Eggs remain in the incubator for 35-39 days, then candled again to check for development of embryo. Eggs showing undeveloped embryo are removed while those showing actively developing embryo are placed in the hatchery into hatching trays.

Egg turning does not occur in the hatchery. Internal pipping where the chick break the air membrane with its beak and lungs begin to function normally, occurs at 42nd day. Candling is done again on this day and thereafter twice a day to determine whether chicks need assistance. If chicks have not started pipping externally (breaking through the egg shell), within 12 hours of internal pipping, they may be assisted by cracking the egg. Do not break the eggshell completely and remove the chicks as this could result in death from external York sac.

MARKING OF BIRDS (OSTRICH) IN CAPTIVITY

It is vital that birds in a farm are marked by use of microchips. A tiny pellet-like object is inserted under the skin of the animal bearing numbers that can only be read by the use of a sensoring device. This enables the farmer to identify his birds once they break and mix with others in the wild. It also helps in breeding purposes. Its use enables the department of wildlife and national parks and the department of animal health and production to clearly identify wild caught and captivity born birds and approve such requests for export and internal trade when appropriate.

ADVANTAGES OF OSTRICH MEAT OVER BEEF

Higher in protein content

Has low fat content

Has less cholesterol level

Fewer risks of heart attack and cardiovascular difficulties

GOVERNMENT EFFORTS TOWARDS GAME FARMING DEVELOPMENT IN BOTSWANA

The government, through her funding agencies and policies such as the national development bank (NDB), Financial assistance policy (FAP), (Citizen Empowerment) CEDA by and SMME provide financial assistance to farmers interested in venturing into any activity, including game farming

Formulation of policies to allow farmers to venture into game farming and ranching as a way of utilizing and conserving wildlife resources

Low prices of animals captured in the wild for stocking or restocking

Technical assistance offered through the department of wildlife and national parks (DWNP) on the capture of wildlife

Provision of advice on breeding, hatchery management and hatchery hygiene through the department of Animal Health and Production (DAHP) officials

Disease control assistance offered through the DAHP

Tracts of land larger than 400m X 430 m can be allocated to farmers upon approval of their projects for the system of game farming to be practiced.

LEGISLATIVE MEASURES ENSURING SUSTAINABLE USE OF GAME

The government of Botswana, through the DWNP, aims at promoting sustainable utilization of wildlife resources in both consumptive and consumptive ways in accordance with the wildlife conservation policy of 1986. Game farming and game ranching are some of the utilization projects geared towards promoting economic development and sustainable game use. Among other things, certain game especially endangered species are conserved and protected within the appropriate areas. This means that there is no shooting in such areas and hunting of such animals is highly prohibited. Examples of these animals are rhinos, ostriches, and giraffes.

Farmers interested in game farming are to register their farms after having been inspected by a DAHP official. Farmers can only sell game animals after approval from DAHP and DWNP. Farmers are required to mark their game by the use of microchips to enable the DWNP and DAHP to clearly identify wild caught and captive born animals and approve requests for export and internal trade when appropriate. This is meant to minimize the misuse of wildlife resources by farmers and poachers.

MODULE 13: RANGE AND PASTURE MANAGEMENT

MODULE 13: RANGE AND PASTURE MANAGEMENT

INTRODUCTION:

Range or Veld (mafudiso in Setswana) is a large area usually not suitable cultivation (i.e., not arable), covered with native vegetation adapted to that environment, and used for grazing by all types of animals- both livestock and wildlife. Pasture is the same as range, but may also have exotic plants introduced into them (i.e., an improved range), and mostly, only use by livestock. Lands covered by range are called rangelands, while those covered by pasture are called pasturelands.

Rangelands (Mafudiso) are the same as range, and are non-arable lands that provide the necessities of life to grazing and browsing animals. The areas are not arable because of any or a combination of the following limitations: low and erratic rainfall, rough topography, and poor drainage. In Botswana, this comprises more than 80% of the land area. However, the 20 % arable lands are also covered by the swamps (Okavango), salt pans (Makgadikgadi), and sand (sandveld). These are found to the East, North East and south east of the country.

SOME DEFINITIONS AND CONCEPTS

Ranch (Polase) is a livestock or wildlife farm with known boundaries, usually fenced, used for grazing and production of livestock and game animals for commercial purposes.
Range/ pasture management is the care and use of grassland to obtain the highest continuous (sustainable) yield without damage to natural resources (soil, water, pasture, etc).
Forage are a herbaceous vegetable material used by browsing animals
Browsers are animals that feed on leaves and branches of shrubs and trees
Palatable means tasty; appetizing; pleasant to eat.
Stocking rate is the amount of grazing land available per livestock unit
Livestock unit (LSU): a mature animal of live weight about 500KG.
Thus 1 calf= 0 LSU; cattle 1-2 years old=0.5 LSU;
6 sheep or 6 goats= 1 LSU; a bull of 1000 kg weight=2 LSU etc.
Carrying capacity (Itekanelo mo phudisong) is the area of land enough to properly maintain 1 LSU without any damage to the land / resources. It is the correct stocking rate for a given area.
Overstocking means putting more livestock units in a pasture than the land is able to maintain; exceeding the stocking rate in a pasture.

Overgrazing (Kgotlhego ya phulo) is a result of overstocking, whereby pastures are grazed beyond their ability to regenerate. Signs include barren lands, and encroachment of unpalatable species.

Figure 122: Ovegrazed land (Mahudiso a a gotlhegileng)

Decreasers (Majang a a monate) – Are desirable plants in a pasture. They decrease in number as grazing becomes excessive. If they are in plenty supply, this indicates good grazing programme and act as an indicator of good range condition. They are more nutritious and palatable and feeding pressure is normally very high in them. They are largely leguminous which are also appetizing because of their higher crude protein level.

Increases - Are plants which increase in numbers as grazing pressure increases. They are usually less attraction or less preference by livestock because they are less palatable, or are not easily seen because they are shorter in size.

Browsers: Animals which feed largely on twigs and other parts of plants located much higher above the ground level- mostly leaves of trees and shrubs. They are not popular grass feeders. Eg goats

Grazers(Phulo) : Animals which largely feed on low-lying vegetation- mostly grass. Include sheep and cattle. Grazers, especially sheep, can cause serious land degradation by eating up most soil cover.

Mixed species grazing: Use of an area by more than one type of livestock and / or game for grazing and / or browsing. Use of sheep, goats and cattle is common in many African rangelands, and Botswana . However, it is also very common to use rangelands where wildlife also feed, for livestock grazing and browsing. This type of range use reduces competition if the different species have different feeding habits, e.g. some are browsers and others grazers. It also improves the efficiency of pasture use, as there is less, if any, waste.

Invaders: These are the undesirable species in a pasture or range. They invade or take over the range where the decreasers and increasers are weakened by either overgrazing or extreme environmental conditions. They may come from other pastures in the dung or attached to the skin (see animal dispersal of weeds), or they may be dormant seeds of annual plants which were growing in the area before it was used as pasture, or they may exist in small undisturbed areas in the range. They are largely grasses with poor feeding value; unpalatable; may come from other areas; may be present on the range.

FACTORS AFFECTING RANGE MANAGEMENT

The following factors affect range management in Africa:

Climate- including rainfall and temperature. (a) Rainfall is scarce and unreliable in most of the country, which is covered by Kalahari Desert. (b) Temperatures are very high and unbearable for most vegetation and wildlife. It is due to these climatic factors that we have rangelands in Botswana. Rangelands are uncultivatable areas that provide the necessity to life to grazing and browsing animals. The areas are uncultivatable because of low and erratic rainfall, rough topography and poor drainage. Only about 20% of Botswana is cultivatable, covering regions such as: the hardveld, the East, North-East and South-East with relatively high rainfall; and is also highly populated. The Okavango swamps and the Makgadikgadi saltpans cover some part in the north. The rest of the country is covered by sand of different depths, and hence is known as the sandveld.

Soil or edaphic factors: soils are loose and sandy, thus not very fertile. They thus cannot support sufficient vegetation to be fed on by wildlife or livestock.

Veld fires: Veld fires are often used to control encroachment and pest (tick, tsetse) control, but sometimes go out of hand and cause damage.

Man and settlement: Increasing population has forced some people to slowly move into pasturelands and practice agriculture. This has interfered with vegetation and wildlife, at times even causing human-wildlife conflicts.

Technical know how: There are not many trained people with skills to manage rangelands effectively. Thus Botswana government at times has to bring in expertise from outside the country.

MAJOR NATURAL VEGETATION TYPES IN BOTSWANA:

Major vegetation types are in Botswana are: The Northern Kalahari tree and bush savanna; the mixed Acacia woodlands; the Mophane woodlands; Kalahari shrub savanna and the aquatic grasslands. Productivity of rangelands of Botswana decrease from North-east to South-west following the rainfall pattern. The Northwest (Chobe) is composed of forest and composition changes to shrub land in Southwest (Bokspits). Productivity decreases from North East to Southwest following the rainfall pattern. The North West (Chobe area) comprises forests whose composition changes to shrub-land in the Southwest (Bokspits).

1 Grassland savanna: Only grass, but once in a while isolated trees and shrubs. Vegetative cover is a mixture of grasses, sedges and herbs e.g. Makgadikgadi pans.

2. Shrub savanna: There are a few trees and many scattered shrubs e.g. savanna of the southwest region.

3. Tree savanna: Comprises mainly the taller woody trees whose upper canopy do not touch each other e.g. Mophane woodlands of the North East.

Figure 123: left-Themeda triandra

Figure 124: The Forest ecosystem

COMPOSITION OF RANGELANDS

1. Indigenous grasses and shrubs or small trees. These include:

(a) Grasses (Bojang): (i) perennials- grow for longer than two years, and are very palatable guinea-guinea grass and finger grass.

Guinea grass

(ii) Annual grasses: complete their life cycle after 1 year or 1 season, are less palatable than most perennials e.g. Phoka (Sugwagaga).

Legumes (Sika lo nawa): Are very few in Botswana rangelands; the few available are often povergrazed because they are the desirable species. (Majang a a monate). Some are also poisonous or unpalatable.

Leguminous feed (Desmond). Others are Vetch, Lucerne etc

(c) Woody plants: These provide browse which is more preferred by goats; e.g. sweet thorn (Mooka), camel thorn (mogotho), umbrella thorn (mosu), Mopane, etc.

Figure 125: The grassland savanna ecosystem popular for Antelopes / springboks
Assignment: Identify 5 other plants found in Botswana rangelands- each representing each group given above).

RANGE PLANTS AND OTHER RESOURCES

There are a variety of resources from the rangelands. The major resource is forage that is used by both livestock and wild animals. Livestock get over 9% of their feed from rangelands, while crop residues, fodder and grain provide the balance. Grain is mostly used in feedlots and in dairy farming. Wood products are used in a variety of ways. Firewood is the most common source of energy in rural households. Fencing materials of wood origin are common in crop fields, ranch and homesteads construction.

A variety of grass species are used in thatching. A high percentage of roofing in rural households is of grass. Important species in this way are Eragrostis pallens (motshikhiri) and Hyperhenia spp. Medicinal plants play an important role in treatment of ailments of different kinds. A number of these plants can be seen in town markets, e.g. sengaparile and monepenepe. Rangelands are known for food plants as fruits, vegetables, roots, bulbs and some can be used in making soup, porridge or beer. Some of these plants are Grewia spp (moretlwa), Azanza garkeana (morojwa), Tylosema esculantam (morama), Mimusops zeyheri (mmilo), Ximenia spp (moretologa), Boscia albitrunca (motlopi), mahupu, mabowa (mushrooms) etc. Other important products of rangelands are phane and honey.

Bee hives in a veld

Aloe as a medicinal veld product

Phane is a seasonal product and it is a source of rural household income. Phane is also used as livestock feed. Rangelands are also watersheds from where rainwater is collected into rivers and dams; and some water infiltrating underground is reached by borehole technology. Borehole technology is a very important system for cattle producers and to some extent the public in general. Minerals are also sourced from rangelands.

MAJOR VELD PRODUCTS

Forage (Phulo): used by livestock and wildlife. Livestock get more than 90% of their feed from these, while the balance is from crop residues, fodder and grains. Grain is mostly used as a concentrate in feedlots and in dairy farms.

Wood products: Firewood is used as an energy source in most rural households. Wood is also used as fencing material in the fields, ranches and homesteads.

Grasses (Bojang): Some grass are used for thatching in rural areas e.g. Eragrostis spp (motshikhiri) and Hyparhaenia spp

Medicinal: Many ailments are treated using Veld products e.g. sengaparile and monepenepe.

Fruits and vegetables: Fruits, vegetables and bulbs are extracted from the rangelands and are used for making soup, porridge or beer. These include Grewia spp, (Moretlwa), Azanza garkeana (morojwa), Tylosema esculentum (morama), Mimusops zeyheri (Mmilo), Ximenia spp (Moretologa) Boscia albitrunca (Motlopi) , Mahupu, and mabowa (Mushrooms).

OTHER VELD PRODUCTS

Phane: Is a seasonal product and a very important source of rural household income. Is also used as livestock feed.

Honey: Is a vital rangeland product.

Enemies of veld products

POISONOUS PLANTS IN THE RANGE

Mogau

USE OF NATURAL VEGETATION TO IMPROVE THE RANGE

Some plants can be introduced into the rangeland to improve both its scope of use and improvement of the environment for grazing. For instance, useful veld products such as Moretlwa (Grewia flava, and G bicolor) provide edible fruits that can be used by wildlife and livestock, as well as man. The natural tees can also be made from Mosukujane and Mosukudu (Lippia scaberrima), which are substitutes to the beverages often used. Even the Sesbania sesbal plant is a natural leguminous plant and can be used to improve the rangelands. If introduced in the rangelands, these plants can improve biodiversity there and widen its ability to accommodate even more wildlife and livestock. The plants also provide shade, which is a useful resource for livestock in the hot Botswana rangelands. Planting of shade trees is a useful management procedure in improving the usefulness of both arable and non- arable lends.

Figure 126: Leguminous crop products

Figure 127: The Sesbania sesban.

EXTENSIVE RANGE MANAGEMENT

Pasture may be utilized using different grazing programs. These may be extensive or intensive. Intensive systems include enclosed grazing, zero grazing, strip grazing, creep and rotational grazing. Extensive pasture management involves a system of free movement of livestock and / or wildlife.

EXTENSIVE RANGE MANAGEMENT

This is the system of grazing where animals graze on well-managed natural pastures. There is some restriction, e.g. there may be a broader fence. The land may be fenced and sometimes herd boys are used to check the movements of livestock. Growth of bushes and trees are also controlled to encourage the grass to make a full cover on the ground. There is more control than in unenclosed system.

Advantages

This system is less costly,

Less labour is involved; and

Animals are left to freely graze.

Disadvantages

Useful pasture may be tramped on, or exhausted beyond recovery.

There is less recovery time for the pasture;

Animals waste energy leading to less milk and meat production

There is high risk of over grazing,

Selective grazing might cause the death of palatable species.

Heavy parasite (ticks and worms) infestation of the land after some time.

Animals may go astray if land unfenced;

INTENSIVE GRAZING

Animals graze on pastures developed by planting good quality grasses and legumes for better production. The pastures are man-established, are fertilized and carefully managed to achieve a high level of production. Grazing land is divided into sections and there is control of livestock movement from one fenced section to another. It may take the forms of creep, rotational, strip and zero-grazing systems described below.

Advantages

Grazing is controlled and pastures are allowed to recover;
Wastage of grass is prevented;
Stocking rate is controlled to avoid grazing; and
There is better utilization of energy for production

Disadvantages

The system requires more skilled labour and,
Is expensive, as more money is required to establish the pasture, fertilise and maintain.

INTENSIVE GRAZING METHODS.

Some intensive pasture management methods include Creep, rotational, strip and zero grazing

CREEP: a feeding method often used in ensuring security of young ones against an angry, temperamental mother or older ones / other adults. The feed for the young one is placed in a creep area where the adults cannot reach. It is commonly applied on pigs, cattle and sheep.

ROTATIONAL GRAZING: dividing the pasture into paddocks (Kampa), and having livestock feeding in each in a pattern. The livestock stay in each for a specific duration matching the carrying capacity, then are moved to the next paddock. The grazed one is top dressed, sprayed against parasites, and irrigated so that it recovers before being used again later. It gives the farmer a chance of cutting and preserving some of the pasture if it overgrows before its time for being used again reaches.

STRIP GRAZING: livestock kept in a narrow strip of pasture whose fence / boundary is movable (it could be electric fence). After the strip is well grazed, the movable boundary is moved again and the livestock are let to use the new area.

ZERO GRAZING: Livestock are kept and fed from a house (also called stall feeding); they move zero distance, and are fed, treated and watered in their shed. They grow faster because less energy is wasted in unwanted movement; also it is easier to control breeding, pests and diseases.

USE OF NATURAL VEGETATION TO IMPROVE THE RANGE

Some plants can be introduced into the rangeland to improve both its scope of use and improvement of the environment for grazing. For instance, useful veld products such as Moretlwa (Grewia flava, and G bicolor) provide edible fruits that can be used by wildlife and livestock, as well as man. The natural tees can also be made from Mosukujane and Mosukudu (Lippia scaberrima), which are substitutes to the beverages often used. If introduced in the rangelands, these plants can improve biodiversity there and widen its ability to accommodate even more wildlife and livestock. The plants also provide shade, which is a useful resource for livestock in the hot Botswana rangelands. Planting of shade trees is a useful management procedure in improving the usefulness of arable and non-arable lands.

The following classes of indigenous vegetation can be used to improve range productivity and use:

1. Indigenous grasses and shrubs or small trees. These include:

Grasses (Bojang):

(i) Perennials- grow for longer than two years, and are very palatable guinea-guinea grass and finger grass. These have soil-borne vegetative matter that can enable the grass to regrow even after a long dry spell.

(ii) Annual grasses: complete their life cycle after 1 year or 1 season, are less palatable than most perennials e.g. Phoka (Sugwagaga). However, they grow very fast to cope with their short life cycle. This makes them able to take chance even in short rains. Thus they can be introduced in areas with persistent short rains. Though less palatable, they offer good feeding material during critical times of drought and famine. However, overgrazing should be avoided, as they have no means of regenerating other than by means of seeds.

Legumes (Sika lo nawa):

Are very few in Botswana rangelands; the few available are often poisonous or unpalatable. However, some that could still be introduced to improve the soil fertility, as they are generally nitrogen fixing. Their adaptation to harsh climatic and soil conditions endear them well to these environments. Once they improve the soil status, more vegetation cn grow in the better soils, thereby increasing the general productivity.

© Woody plants / trees (Setlhare):

These provide browse which is more preferred by goats; e.g. sweet thorn (Mooka), camel thorn (mogotho), umbrella thorn (mosu), Mopane, etc. Some of these can be introduced to provide shade to wildlife and livestock in the dry desert environment of Botswana.

MAN CAN DESTROY RANGELANDS BY

Overstocking, which leads to overgrazing and soil erosion (e.g. donga erosion shown below);

Farming in it even though it is non-arable (as happened in the dust bowl of USA in the early 20th century)

Misuse of fires;

Careless harvesting (e.g. uprooting) of some of the plants for use e.g. medicinal

Figure 128: The extreme case of gully (Donga) erosion

HOW MAN CAN AVOID DESTRUCTION OF RANGELANDS:

Ensuring correct stocking rate

Controlled harvesting of Veld products;

Controlled use of fire (e.g. only at the right time / season and place (;

Introducing useful species e.g. browsers like Leuceana leucocephala etc;
Fertilisation and irrigation if need arises;

INTENSIVE PASTURE MANAGEMENT

INTRODUCTION

Pasture may be utilized using different grazing programs. These may be extensive or intensive. Intensive systems include enclosed grazing, zero grazing, strip grazing, creep and rotational grazing. Extensive pasture management involves a system of free movement of livestock and / or wildlife.

13.1 PRINCIPLES OF PASTURE MANAGEMENT

The following principles are followed in making proper range and pasture utilization. They are: controlled grazing; proper botanical balance; controlled bush burning; replanting; introduction of legumes; fertilizer application; ley farming; conservation; and proper stocking rate. Each is later discussed to details under livestock husbandry.

Figure 129: Extensive grazing field

1. Controlled grazing:

This is meant to protect grasses from destruction, as well as allowing pasture to grow well. It may be done by fencing, rotational grazing, or strip grazing.

2. Proper botanical balance:

This means keeping a good balance between grass and woody, or grass and legume plants. The extra woody plants should be removed so that grasses can have more light and nutrients for their growth. Some woody plants are usually left in, or planted in the range to provide shade for animals, as well as for use as feed by browsers such as goats. The imbalance in a pasture may result from some encroaching plant species growing in sites that seem to have been over-grazed, such as the case of legume species, which are normally selectively grazed.

Controlled bush burning

This helps eliminate the unwanted plant species, especially those that do not readily regenerate (i.e., do not have enough food reserves in their roots). It also encourages early sprouting of grass; kills pests and pathogens (disease-causing agents); kills disease vectors like ticks; as well as burn the hard testa of some pasture seeds such as the red oat grass (Themeda triandra), wattle tree, etc.

Replanting:

The grassland having poor quality grasses should be replanted with better quality grasses such as Rhodes grass (Chloris gayana), Kikuyu grass (Pennisetum clandestinum), Napier/ elephant grass (Pennisetum purpureum), Guatemala grass, Star grass (Cynodon dactylon) etc

Introduction of legumes:

Legumes have a higher crude protein level and are thus more palatable to livestock than the grasses. They therefore improve the appetite. The legumes also fix nitrogen, thus enriching the soil and reducing the need for, or quantities of, top-dressing material to be applied to the pasture. In the long run, it also improves soil fertility. Legumes like Lucerne (Medicago sativa), Vetch (Vicia spp), Desmodiums, Stylo, Sesbania sesban, Leuciana spp among others can be introduced.

6. Fertilizer application:

It provides balanced nutrients to grasses for their proper growth. These plants are then able to supply the nutrients that livestock need. Thus the necessary basal and top dressing should be added. Introducing legumes into the pasture may reduce the need for top dressing. It is normal to apply a top-dress in a paddock from which livestock have just been removed (i.e., after grazing).

7. Ley farming:

This means introducing pasture as part of a crop rotation cycle in a farm. The pasture may be grown in rotation with a field crop after 4-5 years. The legumes in a ley system offer an advantage of more fertile soils. The grazing livestock also throw their dung and urine into the soil, thus making it more fertile.

Figure 130: Sesbania sesban as a pasture and range improvement crop

Proper stocking rate.

This involves using a given pasture for only a limited (right) number of livestock units; i.e. using it to its capacity. The number of livestock units required to feed satisfactorily in a given pasture depends on the yield of the pasture, the species and breed of livestock, its sex (male/female), and adult live weight. The correct stocking rate helps in minimising over-grazing, thus helping conserve the soil, as well as enabling the pasture to have enough food reserved for use in regenerating.

Conservation and preservation (PRESERVATION OF FODDER)

Normally, there are periods of surplus pasture e.g. during and immediately after a rainy season. In the dry season, there is often less pasture, famine and massive deaths of livestock. Cutting and preserving the excess pasture for use during drought can reduce this. If it is cut when still very leafy, i.e., just when there is a correct balance between crude fibre and crude protein levels, it can be packed and compacted in a silo as silage. In the silo, microorganisms ferment the feed, and the resulting lactic acid is useful for preserving it. Thus making a silo needs a skilful compaction such that just the correct amount of air is left. If there is too much air left (Maybe due to inadequate compaction), there may be over heating which may kill the bacteria, thus resulting in incomplete fermentation. The material for silage making should be very leafy, and has a high pasture yield. Often, grasses at or just before flowering are cut and silaged. On the other hand, if the pasture has passed the very leafy stage, and it has long flowered and become hard (due to very high crude fibre content), it may be cut, compacted, and preserved as hay--

in form of cubes. To feed livestock this hard material may require sprinkling some water/ salt solution, and /or even molasses (some form of sugar) to make it more appetising/ palatable. It can be stored in a dry place for a very long time without getting spoilt. The livestock using it need a lot of drinking water.

Baled hay

GRAZING METHODS

These may be intensive or extensive; closed or unenclosed, as discussed below.

ENCLOSED GRAZING

This is the system of grazing in which livestock are kept in fenced areas to graze. They eat up the grass in that area to an acceptable level and then moved to another fenced area to continue grazing. The vacated grazing area is then allowed to recover, and may be fertilized an irrigated to recover faster.

UNENCLOSED OR EXTENSIVE GRAZING

This system of grazing allows the livestock the freedom to graze on a large unfenced expanse of land. Usually herd boys are used to keep an eye on the livestock. It involves mixed species grazing, as even wildlife and animals from neighbours may be grazing in the same area.

Advantages:
It is cheap and simple to manage.
There is free breeding, thus qualities of good bulls easily spread across the whole herd in the community;

Disadvantages
It easily leads to disease transmission and spread,
It is labour intensive,
Livestock easily go astray, or can get injured or attacked by wildlife,
It leads to low livestock production e.g. growth rate and milk yield etc.
Difficult to monitor how much the livestock have fed on;
Difficult to control breeding;
On the whole, this method is not very efficient because the livestock usually select the best grasses first leaving the less palatable ones; useful pasture; and animals waste a lot of energy moving about in search of pasture.

EXTENSIVE GRAZING
This is the system of grazing where animals graze on well-managed natural pastures. There is some restriction, e.g. there may be a broader fence. The land may be fenced and sometimes herd boys are used to check the movements of livestock. Growth of bushes and trees are also controlled to encourage the grass to make a full cover on the ground. There is more control than in unenclosed system.

Advantages
This system is less costly,
Less labour is involved; and
Animals are left to freely graze.

Disadvantages
Useful pasture may be tramped on, or exhausted beyond recovery.
There is less recovery time for the pasture;
Animals waste energy leading to less milk and meat production
There is high risk of over grazing,
Selective grazing might cause the death of palatable species.
Heavy parasite (ticks and worms) infestation of the land after some time.
Animals may go astray if land unfenced;

INTENSIVE GRAZING

Animals graze on pastures developed by planting good quality grasses and legumes for better production. The pastures are man-established, are fertilized and carefully managed to achieve a high level of production. Grazing land is divided into sections and there is control of livestock movement from one fenced section to another. It may take the forms of creep, rotational, strip and zero-grazing systems described below.

Advantages

Grazing is controlled and pastures are allowed to recover;
Wastage of grass is prevented;
Stocking rate is controlled to avoid grazing; and
There is better utilization of energy for production

Disadvantages

The system requires more skilled labour and,
Is expensive, as more money is required to establish the pasture, fertilise and maintain.

CREEP GRAZING

Involves use of boundaries to control free movement of some selected animals, while allowing others free passage. This selection is mostly based on size difference. For instance some high quality feed may be placed in a compartment that can only be accessed by smaller animals, e.g. calves, lambs, piglets or kids. The larger animals, the mothers cannot reach the creep area. This method is commonly used to provide supplementary feed to piglets and lambs. The special feed placed in the creep area is called creep feed. In pigs, creep area also serve as a resting place for the piglets, as well as a n escape point from the mother when it is angry. Angry lactating pigs can normally even crush their young ones to death.

Advantage

Good quality feed (creep feed) given to young delicate animals;
Creep area may provide an opportunity for capturing young ones for various reasons, in cases where the mother is too harsh;
Selective grazing possible where sheep have an access to areas where cattle cannot access.

Disadvantage

Expensive to build

ROTATIONAL GRAZING

Involves letting livestock graze in an area for some time, then remove it and transfer it elsewhere once the resources in the first area are exhausted. It may be done by strip or paddock systems.

Advantage:

Diseases, pests and parasites easily controlled;
Livestock feeding controlled, i.e. they get enough feed;
Overgrazing is avoided;
Pasture have time to recover;
Avoids wastage of pasture- e.g. reduces the selectivity of grazing among livestock, as they are not taken to a new area until the first one is sufficiently grazed;
Excess pasture can be conserved as standing hay, hay or silage.
Breeding easily controlled.

Disadvantages

Expensive to lay the paddocks and related structures;
Needs skilled management;

ZERO GRAZING

It is the type of grazing in feed and water is given to livestock in an enclosure called zero-grazing unit. The livestock do not go out to graze (i.e., they move almost zero distance for food).
Fresh or preserved feeds may be given.

Advantages

Less land required;
Livestock are more productive – e.g. milk yield;
Livestock grow faster as less energy is wasted in movements;
Good control of diseases;
Good control of pests and parasites;
Controlled breeding possible;
Quality of management e.g. feeding, milking etc maintained.

Disadvantages

Animals may develop limb problems because of lack of exercise;
Expensive;

Needs a lot of labour (full time management);
Needs a lot of skills.
Animals kept under a very artificial system and may suffer from stress.

STRIP GRAZING

This is a form of rotational grazing whereby a thin strip of pasture is left for livestock to feed on, and then transferred to a new strip when the first one is exhausted. The strip is mostly made by use of electric wire lines that the livestock cannot pass. Once they get in contact, they get shock and retreat. The advantages and disadvantages are as those under rotational grazing.

SOME POISONOUS PLANTS FOUND IN PASTURES:

Thorn apple- poisonous to livestock
Phoka

SOME PASTURE SPECIES

GRASS: PANICUM MAXIMUM (GUINEA GRASS)

Growth habit: there are short and tall varieties, but average height is 50-100 cm. May reach 3 m tall. They are tufted.
Persistence: perennial
Soil requirements: fertile soil preferred, but may also grow in soils of moderate fertility.
Climatic requirements: warm places with moderate to high rainfall preferred. But can establish very fast in short rains and produce seeds within 2-3 months.
Fertilizer requirements: Basal dressing at establishment; can top dress after cutting or grazing.
Planting time and method: beginning of rainy season, either from seed or from splits
Nutritive value: high crude protein when young, but this reduces and it becomes less palatable after flowering. It is rich in carbohydrate.
Other grass pastures: Rhodes grass, Kikuyu grass, weeping love grass, teff grass, star grass, and buffel grass

LEGUME: LUCERNE / ALFALFA (MEDICAGO SATIVA)

Growth habit: Deep rooted bushy plant
Persistence: perennial
Soil requirements: well drained
Climatic requirements: Does well even in dry places once established

Fertilizer requirements: basal dressing only

Planting time and method: Onset of rainy season by broadcasting the seeds; can plant in an established pasture to improve it.

Nutritive value: Rich in crude proteins; very palatable.

Other legumes: Stylosanthes, Leuceana leucocephala (Leuceana), Sesbania sesban (sesbania), and Lablab (eg Dolichos lablab).

Whatever crop is established, it is always better to establish a mixed pasture with grass and legume species.

Table 26: Characteristics of some livestock feeds

Name	Growth habit	Palatability	Nutritional value	Comments
Clover (Trifolium spp)	Bushy	Very palatable, thus tend to be overgrazed	High, as it is a good protein source	Can be annual or perennial; Better if mixed with hay grasses; Prefer moist soils; Can withstand grazing; Inoculated seed recommended; Inoculated seed recommended.
Green leaf (Desmodium spp)	Trailing stem with roots at nodes	V Palatable; tends to be overgrazed	High, as it is a good protein source	Perennial; Compatible with grass pastures; Prefer moist soils; Inoculated seed recommended.
Rhodes grass (Chloris gayana)	Tufted; spreads by stolons	V Good; likes even when mature	High before flower ing	Annual and perennial grasses Can stand grazing well Should be allowed to seed in first year.

OTHER GRASS AND LEGUME SPECIES

The Emba Rhodes grass

MODULE 14: CASE STUDY: TICH APPROPRIATE POULTRY FEED AND HOUSING PROJECT

Established: June, 2004; Developed and implemented in June-July, 2004; DR AFULLO, Augustine

INTRODUCTION

In a series of situation analyses and interactions with its partner communities, TICH has identified improved poultry farming as a possible key income and food security intervention. This arises from the fact that the communities already have in-depth knowledge about poultry husbandry. Except it has not been done as a business. The farmers have also recognized poultry as fast growing, thereby rendering them feasible investment, food security and quick poverty alleviation / eradication assets. With incremental management, this is a sector that has all the potential to turn round the household economy, which is the foundation to the local and national economy. In view of this, the TICH food security and environment faculty was challenged by a TICH technical committee meeting dated 7th June, 2004, to come up with:

An optimal feed formulation that is both nutritionally adequate and affordable by the partnership communities;

A housing design that caters for the key problems facing poultry farmers in the partnership area.

KEY BOTTLENECKS AND CHALLENGES IN THE POULTRY SECTOR IN NYANZA PROVINCE, KENYA

The following are the key problems associated with poultry production in Nyanza province, Kenya:

Housing: None or inappropriate.

Feeding: No particular feeding plan; chickens fend for themselves within the context of extensive management.

Predation: Cats, mongoose, dogs etc commonly kill and / or eat both small and adult chickens.

Eagle: Is a single greatest danger to chicks. Reduce chick population by more than 70% in some places.

PICA: Imbalanced feeding leads to odd feeding habits, which also cause the chicken to catch diseases and other infections.

Diseases: Often wipe a whole flock, especially New castle disease (NCD). This results from lack of both knowledge about the disease, and/ or lack of means to prevent it.

Livestock-crop conflict: There are cases where the extensively produced chickens become pests. They eat up the farmers' crops especially beans.

KEY PROBLEMS IDENTIFIED IN THE POULTRY SECTOR AND INTERVENTIONS USED

Problem	Cause	Proposed Intervention
Mass death of adult chickens	Diseases (especially New Castle Diseases (NCD))	Mass NCD vaccination followed by regular preventive vaccinations by age
Loss of adult chickens	Mongoose, theft (man), wild cats, alligator etc	Housing; Restricted movement of chickens.
Morbidity (diseases) of poultry	Poor housing Poor Feeding Unhygienic conditions Lack of preventive measures	Training on poultry requirements on: Improved housing; Feeding requirements; Minimum hygiene; Vaccination.
Low survival rate of chicks	Predation (eagle); Death from unfavorable environmental conditions (heat / cold); Trapped by an object e.g. metal wires, grass etc;	Make fold for demonstration; Encourage each household to make their own fold; Provide feed and water; Vaccinate; Provide medicine;
Low hatching rate		Training on: The correct number of eggs per brooding hen; Correct storage of and selection of eggs

		for a brooding hen; Correct cork: hen ratio.
Low growth of chickens	Improper feeding / not feeding at all; Diseases; Inappropriate or no selection;	Encourage selection of breeding stock for fast growth; Train on feed requirements and identify locally available and appropriate food substitutes;
Low laying rate and low quality eggs (small, weak-shelled, malformed, blood spotted etc	No records; No or improper selection / culling practices; Poor or no feeding; Diseases;	Encourage selection of breeding stock for fast growth; Train on feed requirements and identify locally available and appropriate food substitutes; Provide Training on and demonstrate disease control;
Vices (Egg pecking, cannibalism etc)	Improper feeding; Improper selection of breeding stock.	Improved feeding (identify balanced feed rations via training); Train on feed requirements for poultry (by age); Training on selection and culling (breeding management).

COMMONLY ASKED QUESTIONS: EXPECTATIONS OF MOST TRAINEES

Why do some chicken eat their eggs?

What are the housing space requirements for (i) chicks; (ii) layers (iii) Broilers

What is the problem of putting a lot of chickens in a small space / house? (i.e., maximizing space use)?

Why are grade chickens uniform in colour, size etc unlike local chicken?

Why do some chickens lay soft eggs?

Is it wrong for a cock to mate with its relatives such as mother, or sisters?

How can we protect the young chicken from enemies such as the eagle, alligator, or mongoose?

What materials can we use for making a modern poultry house?

Why do some chickens have diarrhoea?

Why do some chickens have bloodstained feaces?

What are the signs and symptoms of poultry diseases?

Why do chickens die on regular basis, once or twice per year?

What medicine can be used to treat sick poultry?

What methods can be used to prevent poultry diseases?

What feeds should be given to chickens?

Are there different feeding requirements by age of chickens?

How can local poultry be reared on business / commercial basis?

Why do local chickens grow slowly and produce fewer eggs?

Why do some eggs never hatch?
Why do some chicken run away from their eggs?
What age do chickens start laying?
What is the right cock: hen / pullet ratio?
What is the best culling age for an adult chicken?
How many tines can a hen produce in a year?
What can happen if one eats a sick chicken?

ACTIVITIES:
MANDATE I: POULTRY HOUSING PROJECT
Objective 1: To design the most optimal housing for chicken that caters s for all housing needs;

THE TICH IMPROVED FOLD (OSERA OR OSERO)
This adopts the fold system, but the arks are made larger than the standard folds, using lighter material. This can then contain more birds. 1/3 of the fold is covered/ roofed to provide shelter and laying place. The rest is open and covered with wire net.

SPACE REQUIREMENTS BY AGE
Most convenient fold size is 3.5m x1.5m x 1.5m for adult 10 birds: This gives a floor space of 5.25 m2/10 birds, which equals 0.525 m2 / bird, i.e. about 2 birds / m2. For young chicks, each of which requires 50cm2, (0.005 m2) up to 1050 chicks can be kept in one standard fold. Thus one such a facility may cater for up to 1050 0-5 weeks' old chicks. For chicks of 6-10 weeks old, we can assume broiler density requirements of 6 birds/ m2. Thus, up to 30 growers (5-10 weeks) can be kept in a standard fold. For older categories (10-12/16 weeks), a density of 5 birds/m2 (slightly less than the layers specifications of 3-4 / m2) may be necessary. This may contain a total of 26 pre-laying birds aged 10 up to 12 or 16 weeks. The same standard fold is recommended for 10 layers. This is as shown in the table below.

Thus, the maximum number of folds a farmer may decide to have if s/he chooses to go fully commercial is 8. The last 3 folds are those used to keep those for sale (i.e. those which have attained slaughter weight). In such a fold, non-fish ration is essential, to avoid a fishy smell of the chicken being sold. Use of plain grains may be necessary, or perhaps supplemented with small supply of proteins to enable the animals maintain their weight. Alternative sources of proteins here may be white ants.

TABLE 27: FOLD CAPACITY AND NUMBER OF FOLDS NEEDED BY AGE-GROUP

**Comment	Brooding chicks; DCP-rich foods required	CHO-rich foods; need much energy	Finishing; DCP-rich feed	SALE	Losing
Age (weeks)	0-5	5-10	10-12	12-16	(17+)
Minimum Space requirement per bird (m2)	.005	0.17	0.2	0.3	0.5
Max capacity of a standard fold	1050	31	26	17	10
# of Standard folds required for 1050 birds	1	34	41	62	105
# of folds needed for 500 birds (Annual target per HH)	1	17	20	31	53
# of folds needed for 45 birds (Monthly target per HH)***	1	2 (1.5)	2 (1.75)	3 (2.66)	5(4.5)
Cumulative number of FOLDS	1	3	5	8	

HH means household.

Trapping white ants:
Make a small hole of dimensions 50-100 cm long x 30-60 cm wide x 20-50 cm deep.
Put some dry grass bedding into the hole.
Put more grass and / or dry maize / sorghum straw well arranged in layers.
Cover with grass (dry or fresh);
Cover with a little soil.
Leave this overnight, or for 2 days.
The ants will accumulate in the hole, and can be harvested.
Harvesting white ants
Put on a hand glove
Open up the hole by first removing the soil cover, then grass.
Once you reach the straws, there is likely to be a swarm of ants. Hold the straws together and hit the bundle onto your harvesting a basket to transfer the ants attached to the straw into the basket.
Once all these are removed, remove the next layer of dry grass. This is where the majority of the ants will have accumulated. Remove the dry grass and put these into your harvesting basket.
Take the basket, and transfer its contents onto a surface from where the chicken can eat up the ants.
Advantages of the fold system:

The arc is easy to make; it can be made by the local jua kali artisans, and even by the farmer him/herself.

Uses local materials e.g. Poles-Bamboo and / or bush poles; Fold cover: Plastic or papyrus reeds made like a mat.

Plastic material can be used to cover the shaded section of the fold to keep the inside dry even when it is raining;

The handle used to move the ark to fresh ground every few days such that the chicken inside can graze and find some of their own food in form of insects and worms.

Feeder can be made from split bamboo, while the waterer can be made from split motor tyre. The feeder and the waterers can be installed inside the fold.

CONTROLLING BROODINESS:

This strategy can help ensure that the target of 450-500 young chicks is produced per year. To ensure that each hen sits on the right number of eggs at brooding, all eggs laid within a given time can be collected and given to only one of the layers to lie on, while the others are kept laying. Use of trap nest can lengthen the laying duration. A target of 4 chickens can be allowed to brood per month, in a type of relay system. This guarantees regular supply of at least 42 chicks per month, which should be available for sale 12-16 weeks later.

TRAP NEST

A hen suspected of not laying is put in a trap nest for 36 hours to see if she does not lay. Broody hens do not lay, and tend to want to sit on eggs, and being protective if one goes near it. If she does not lay, perhaps it is due to broodiness. She can be brought back to lay by putting it in a trap nest for 4-5 days. Cool air under their bodies stops them from brooding. The trap nest can be raised above then ground to allow air circulate under

MANDATE II: THE POULTRY FEED FORMULATION PROJECT

Theoretical Background: Principles of Poultry feeding:

Generally, chickens are regarded as grain feeders such that basic ingredients of poultry feed the world over are a form of grain or seed. The overall protein levels in feeds can be kept relatively low provided a full range of amino acids is present in approximately correct amounts. Whenever poultry are fed in diets with a quantity of ungrounded cereal and seeds, it is important that they be provided with insoluble grit ad lib. Improperly fed poultry develop malnutrition diseases and conditions such as pica. This is an odd feeding habit resulting from nutritional deficiency e.g. egg pecking resulting from calcium and phosphorous deficiency in layers;

Younger chicks can be fed ad lib (i.e. as much as they want, for quick growth). Adult chicken need about 120 g feed per day. This gives 10.8 kg for three months (i.e., up to 12 weeks), and 21.6 kg per 6 months, or 44 kg / year. Taking care of losses (assume 10%), this gives 11.88 kg / 3 months and 23.76 kg / 6 months or 47 kg / year per bird. Therefore with 10 birds, a farmer needs about 115 kg feed/ 3 months or 230 kg/6 months, 470 kg / year. These can be set aside in advance before the production process starts, so as to be sure of constant uninterrupted supply throughout the production cycle. At 8 weeks, birds produce 1 kg of meat from every 1.93 kg of feed taken. Broiler cockerels reach 2 kg in 8 weeks, while pullets reach 1.65 kg in 8 weeks. Where they are mixed, the average mass of birds at 8 weeks is 1.85 kg. Poultry need green leaves for:
Keeping chicken busy, thus avoiding vices;
Source of vitamin A (especially for layers);
Appropriate technology approach: Poultry feeding in the local context

Generally, home mixed rations are usually cheaper by 10-15% and the farmer can vary the energy or protein content of the ration depending on the needs of the animal. Locally available material (cereals) can be used as poultry feed such as sorghum, millet, maize and other grains. These can be mixed with Ochong'a (remains from Omena fish sorting / miniature omena), or whole omena. Some little sand can be added to help in digestion (proper functioning of the chicken organ called the gizzard). Alternatively, the mixture can be ground in a posho mill, and then mixed with some sand. Alternative sources of proteins are white ants (Biye). Requires some skills to trap and collect for use by chicken.

Objectives:
To identify different raw materials in the area that can be part of the formulation;
To get market prices of different possible raw materials for the formulation;
To formulate the most optimal feed ration that is nutritionally adequate / superior and cheap;

Background targets
To accomplish this mandate, a target was set as follows:
Have a system which can accommodate 450 birds / year (495 / 500 birds per year, corrected to take into account 10% mortality and other losses). This translates to at least 42 birds per month (minimum 4 adults with their young ones numbering at least 11 at any time). The adults are not counted here because they become part of the breeding stock. The young ones can be sold on relay basis (i.e., have about 45 chicken per month

for sale). This gives a figure of 540 per year, but assuming 10% mortality, it boils down to about 490 live chicks per year.

METHODS

Market survey by visiting Kibuye and Chiro mbero to get prices for different grains and fish products (such as ochong'a);

Literature review to get the nutritional value of possible raw materials for the formulation;

Use of pearsons square method to formulate the ratio of key inputs of the formulation;

Use of linear programming to identify feasible mixes of the different raw materials

RESULTS:

The possible raw materials for use in feed formulation were:

1.Sorghum 2.Maize 3.Omena / Ochong'a (Omenalets)

These materials were selected based on their presumed richness in carbohydrates and proteins, the major feed constituents. The energy and protein composition of these materials as shown in the table below:

Table 28: Possible raw materials available for feed formulation

RAW MATERIAL	Crude protein%	Energy Kcal / kg	Energy / CHO %
Large fish	47	255	
Ochonga / omena- small fish	42-44 (43)	320	58***
Maize	10.6	345	70-80
Sorghum	10.5	335	74

These figures were then compared with standard feeds in the market used for specific purposes. The layers and broilers marsh acted as guidelines for the formulation. Protein was particularly important, as it was used as the benchmark in pearsons square analysis.

Table 29: raw materials actually used in the feed formulation

	Protein content (kg/ton of dry feed)	Energy content (Mcal/ton of dry feed)
Sorghum	105	335
Ochong'a	430	320
Protein requirement for chicks (0-8 weeks)	200	
Energy requirement for chicks (0-8 weeks)		2970
Protein requirement for growers (9-18 weeks)	150	
Energy requirement for growers (9-18 weeks)		2700
Energy requirement for layers (18 weeks+)		3130

Protein requirement for Layers / breeding stock (18 weeks+)	175	

Method for determining the appropriate feed mixes (Ratios) using different raw materials include the Pearson Square and linear programming.

Figure 131: USING PEARSON'S SQUARE TO FORMULATE GROWERS FEEDS (9-18 weeks)

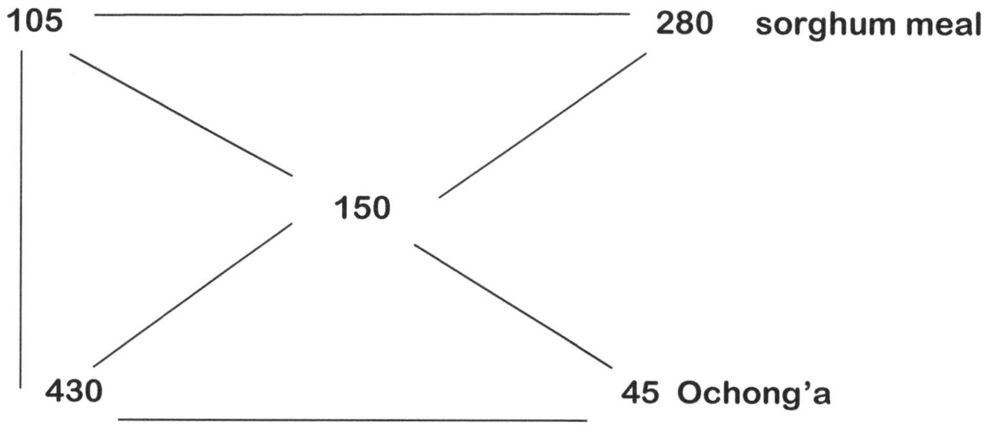

Therefore 280 parts of sorghum meal mixed with 45 parts of ochong'a gives a feed formulation with 15% protein. This is sufficient for feeding the growers aged 9-18 weeks. Thus 13.846 kg (13.85%) of ochong'a mixed with 861.53 kg (86.15 %) of sorghum grains give 1 tone of appropriate chick feed. The formulation has (i) 15% protein content (ii) 332.92 Mcal/ton

Figure 132: PEARSONS SQUARE FORMULATION OF CHICK MARSH (0-8 weeks)

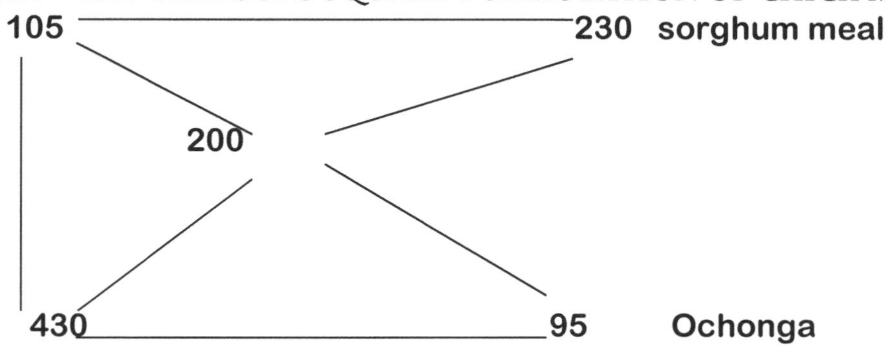

Therefore 230 parts of sorghum meal mixed with 95 parts of ochong'a gives a feed formulation with 20% protein. This is sufficient for feeding the chicks aged 0-8 weeks. Thus 292.3 kg (29.23%) of ochong'a mixed with 707.7 kg (70.77%) of sorghum grains

give 1 tone of appropriate chick feed. The formulation has (i) 20% protein content (ii) 330.62 Mcal/ton. Once the ratios were determined using Pearson's square method, the cost elements were assessed considering the market price of each ingredient, and also the feed requirement per bird per age. These are as shown in the tables below.

Table 30 :Weekly requirement of dry feed by poultry at different live weights

Age (weeks)	Liveweight kilograms	Total weekly dry feed Requirement (kg)	Total daily dry feed requirement (g)
1.25	0.23	0.2	28.4
3	0.45	0.37	52.7
4.75	0.68	0.54	77.0
6	0.91	0.65	93.2
7	1.36	0.77	109.5
9	1.82	0.82	117.6
11.5	2.27	0.88	125.7
13+	2.72	0.91	129.7

Table 31: Age against weight and weekly feed requirement of chicken

Age (weeks)	Weight (kg)	Approx weekly feed requirement per bird (kg)
0	.05 (approx)	
1	.15	0.1
2	.4	0.25
4	.5	0.45
5	.75	0.55
7	1.5	0.7
9	2.05	0.9
11	2.25	0.9
12	2.3	0.95

230 parts of sorghum meal mixed with 95 parts of ochong'a gives a feed formulation with 20% protein. This is sufficient for feeding the chicks aged 0-8 weeks. Thus 292.3 kg (29.23%) of ochong'a mixed with 707.7 kg (70.77%) of sorghum grains give 1 tone of appropriate chick feed. The formulation has (i) 20% protein content

Table 32: Feed requirements and projections

Age /Week	Feed requirement per bird (kg)	Cumulative feed requirement/ bird (kg)
1	0.1	0.1
2	.25	0.35
3	.35	0.70
4	.45	1.15
5	.55	1.70
6	.6	2.30
7	.7	3.00
8	.8	3.80
9	.9	4.70
10	.9	5.60
11	.9	6.50
12	.95	7.45
13	.95	8.40
14	.95	9.25
15	.95	10.20
16	1.0	11.20

COST-EFFECTIVENESS OF THE FORMULATION

Price of different raw materials use in the feed formulation are as follows (as got from the traders in Kibuye and Chiro Mbero markets in Kisumu between 7th and 11th June, 2004).

Table 33: prices of raw materials for chich feed (June 2004)

Item	Price per gorogoro (Shs)	Assumed price per kilogram (Shs)	Amount of raw material per tonne of mix (kg)	Total input price per tonne of formulation
Sorghum	40	20	707.7	4384.50
Ochong'a	15-20 depending on season	15	292.3	14154.00
Price of formulation per ton (Shs)				18,538.50

Given 280 parts of sorghum meal mixed with 45 parts of ochong'a gives a feed formulation with 15% protein, giving feed for the growers aged 9-18 weeks, 138.46 kg (13.85%) of ochong'a mixed with 861.53 kg (86.15 %) of sorghum grains give 1 tone of appropriate chick feed. The formulation has (i) 15% protein content

Table 34: Price comparison among different poultry feeds.

Item	Price per gorogoro	Assumed price / kilogram (Shs)	Amount of raw material /ton of mix (kg)	Total input price per ton of formulation (Shs)
Sorghum	40	20	861.53	17230.60
Ochong'a	15-20 depending on season	15	138.47	2077.05
Price of formulation per ton (Shs)				19307.65

Using the above formulation model and table, the unit price of formulation per kilogram is Shs 18.55. This can be used alongside other data to give projections in terms of requirements and profits. This is as shown below:

Table 35: Financial projections for poultry business

Selling Age (weeks)	Total cumulative Feed requirement (kg)	Cumulative cost of feed Shs) A	% of other costs	Other costs (Shs) B	Total Production costs (Kshs) A + B	Projected Weight of chick (kg)	Projected Sale price per chicken (Kshs) *	Expected profit (Shs) per bird
8	3.80	70.50	15	10.60	81.10			
9	4.70	87.20	16	14	101.20			
10	5.60	103.90	17	17.70	121.60			
11	6.50	120.60	17	20.50	141.10			
12	7.45	138.20	18	24.90	163.10	2.0	240	80
13	8.40	155.80	20	31.20	187.00	2.1	250	63
14	9.25	171.60	23		211.10	2.2	260	48.9
15	10.20	189.20	25		236.5	2.3	280	43.5
16	11.20	207.80	25		259.75	2.35	2822	22.25
17	12.20	226.30	25		282.90	2.4	288	5.10
18	Point of lay							
19								
20								
21								
22								

Assume sale at the rate of Shs 120 / kg (standard for red meat; white meat normally costs higher)

OTHER COSTS IN POULTRY PRODUCTION

After the structure and equipment are laid, Feeding comprises 70-85% of all other inputs in poultry management. Other costs arise from Vaccination, medicine (especially dewormers and antibiotics), routine management e.g. cleaning, heating for young chicks under 5 weeks, etc. Vaccination schedule is shown at the appendix. If properly followed, it may reduce other costs by as much as 25-50% since disease management is very expensive, and vaccination keeps diseases to a mere minimum.

ASSUMPTIONS IN THE FEEDING AND HOUSING DESIGN:

All chicken will have been sold by the age of 16 weeks (4 months) since experience shows that 4 months' old chicks without improved feeding can acquire a live weight able to fetch Shs 200 each. Thereafter, those wanting to keep layers for eggs can buy them and use them only for a short duration before they start laying. Those on improved feeding programs therefore can attain the selling weight up to 4 weeks earlier (by 12-16 weeks of age).

For commercial broilers, a live weight of 2.5-3kg (about 2-2.5 kg slaughter weight) is normally achieved by 10-12 weeks of age.

After 16 weeks, the farmer can use the facilities of the first batch (which has just been sold), i.e. the cycle begins after every 4 months. Thus, one can plan for one-third the number of annual bird requirements;

Every month, at least 45 eggs are being incubated (under natural system) so that 12 – 16 weeks later, there is a flock for sale / slaughter.;

It is important to keep the different age groups in different folds/ enclosures due to varying feed requirements (quality and quantity);

POULTRY PROJECT MONITORING AND EVALUATION: Record keeping.

We expect these initiatives to be easy to adopt by the farmers. Records to be kept will be weekly, monthly and annual Records as shown below are recommended during the poultry production programme. These can then be used to make graphs such as:

Month (X-axis) VS Progressive average monthly value of sales (in Shs)

Month (X-axis) VS Progressive Total monthly value of sales (in Shs)

Month (X-axis) VS Flock size

Month (X-axis) VS Progressive average monthly value of Profit (in Shs)

Month (X-axis) VS Progressive Total monthly value of profit (in Shs)

Table 36: CHICKS PRODUCTION RECORD FOR YEAR 2...........

Month	Week	Number of chicks hatched	Number of chicks dead	Number of chicks alive by end of week
Jan	1			
	2			
	3			
	4			
	TOTAL			
Feb	1			
	2			
	3			
	4			
March	1			
	2			
	3			
	4			
	TOTAL			
APRIL	1			
	2			
	3			
	4			
	TOTAL			
MAY	1			
	2			

Table 37: POULTRY SALES RECORD FOR YEAR 2...........

Month	Week	Number of chickens sold	Number of chickens eaten at home (bought at homestead)	Number of chickens sold by end of week	Income from sale Shs
Jan	1				
	2				
	3				
	4				
	TOTAL				
Feb	1				
	2				
	3				
	4				
	TOTAL				
March	1				
	2				
	3				
	4				
	TOTAL				
APRIL	1				
	2				
	3				

Table 38: SALE RECORDS FOR THE YEAR 2............

Month	Jan	Feb	Mar	Apr	May	Jun	Jul	Aug	Sept	Oct	Nov	Dec
Total monthly												
Cummulative												
Total monthly												
Total cumulative												

POULTRY HEALTH RECORDS: VACCINATION RECORD FOR POULTRY (CHICKEN)

BATCH ....

Table 39: The poultry vaccination schedule

Age	Vaccination	Comment
DAY 1	Marex Disease	Marex vaccination
Day 1-14	New castle disease (NCD)	NDV intraoccular
14th day	Gumboro disease	Gumboro vaccination
3rd-4th weeks	Fowl pox	Fowl pox vaccine
5th week	Gumboro disease	Gumboro vaccination
6th week	NCD	NDV Komorov or lasota vaccination
16th week	NCD	NDV Komorov for point of lay
18th week	Gumboro disease	Gumboro disease for parent stock

Table 39b: Different raw materials for poultry feed

	Protein content (kg/ton of dry feed)	Energy content (Mcal/ton of dry feed)
Sorghum	105	335
Ochong'a	430	320
Protein requirement for chicks (0-8 weeks)	200	
Energy requirement for chicks (0-8 weeks)		2970
Protein requirement for growers (9-18 weeks)	150	
Energy requirement for growers (9-18 weeks)		2700
Energy requirement for layers (18 weeks+)		3130
Protein requirement for Layers / breeding stock (18 weeks+)	175	

MODULE 15: CROP AND LIVESTOCK IMPROVEMENT

GENETICS BREEDING BIOTECHNOLOGY

15.1 GENETICS
15.2 BREEDING
15.3 BIOTECHNOLOGY

15.1 GENETICS

DEFINITIONS OF CONCEPTS

Genetics: the study of how characteristics are passed from parents to offsprings.

Trait: characteristic such as height, eye colour, seed colour, seed size, etc

Chromosomes: are thread-like structures or strands found inside a nucleus that contain units of inheritance called genes. They exist in pairs called homologous chromosomes.

Genes:

Are the units of inheritance that control a characteristic in an organism; a hereditary determinant of a trait; they carry the trait from one generation to the next. They are arranged within chromosomes, and therefore also exist in pairs.

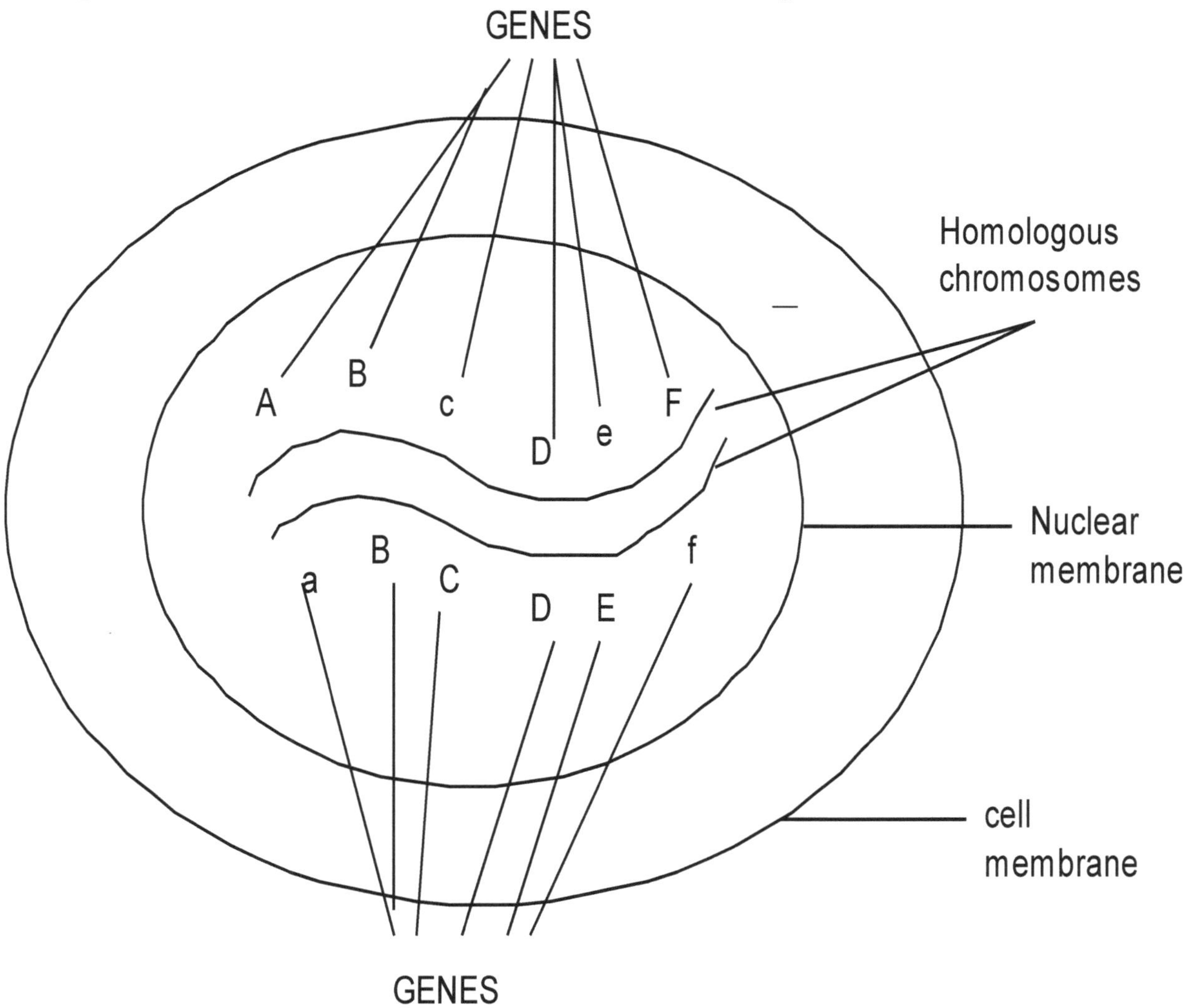

Figure 133: The details of a cell structure showing the genes and the chromosomes

Some concepts in Genetics and Breding

Each species has a fixed number of chromosomes (all in pairs); cattle have 30 pairs; sheep has 27; pigs have 19 pairs; chicken has 39 pairs, and man has 23 pairs. Any deviation leads to an abnormality of the organism.

Allele (or allelomorphic pair): Are the different forms of a particular gene OR a pair of genes on homologous chromosomes that control a specific trait e.g. length of nose (thus one form of the gene may be associated with short nose, and the other with long nose), colour of hair etc.

Genotype: The genetic make-up or constitution of a cell or individual; it may be TT for homozygous dominant and tall plant; Tt or tT for a heterozygous but tall plant; and tt for homozygous but short plant.

Phenotype: the physical outlook or the observable properties of an organism e.g. tall, short, round seeded, wrinkled seeded etc. The maize plant below is a typical undesirable one since it is prone to pests e.g. birds, among other crop enemies since the cob and grains are exposed. Geneticists and plant breeders would want to come up with varieties whose entire cobs are covered.

Homozygous: A genotype in which the 2 alleles are alike; a genotype in which the members of a pair of alleles are the same; a case where the 2 forms of the gene controlling one characteristic are exactly alike. The 2 genes in the paired chromosomes are exactly alike; the allele is exactly the same. e.g. RR, TT, nn, gg etc. An individual or cell with a homozygous genotype is called a homozygote.

Heterozygous: A genotype whereby members of a pair of alleles are different; a genotype in which the members of a pair of alleles are different; If the 2 genes on homologous chromosomes are different- thus there are 2 different forms of the gene on the same locus in strand 1 and in 2. Example Rr, Ss, qQ, xX etc. Such an individual is called a heterozygote.

Somatic cell: these are body cells, and have the full number and paired chromosomes. They are diploid (have paired chromosomes). They are found in all non-reproductive (vegetative) parts of the body. They are thus also called body cells. Multiplication of body cells causes growth of individuals, and the process of their multiplication is called mitotic cell division.

Mutations: Are sudden changes in the genotype of an organism, resulting in changes in the phenotype of the offsprings. These may be caused by chemicals (mutagens), strong radiations e.g. UV, etc

Crossing over: this means the exchange of some genetic material between the 2 homologous chromosomes during their stages of cell division. Sometimes they come too close, with the risks of some part of one chromosome breaking. This may affect the phenotype.

Gamete: These are reproductive or sex cells

Dominant gene: One that always shows or exhibits its characteristics regardless of the nature of the other member of the pair; e.g. if gene T controls tallness in a plant and t controls shortness, then genotypes Tt, TT and tT- will show tallness- a characteristic of the gene for tallness. The gene whose characteristic is not exhibited (shown) as long as T has appeared in the genotype is said to be recessive.

A recessive gene: is that suppressed by the dominant one, and its trait is only shown/ exhibits itself/comes out if the dominant gene does not appear anywhere. Thus in the case above, t is the recessive gene, and since it controls shortness, the plant only becomes short if the genotype is tt.

Figure 135: A maize cob exposed to pests eg birds

Gamete: These are reproductive or sex cells. They are found in the ovaries, testes and anthers. They have half the number of chromosomes found in a body (somatic) cell, and often reunite in fertilization to give rise to a full diploid cell called a zygote. Gametes are said to be haploid because they have HALF the number of chromosomes found in a body or somatic cell. They are produced by the process of meiosis.

CELL DIVISION:

Two types of cell division exist- mitotic and meiotic. Mitotic cell division is also called mitosis, and is a kind of cell division whereby the daughter cells are exactly like the parent cell, diploid in nature. It takes place in all body cells, and is meant for growth. Meiotic cell division is also called meiosis, and is a kind of cell division whereby the daughter cells are having half the number of chromosomes as the mother cell (i.e. are haploid). Meiosis occurs in reproductive organs such as ovaries, anther etc, and is meant to produce reproductive cells (gametes which are haploid. Upon fertilization, the haploid female and male gametes unite to give rise to a new individual, which is diploid.

Mitotic cell division (mitosis): This is a kind of cell division whereby the daughter cells are exactly like the parent cell, diploid in nature.

Meiotic cell division (meiosis): This is a kind of cell division whereby the daughter cells are having half the number of chromosomes as the mother cell (i.e. are haploid).

Table 40: Differences between meiosis and mitosis

	MEIOSIS	MITOSIS
# of products	4 daughter cells	2 daughter cells
Nature of products	Haploid products	Diploid results
Purpose	Reproduction / production of gametes	Growth / multiplication of cells
Where it occurs	Reproductive organs (testes, anther, ovary	All body parts

EXAMPLE OF SIMPLE GENETIC CROSSES: PREDICTING THE NATURE OF OFFSPRINGS (F1 AND F2).

Case 1:

A red seeded plant is crossed with a white seeded plant. Assuming that red is dominant, predict the nature of F1.

Solution:
Red seeded plant- male; and white seeded one female
Red represented by R, while gene for white is r
Male plant may be heterozygous (and therefore has genotype Rr), or homozygous (and therefore has genotype RR)
Female plant is of genotype rr

1a (Scenario 1) - if male parent heterozygous:

PARENTS	**Male**		**Female**	
Genotypes of parents	**Rr**		**rr**	
Meiosis (Gamete)	**R**	**r**	**r**	**r**
Fertilisation				
F1	**Rr**	**Rr**	**rr**	**rr**
F1 Phenotype	Red	Red	White	White

Genotypic ratio50% Heterozygous; 50% homozygous recessive
Phenotypic ratio:50% red seeded; 50% white seeded

Use of a Punnet Square

	Female parent			
Male parent	X	rr		
		Gametes	r	r
	Rr	R	Rr	Rr
		R	rr	rr

Genotypic ratio:50% Heterozygous; 50% homozygous recessive
Phenotypic ratio:50% red seeded; 50% white seeded

1b SCENARIO 2: MALE PARENT HOMOZYGOUS

PARENTS	**Male**		**Female**	
Genotypes of parents	**RR**		**rr**	
Meiosis				
Gametes	**R**	**R**	**r**	**r**
Fertilisation				
F1	**Rr**	**Rr**	**rR**	**rR**
F1 Phenotype	**Red**	**Red**	**Red**	**Red**

Genotypic ratio 100% Heterozygous (Rr) Phenotypic ratio:100% red seeded

Punnet Square.

	Female parent			
Male parent	**X**	**rr**		
		Gametes	**r**	**r**
	RR	**R**	**Rr**	**Rr**
		R	**rR**	**rR**

Genotypic ratio: 100% Heterozygous (Rr); Phenotypic ratio:100% red seeded

3. Predict the nature of F2 if the above offsprings were selfed, and their male parent was homozygous.

If male parent of F1 was homozygous, it means all F1s were heterozygous; selfing means mating a brother and a sister (F1 products). Thus

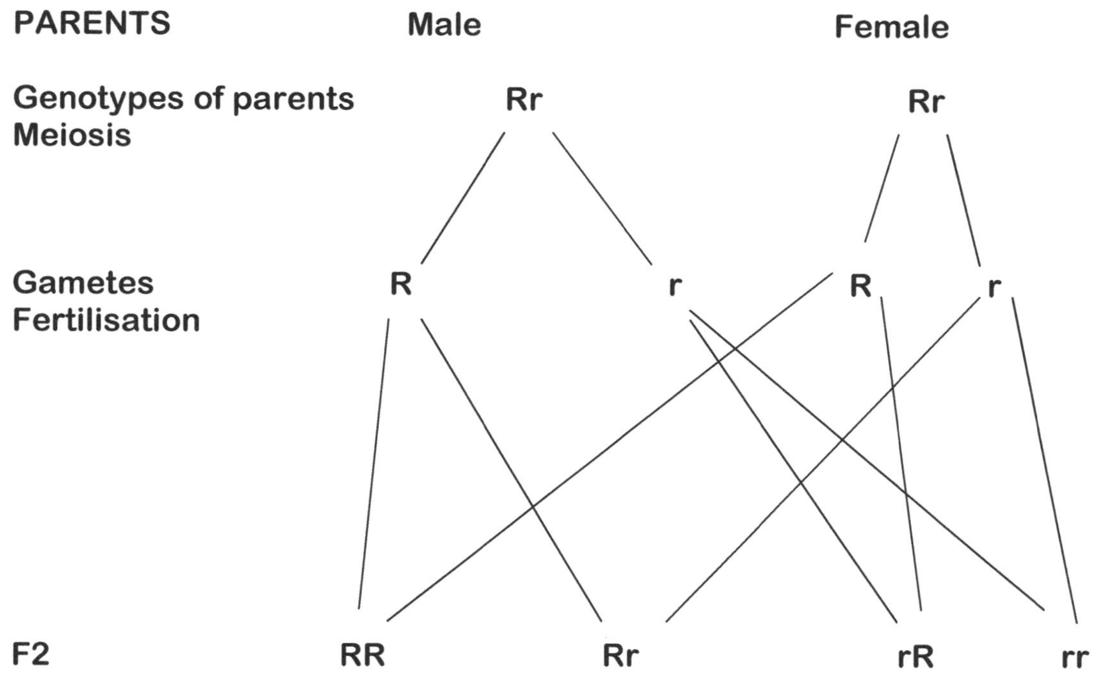

F2 Phenotype Red Red Red White

Genotypic ratio
25% RR (Homozygous dominant);
50% Heterozygous (Rr);
25% homozygous recessive (rr)
Phenotypic ratio: 75% red seeded; 25% white seeded

Use of a Punnet Square:

	Female parent			
Male parent	X	Rr		
		Gametes	R	r
	Rr	R	RR	Rr
		r	rR	rr

Genotypic ratio
25% RR (Homozygous dominant); 50% Heterozygous (Rr); 25% homozygous recessive (rr)

Phenotypic ratio:75% red seeded; 25% white seeded

4Predict the nature of offsprings if a heterozygous parent was crossed with a homozygous dominant parent

	Male	Female
PARENTS	Male	Female
Genotype of parents	Homozygous dominant	Heterozygous
Genotypes of parents	RR	Rr

Meiosis (Gametes)
Fertilisation

R **R** **r** **R**

F2 Rr RR Rr RR

F2 Phenotype Red Red Red Red

Genotypic ratio: 50%Heterozygous (Rr); 50% Homozygous dominant

Phenotypic ratio: 100% red seeded

Punnet Square:.

	Female parent			
Male parent	**X**	**Rr**		
		Gametes	**R**	**r**
	RR	**R**	**RR**	**Rr**
		R	**RR**	**Rr**

Genotypic ratio: 50%Heterozygous (Rr); 50% Homozygous dominant

Phenotypic ratio: 100% red seeded

5Predict the nature of offsprings if red and white were co-dominant (i.e., the partially dominant, i.e. there would be a third phenotype in case of heterozygosity- pink) if
Male is red and female white; (b) Male is pink and female white
Male is white and female pink
6.Crossing between a red fowl and white fowl results in red fowls. Explain this. Show by use of genetic crossing how this takes place.
Smooth – SS; Shrunken ss; All smooth seeds –1; Self crossing Ss –1
Explanation: 3:1; 75% smooth; 25%shrunken

DEVIATIONS / VARIATIONS

Sometimes, the offsprings may result with characteristics not exactly, or completely not in agreement with the predictions or expectations. This could result from either mutations or crossing over.

1. Mutations: Are sudden changes in the genotype of an organism, resulting in changes in the phenotype of the offsprings. These may be caused by mutagens, strong radiations e.g. UV, etc

2. Crossing over: this means the exchange of some genetic material between the 2 homologous chromosomes during their stages of cell division. Sometimes they come too close, with the risks of some part of one chromosome breaking. This may affect the phenotype.

BOX 31: HAIRY CELL LEUKEMIA

SOURCE:http://encarta.msn.com/media_461570368_761578429_-1_1/Hairy_Cell_Leukemia.html

BREEDING

Breeding is the manipulation of the genotype of organisms with the aim of producing crops and/ or livestock of desirable characteristics such as high yield, high butter fat content, disease resistance, etc. It is done by using selection and controlling the pollination/ breeding. A villager unconsciously does breeding by selecting the largest cobs from his field and uses these as seed in the next planting season. Livestock breeding is done by selecting what animals should be mothers of the future herd (dams), and which bull as future fathers of the herd should be left when others are being castrated.

OBJECTIVES OF (IMPORTANCE OF) BREEDING

Aim is to produce crop and livestock types with the maximum number of desirable traits, and the minimal number of undesirable traits.

Objectives/ Importance of breeding:

1. Increase yield e.g. of crops, milk, egg, etc;
2. Increase growth rate e.g. in pigs, broiler chicken, beef cattle;
3. Improve quality e.g. tenderness of meat, thicker milk with higher butter fat content, maize grains with higher protein content, beans with less concentration of stomach upset chemicals, etc;
4. Produce disease resistant breeds and varieties;
5. Produce pest resistant breeds and varieties;
6. Produce drought resistant varieties;

7 Produce early maturing crops or livestock.

High Milk producing livestock

Good meat producing livestock

SOME CONCEPTS IN BREEDING (METHODS OF BREEDING).

1. Selection 2. Crossbreeding 3 inbreeding 4. Upgrading

SELECTION:

Means choosing individuals to be parents of the future herd, i.e., for continued reproduction.

Selection involves giving priority / preference to some seeds, plants or livestock to be parents of a herd by favoring them in mating (livestock/ males), pollination (flowers), planting (seeds) etc at the expense of others which may be culled, slaughtered, castrated etc. This denies the latter a chance to breed / mate, and therefore are less likely to have children / offsprings.

LINE BREEDING:

This process involves producing a pure line / pure breed. The parents are related but not as closely as in inbreeding. This could be for instance mating 1st, 2nd and 3rd cousins. A pure line thus produced will always breed true to type- i.e. all parents, offsprings (F1, F2, F3... Fn) will be the same in one or more characteristics that make that line or breed unique (as long as there is no crossbreeding). (Examples are black/ white stripes in Friesian dairy cow, colour of seed associated with a particular crop variety, etc). Producing a pure line, however, involves tiring effort of repeated selection and controlled breeding / mating (pollination, insemination, etc).

INBREEDING:

The mating of close relatives (brother/ sister, mother/ son, father/ daughter). This method may be used if the breeder wishes to concentrate a particular good trait in his herd or crop. If both a bull and his daughter show a particular desirable quality, then their offspring is most likely to show it even more conspicuously. However, the repeated crossing / mating of these close relatives may also increase the undesirable qualities associated with the family. Therefore inbreeding must be accompanied by very strict selection and culling. It increases homozygosity. Offsprings do not benefit from hybrid vigour, and they get bad traits from parents.

CROSSBREEDING OR HYBRIDIZATION:

Crossing unrelated parents from different lines / breeds. It is the crossing or mating of individuals from two different pure lines / breeds. One line/ breed may have some good qualities (that the second one is poor in), and bad qualities (that the 2nd one is good in). Thus crossing them would be meant to combine all the good qualities from both the lines or breeds into one organism (product). The process of crossing these pure breeds / lines is called hybridization, because a hybrid results; the product (offspring) is called a hybrid.

Since the hybrid (also called cross-breed) has almost all the good qualities from both parents, it is very superior and is better than both the parents, it tends to grow fast, pest resistant, etc- thus has hybrid vigour. Example is Tswana bull crossed with a Simmental cow. The offsprings are heterozygous. When hybrid offsprings are mated, hybrid vigour is lost. Due to the phenotypic ratio associated with cross breeding, the process of selection is not necessary since all offsprings (F1) are heterozygous. However, election may still come in if there are other selection criteria being considered e.g. eye colour, milk yield, size of udder, disease resistance etc, which may be individual traits.

HYBRIDISATION

Crossing unrelated parents from different lines / breeds is called crossbreeding or hybridization. This is the crossing or mating of individuals from 2 different pure lines / breeds. One line/ breed may have some good qualities (that the second one is poor in), and bad qualities (that the 2nd one is good in). Thus crossing them would be meant to combine all the good qualities from both the lines or breeds into one organism (product). The process of crossing these pure breeds / lines is called hybridization, while the product (offspring) is called a hybrid. Since the hybrid (also called cross-breed) has almost all the good qualities from both parents, it is very superior and is better than both the parents, it tends to grow fast, pest resistant, etc- thus has hybrid vigour. Eg Tswana bull X Simmental cow. The offsprings are heterozygous. When hybrid offsprings are mated, hybrid vigour is lost.

UPGRADING

Under this scheme, Lower performers are graded up. It is a form of crossbreeding in which the male is of the breed whose characteristics the farmer wishes to establish in his herd. The herd may already be excellent, but are not very good in one trait that the bull's family is excellent in. The bull is then mated with the cows in the herd, and their products are expected to be better than the cow's parents' performance. An example is whereby a bull' s family produces 30 litres of milk / day, while the herd produces 20. With repeated mating of parent, then daughter, then granddaughters..... etc, eventually the herd will give 30 litres of milk/ day. The herd is then said to have been upgraded. An illustration of this is given below.

GENETIC UPGRADING PROCESS.

This is an explanation of the illustration below. At every stage in the mating process, a new generation results whose Friesian qualities increase in the order: 50%- 75%-87.5%-93.75%-96.875%-98.43875%-99.219375%-99.6096875%-99.804843875-........ 100%, and whose Afrikander qualities decrease by 50%-25%-12.5%-6.25%-3.125%-1.56125%-0.780625%-0.3903125%-0.195156125-….0%. This milk yielding quality will have been imparted to the new upgraded breed, while it would retain most of its other good qualities. This is as shown in the figure below.

UPGRADING: 5 serial crosses with an upped grade bull (Friesian)

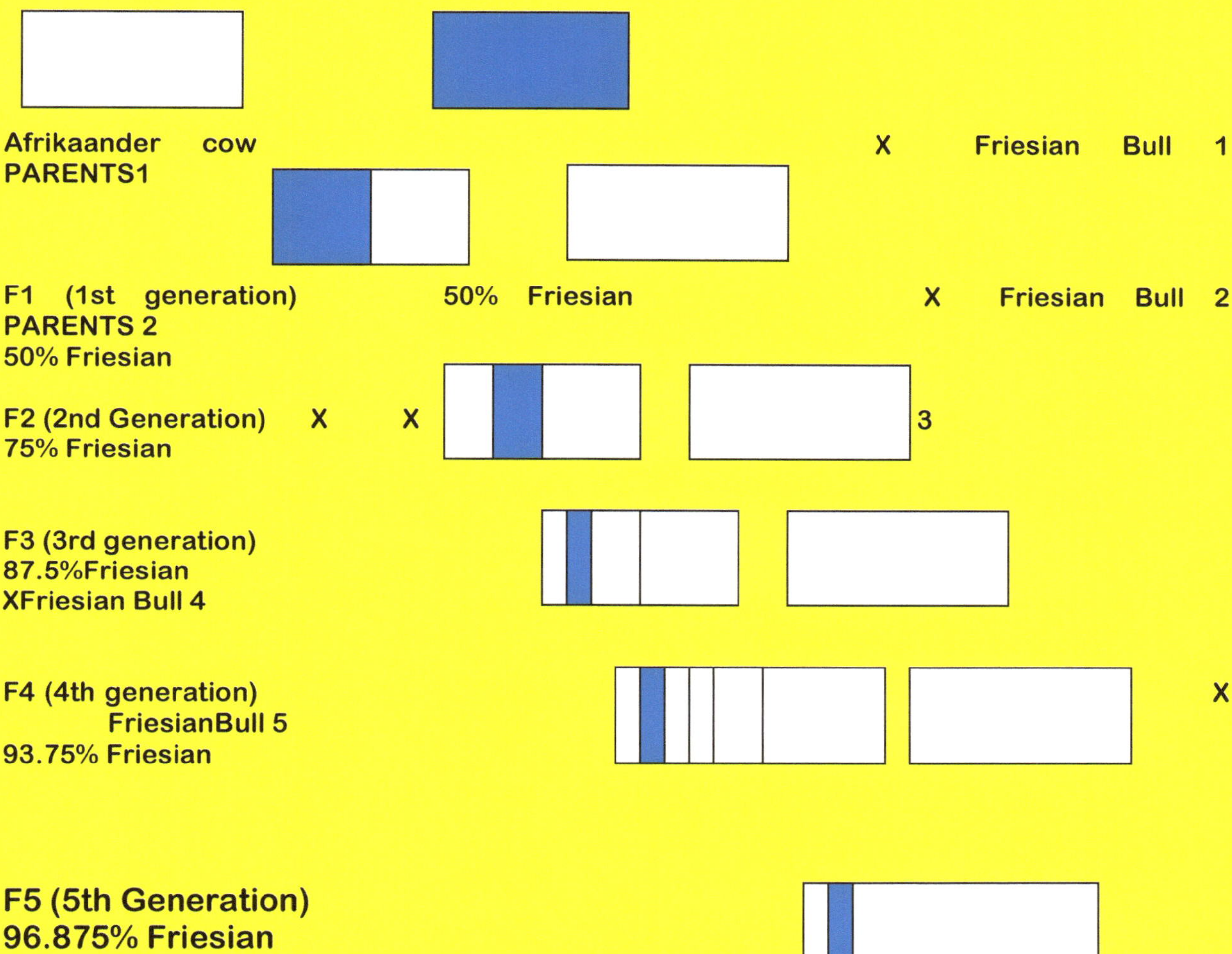

Figure 136: Genetic Upgrading process.

15.2 GENETIC ENGINEERING AND BIOTECHNOLOGY

BIOTECHNOLOGY:

Biotechnology refers to the judicious manipulation and utilization of the microbiological processes for man's own good especially in producing useful products.

OR The application of basic biological (life) principles in production of new products such as food;

OR is the use of microorganisms to make products that are useful to man.

The products may be food, energy, medicinal, or waste-management related. Biotechnology can be modern or traditional. The new products may have better taste, smell, acceptability, concentration, look and feel (texture), or may keep longer. Examples are fermented milk from a factory, or farm inputs such as multiplication and packing of strains of Rhizobia so that they can be used to inoculate planting materials of leguminous crops e.g. beans to facilitate nitrogen fixation in the plant's root nodules. It involves the use of applied molecular biology, biochemistry, microbiology and related physiological processes in a production process. There are different forms of biotechnology: traditional and modern

The traditional Biotechnology systems are those that have been used by rural communities for many centuries, and are largely not documented, i.e. nothing or very little is recorded about them. These include milk processing and preservation, brewing etc. Simple common sense science, ideas and knowledge were used to develop them, and have been tested and improved overtime from generation to generation. They have a lot of scope for development, and perhaps in solving some of the current world problems. Are largely assumed by modern scientists, regarding them as primitive and outdated. Are knowledge systems or ideas that man e.g. farmers have been using for centuries, mostly unconsciously. It is done without knowing because they are part of life.

Modern biotechnology is the current complex industrial-based knowledge systems. They are well documented, i.e. a lot has been written about them. They are mostly developed from traditional technologies either directly or indirectly. They have been researched on for some time, and are highly improved, and use high level of skills, labour and technical science.

Examples of biotechnology:

Making of traditional and modern brews e.g. chibuku, beer, wines and spirits; Making of bread; Preservation of milk in forms such as madila; Production of biogas; Making of vaccines; Making of hormones such as insulin; Making of bio-fertilizers, which are used in inoculation of seeds of legumes before planting to enhance the nitrogen fixation process.

15.2 GENETIC ENGINEERING (GE)

Genetic Engineering (GE) is a very refined manipulation of the genetic make-up of organisms by artificially inducing mutations so as to produce an organism with some desirable traits, or is the manipulation of the genotype of an organism to maker it and / or its characteristics more desirable to man. This is done by artificial mutations. It is a more specialized form of breeding- since it involves more refined deletions and transfers at gene level within the nucleus. GE may involve the removal of a DNA segment from one organism's nucleus and inserting it into another. It mimics natural mutations and the nature and behavior of products are often difficult to predict. Genetic engineering enables scientists to produce clones of cells or organisms that contain the same genes. An illustration of GE is shown above, with the explanation below:

Figure 137: Genetic Engineering Illustration:
SOURCE: http://encarta.msn.com/media_461516631_761557775_-1_1/Genetic_Engineering.html

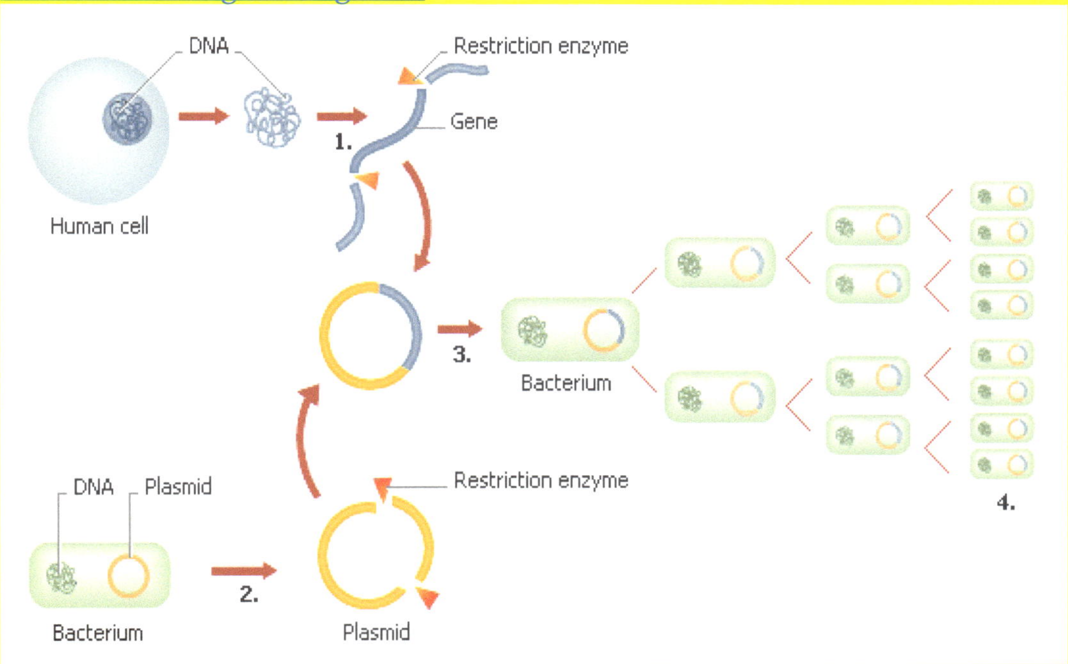

1. Scientists use restriction enzymes to isolate a segment of deoxyribonucleic acid (DNA) that contains a gene of interest—for example, the gene regulating insulin production.
2. A plasmid removed from a bacterium and treated with the same restriction enzyme binds with the DNA fragment to form a hybrid plasma.

3. The hybrid plasmid is re-inserted back into the bacterium, where it replicates as part of the cell's DNA.
4. A large number of identical daughter cells (clones) can be cultured and their gene products extracted for human use.

In the creation of a monoclonal

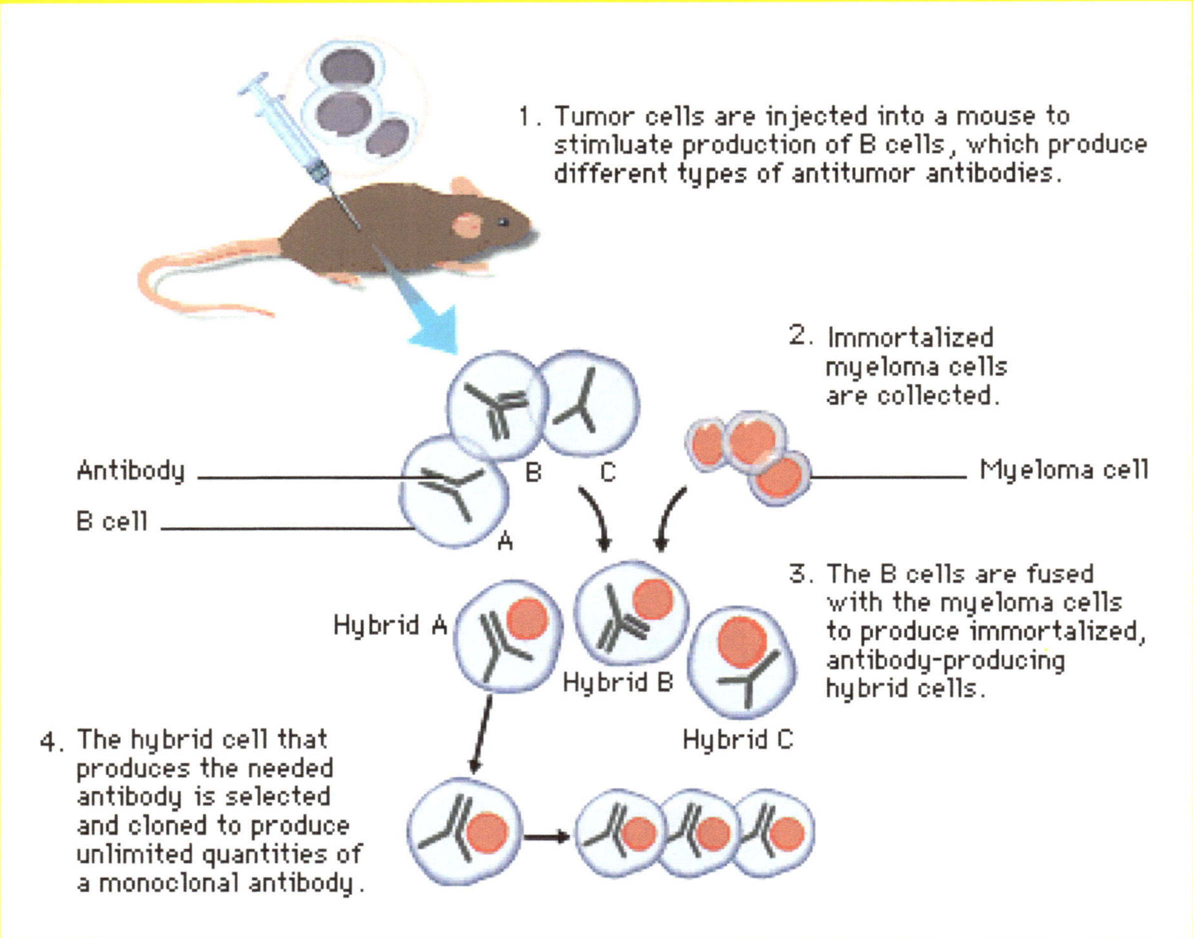

Antibody, a normal B cell (a type of lymphocyte, or white blood cell) is united with a myeloma cell (a type of cancer). This union results in the formation of cloned hybridomas (hybrid cells) that have the cancer cell's trait of dividing endlessly and the B cell's ability to produce a specific type of antibody. Monoclonal antibodies are designed to target specific molecules in the body. They have a variety of uses from pregnancy testing to diagnosing and treating certain diseases.

SOURCE: http:// encarta.msn.com/media_461540240_761571643_- 1/Producing_ Monoclonal_Antibodies.html

GENES

Genes are individual units that control characteristics of organisms, and are found on the chromosomes. Man may remove a gene from one organism's chromosome and plant it in the chromosome of another organism that he wants to have similar trait controlled by the transferred gene. Often, the behaviour of the products of such genetic manipulations is unknown and unpredictable. It is believed that some laboratories in the developed world created the HIV / AIDS virus by such systems and injected it into other primates such as monkey, baboons, chimpanzees in the rainforest of the great Lakes region of Africa (Zaire / DRC). Later, these were injected into man, with the current disastrous results- no treatment for HIV/ AIDS. The same manufacturers then claimed that HIV / AIDS originated in Africa.

THE BENEFITS OF BIO TECHNOLOGY IN THE PROCESS OF MAKING BREAD

When making bread, flour is mixed with water dough. Flour contains starch and some of this break down to maltose when there is moistened. Yeast is added to the dough and breaks down the sugar (maltose) as it respires. There is air in the dough, so the yeast respires aerobically first until the oxygen is used up. It produces carbon dioxide and bubbles of this gas get caught in the dough, making it rise. The yeast is killed when the bread is cooked.

THE PROCESS OF MAKING MADILA

The milk from cow, sheep or goats may be used to prepare madila (Botswana special milk). This is made naturally by the fermentation of milk by bacteria. Initially these bacteria came from natural sources e.g. from the animals themselves or from the atmosphere. The fresh milk is collected and left in a warm place, and after a day or two the milk has fermented and turned into madila. Here the bacteria act on the milk sugar lactose and convert into lactic acid which acid gives madila its slightly sour taste. In the industries the milk is first pasteurized to culture bacteria. After madila is ready it is cooled to 50C to stop the bacterial process. Throughout the process, the materials and the containers are kept in a sterile condition e.g. free from any harmful pathogens.

THE PROCESS OF MAKING CHIBUKU

This process of making chibuku dates thousands of years. Microorganism called yeast brings about the fermentation. Chibuku is made from sorghum grains. These grains are soaked in water for two days at room temperature (200C) and then allowed to germinate for four days. During germination the starch–digestive enzymes in the grain are activated. The enzymes in the sorghum grains are then dried and crushed (milled) and mixed with water. The enzymes in the sorghum grains now digest the starch to maltose and glucose. The sugary solution (worst) is filtered and kept for two or three days to get the beer naturally.

Benefits of Biotechnology in Agriculture:

Some medicinal products such as vaccines can be made from biotechnology, thereby helping improve livestock health;

Some food products can be kept longer in good condition e.g. milk in the form of madila;

Some products have improved acceptability, i.e. they have better taste and form, e.g. madila. Some people cannot take fresh milk as it causes stomach upsets in them. But they can use preserved milk in the form of madila.

Some farm products are converted into usable forms e.g. wheat floor made into bread by fermenting using yeast;

May help reduce environmental pollution by production of micro-organisms which can directly use air-borne or sewage pollutants.

Has helped reduce reliance on fossil fuels by its use in producing biogas for use in the farm.

Disadvantages of Biotechnology in Agriculture

Some biotechnology products cause stomach upsets to some people;

Some products have very artificial taste and look;

Some have been misused as biological warfare by rich industrialized countries against enemy and weak countries e.g. the anthrax powder put inside envelopes and posted to enemies used during and after the Osama-Bush war in 2001, among others.

Some may be preserved with chemical which may have side-effects on the body;

The biological agents used may be dangerous- either as direct pathogens, or may produce some toxic products and by-products which endanger man and livestock life;

MODULE 16: AGRICULTURAL ECONOMICS

16.1.1 PRINCIPLES OF AGRICULTURAL ECONOMICS
16.2 FARM MANAGEMENT
16.3 AGRIBUSINESS AND ENTREPRENEURSHIP
16.4 MARKETING

16. PRINCIPLES OF AGRICULTURAL ECONOMICS

Introduction:

Agricultural economics is the study of how scarce agricultural resources are allocated to produce various farm products and their distribution to consumers. The central problem of economics is the fact of scarcity. Human beings, except for the fortunate few, cannot have all they want, and what is true of individuals is true of governments. Human beings cannot get all they want because resources are limited e.g. there is not enough arable land in Botswana for everyone to have 10 hectares. The resources that are available enable people to make a certain amount of each good and not a big amount of all goods. Given that people cannot produce all they want of each good, therefore they have to choose what they want to produce because of the scarcity of resources e.g. on the available piece of land, a farmer can grow either maize or sorghum. Here the farmer has to choose whether to produce more maize and less sorghum or less maize and more sorghum or no maize and no sorghum but put all the land in dairy farming.

DEFINITIONS OF SOME TERMS

1.Inputs -anything used to produce goods and services e.g. some of the inputs of farm producing maize might be seed, land, labour, water, fertilizer, and machinery

2. Resources- are inputs used to produce goods /services

3.Firm -an organization that produces goods or service for sale

4. Supply: Is the quantity of commodity which sellers offer for sale at a given Price and time

5.Demand: Is the quantity of commodity which a buyer is willing and able to purchase at a given Price and time.

6. Output: final result of a production process

7. Production cycle: this is a period from which a farmer puts inputs together until he gets an output and sells or uses it.

8. Scarcity: it is a relative term and refers to a situation which arises when there is less of something e.g. an economic good, service, resource than people would like to have.

9. Opportunity Cost: Opportunity cost refers to the value of a resource in its alternative use. Opportunity cost is not a real cost e.g. If a farmer is having P40000 and having two alternatives, either to buy a car or to invest in agriculture, but he finally buys a car with P40000, and he does not invest in agriculture. In this case the farmer has lost the opportunity to invest P40000 in agriculture. Therefore the opportunity cost for not investing in agriculture is the P40000.

ECONOMIC SYSTEMS

1. Free market economy
2. Mixed economy
3. planned economy

PLANNED ECONOMY

In a planned economy, all economic activities such as production of goods and services are controlled by government. Here the government takes responsibility for all economic resources/factors of production (labour, capital, land & entrepreneurship / management)

FREE MARKET ECONOMY

Production is mostly controlled by private sectors. People are able to own resources as a means of production. Individuals are free to choose what to produce, when to produce and how to produce. Firms make choices that maximize their profits and they produce goods and services that consumers want. In other words the forces of demand and supply determine the production of goods and services.

MIXED ECONOMY

The economy is partly controlled by the government and partly by private sectors e.g. the state might control the mining industry, but consumers good are manufactured by private industries. Individuals are free to produce goods of their choices and use method of production that seems appropriate to them but government is able to set standards of the products as well as the pricing systems

PRODUCTION ECONOMICS

Production involves the making or creation of goods and services.

THE PROCESS OF PRODUCTION

Extraction (e.g. mining, farming, ranching, forestry)
Manufacturing e.g. shoes, beer, cigarette, books, and fertilizers.
Service industries e.g. banking, insurance, transport, and health services. Provide services at all stages of production
Distribution e.g. wholesalers, retailers salesmen
Consumers

CLASSIFICATION OF PRODUCTION

Primary- extractive industries e.g. agriculture, mining, fishing, forestry

Secondary-Manufacturing industries e.g. motor cars, canning, fertilizers
Tertiary -services industries e.g. banking, insurance, education, transport, health. These provide Services at all stages of production.
Production is never complete until a commodity or service reaches a final consumer.

FACTORS OF PRODUCTION

A factor of production is anything that contributes directly to output. They are economic resources required for the production of goods and services. These include land, labour, capital and management (entrepreneurship).

LAND

This refers to the physical nature of land without any improvement made on it by man. Land in economics includes all sorts of natural resources i.e. all the gifts of nature as distinct from man-made resources. It includes all the minerals, forests, farmlands, fishing grounds and the space on which productive activities take place.

LABOUR

Labour can be defined as all human effort of a body or mind that is undertaken for reward e.g. the work of teacher, doctor or administrative officer. Any work undertaken for mere pleasure or amusement and not for economic reward is not a labour e.g. a friendly game of tennis, the voluntary care of the sick, performance of students at assembly ground. The effort of musicians becomes labour when they play for money.

CAPITAL

Capital resources that are created to assist in the production of goods and services. This includes all man-made assets that help land and labour to produce e.g. buildings, machinery, power stations, irrigation furrows, roads, dams, boreholes, factory, raw materials, etc. Money is considered capital (liquid capital) only when it enables us to buy assets that can be used in production.

CLASSIFICATION OF CAPITAL

1. Fixed capital: Are items whose quantity cannot change during the production time. They include farm buildings, machinery or equipment, permanent improvements on the land such as fences, irrigation furrows, roads etc. These items are not diminished by production but they are affected through depreciation i.e. wearing/decreasing in their value slowly as they are being used.

2. Working/variable capital are items whose quantity can be changed during production process. They include all the raw materials used for production e.g. fuel, fertilizer, seed, pesticide, foodstuff etc. These items are diminished (used up) during the production process.

MANAGEMENT (ENTREPRENEURSHIP)

We have now discussed land, labour and capital and indicated their importance as factors of production. No productive activity can take place without the cooperation of all the three factors. These three factors however cannot do anything on their own without someone to organize them. That person is called the entrepreneur. For example, a farmer who produces maize will involve some direction and organization so that he may obtain the best results. Although the farmer may own other factors of production like land, capital (machinery, buildings, dams, irrigation furrows, spraying pumps etc) and labour, he must perform the important task of combining these factors in a way that will yield him the best maize-both in quality and quantity. He must develop fully the various sections of his land; he must provide machinery, and other tools to his laborers and assign them with different pieces of work. The farmer must arrange when to prepare, the planting and the harvesting will be done depending on the season.

DIMINISHING RETURNS

This term means that it does not always pay to keep on putting more money into producing any commodity. After a time it stops being worthwhile. The returns diminish (get smaller). For example, if you buy fertilizer and put it on a crop, you might double the yield. Then you might think, why not put on twice as much fertilizer the next time, and double the yield? In practice this would not work, because there is a limit to how much the crop can respond to fertilizer treatment. In terms of money it is uneconomic to keep on adding more fertilizer to increase the yield. It will not pay. The increased cost of fertilizer will exceed the value of the increased yield. What the farmer looks for is the optimum return and not the maximum yield. The law of diminishing return can be summarized as follows:

A variable input is increased while all other inputs remain constant; a point is eventually reached where the additional output for each additional unit of input will decline The law of diminishing returns is presented in a graph in a later section. The law of diminishing returns helps the farmer to know that he should not try to aim for maximum output without consideration of the input cost because by producing the maximum output, he may not be getting high profit. He must aim at producing optimum output and not more output at very high cost.

DEMAND, SUPPLY AND PRICE

History of supply and demand

Attempts to determine how supply and demand interact began with Adam Smith's The Wealth of Nations, first published in 1776. In this book, he mostly assumed that the supply price was fixed but that the demand would increase or decrease as the price decreased or increased. David Ricardo in 1817 published the book Principles of Political Economy and Taxation, in which the first idea of an economic model was proposed. In this, he more rigorously laid down the idea of the assumptions that were used to build his ideas of supply and demand. During the late 19th century the marginalist school of thought emerged. This field mainly was started by Stanley Jevons, Carl Menger, and Léon Walras. The key idea was that the price was set by the most expensive price, that is, the price at the margin. This was a substantial change from Adam Smith's thoughts on determining the supply price. Finally, most of the basics of the modern school theory of supply and demand were finalized by Alfred Marshall and Léon Walras, when they combined the ideas about supply and the ideas about demand and began looking at the equilibrium point where the two curves crossed. They also began looking at the effect of markets on each other. Since the late 19th century, the theory of supply and demand has mainly been unchanged. Most of the work has been in examining the exceptions to the model (like oligarchy, transaction costs, non-rationality). (Source, unknown author)

DEMAND:

Demand is the amount of commodity which an individual consumer wishes to purchase at a given price and time. Demand is different from want. Want is the mere wish to have something while demand is the willingness and ability to buy something. We all want or even need so many things. We want clothes, television sets, food, cell phones etc. If you enter a shop where cell phones are sold and you only ask for the price but you have no money with which to buy, you only want the cell phone, you have not yet demanded it. You will be demanding it the very moment you are in a position to pay for it. We can also learn something more about the shape of this demand curve. The curve clearly slopes downwards from the left to right. This means that the lower the price, the greater the quantity demanded. This is the law of demand that states that: The lower the price of any commodity the greater the quantity demanded.

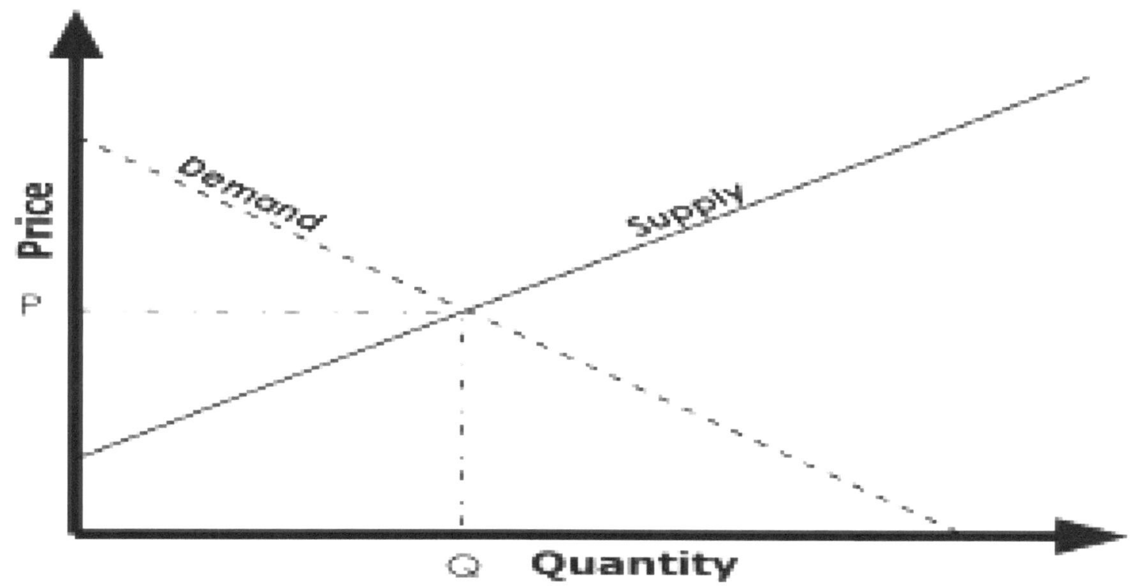

Graph of simple supply and demand curves: Demand CURVE: Source: http//en.Wikipeadia.org/wiki/image

The slope of the demand curve (downward to the right) indicates that a greater quantity will be demanded when the price is lower. On the other hand, the slope of the supply curve (upward to the right) tells us that as the price goes up, producers are willing to produce more goods. The point where these curves intersect is the equilibrium point. At a price of P producers will be willing to supply Q units per period of time and buyers will demand the same quantity. P in this example, is the equilibrating price that equates supply with demand.

DEMAND SCHEDULE

It is a tabular presentation of the quantity of a commodity that would be bought over a range of prices. The following table shows the demand of maize in Maun, Botswana.

Table 41: demand schedule

Price/kg (pula)	20	30	40	60	100	150
Quantity demanded/day (kg)	14	10	7	5	3	2

We can draw a graph from above demand schedule, having price on the Y-axis and quantity demanded on the X-axis and we obtain what is known as demand curve.

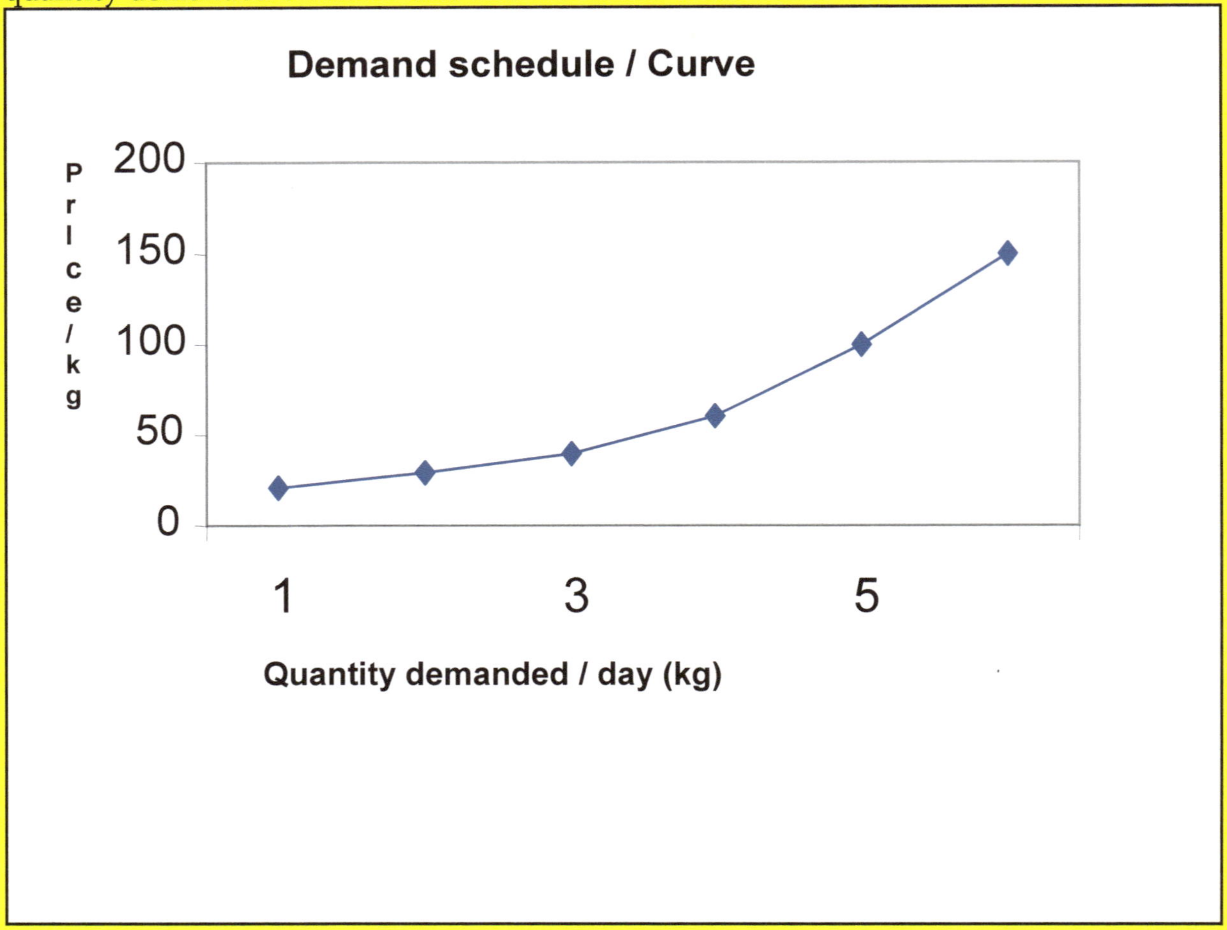

Figure 138: Demand curve

FACTORS AFFECTING DEMAND OR DETERMINANTS OF DEMAND

1. Price

The higher the price of the commodity the lower would be the quantity demanded

2. Income of consumers

If the income of buyers/consumers were high, the amount of a commodity that will be purchased would be higher than if their income was lower.

3. The population of consumers

If the number of people who want to purchase the same product were higher, then the amount of the commodity being demanded would be higher.

4. Price of alternative/substitute goods

Substitute goods such as spaghetti and macaroni are similar and therefore can easily replace each other in a diet. In this case, if the price of spaghetti goes up while that of macaroni remains the same, people will reduce the amount of spaghetti they buy and increase that of macaroni.

5. Taste and preference

This refers to what a consumer likes, if a customer likes rice, then there may still be a high demand for it even at higher prices. i.e. people tend to still buy a product they like even at higher price as long as they can afford it.

6. Price expectations or speculations

Often at times, there may a sudden increase or fall in demand of a good simply due to fears of prices. Example when budget day is approaching and there is a feeling that the price of a product may be increased on the budget day, there may be a general rush for those products before the said day.

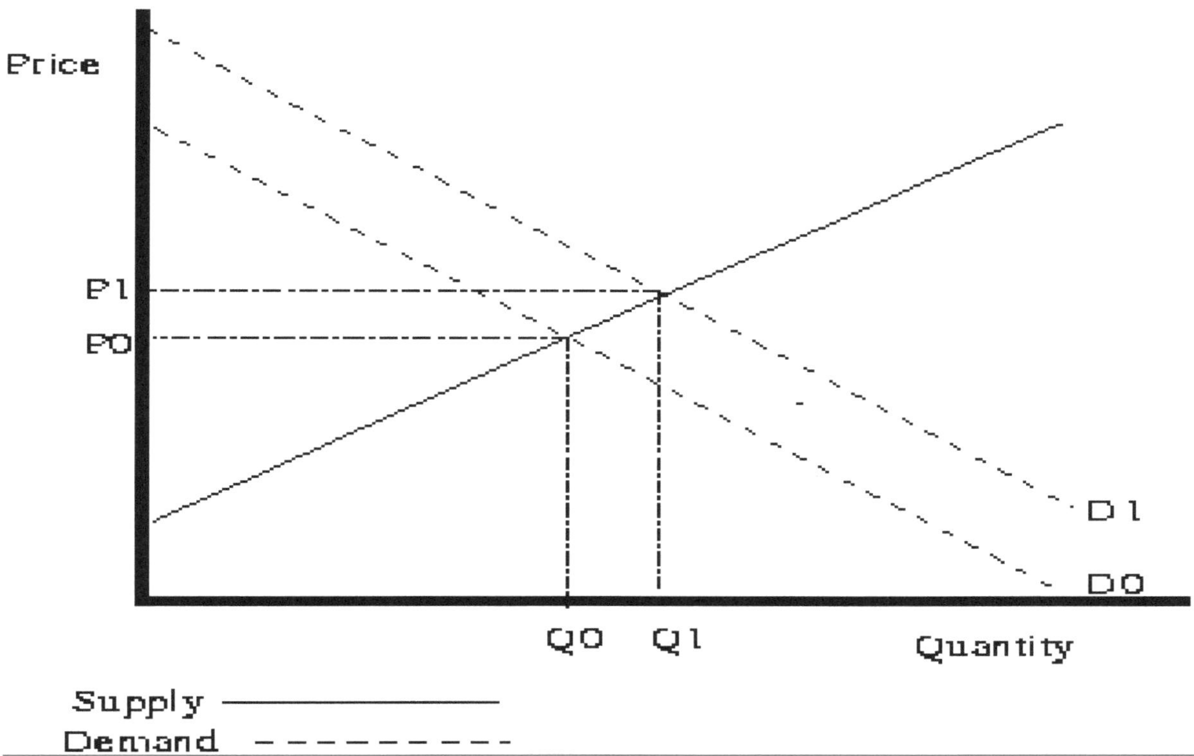

Demand CURVE: Source: http//en.Wikipeadia.org/wiki/image

OTHER MARKET FORMS

In a situation in which there are many buyers but a single monopoly supplier that can adjust the supply or price of a good at will, the monopolist will adjust the price so that his profit is maximized given the amount that is demanded at that price. This price will be higher than in a competitive market. A similar analysis using supply and demand can be applied when a good has a single buyer, a monopsony, but many sellers. Where there are both few buyers or few sellers, the theory of supply and demand cannot be applied because both decisions of the buyers and sellers are interdependent—changes in supply can affect demand and vice versa. Game theory can be used to analyze this kind of situation. (See also oligopoly.) The supply curve does not have to be linear. However, if the supply is from a profit-maximizing firm, it can be proven that supply curves are not downward sloping (i.e., if the price increases, the quantity supplied will not decrease). Supply curves from profit-maximizing firms can be vertical, horizontal or upward sloping. While it is possible for industry supply curves to be downward sloping, supply curves for individual firms are never downward sloping. (Unknown author)

DEMAND SCHEDULE

It is a tabular presentation of the quantity of a commodity that would be bought over a range of prices. The following table shows the demand of maize in Maun, Botswana.

Table 42: demand schedule

Price/kg (pula)	20	30	40	60	100	150
Quantity demanded/day (kg)	14	10	7	5	3	2

We can draw a graph from above demand schedule, having price on the Y-axis and quantity demanded on the X-axis and we obtain what is known as demand curve.

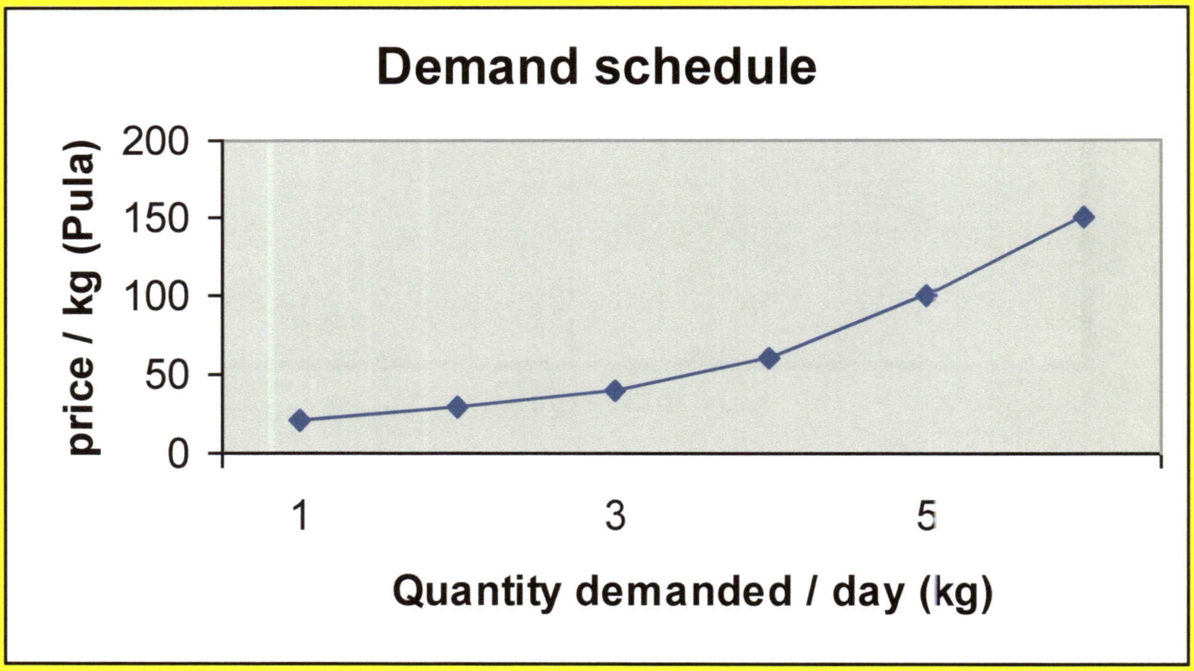

Figure 139: Demand schedule

We can also learn something more about the shape of this demand curve. The curve clearly slopes downwards from the left to right. This means that the lower the price, the greater the quantity demanded. This is the law of demand that states that: The lower the price of any commodity the greater the quantity demanded.

ELASTICITY OF DEMAND (ED).

A small change in price of a commodity makes a demand to change. Elasticity of demand shows how prices changes affect quantity of a commodity demanded. It is a rate at which the quantity of a commodity demanded changes as prices changes. Mathematically y can be stated as:

Elasticity of Demand (ED) = %change in quantity demanded / %Change in price

The demand of a commodity is elastic when it expands and contracts as a results of a change (fall or rise) in prices. On the other hand demand is inelastic when it is not affected by a small change in price.

PRICE Elasticity OF DEMAND AND SUPPLY

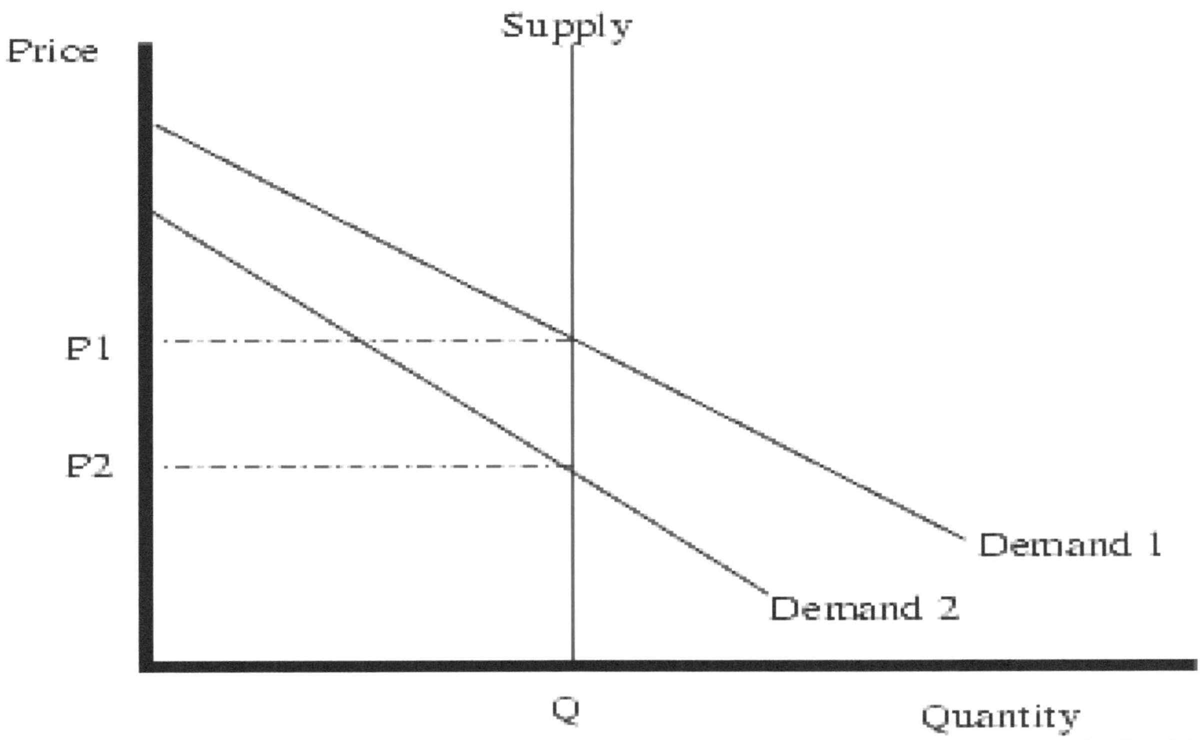

Supply curve shifts

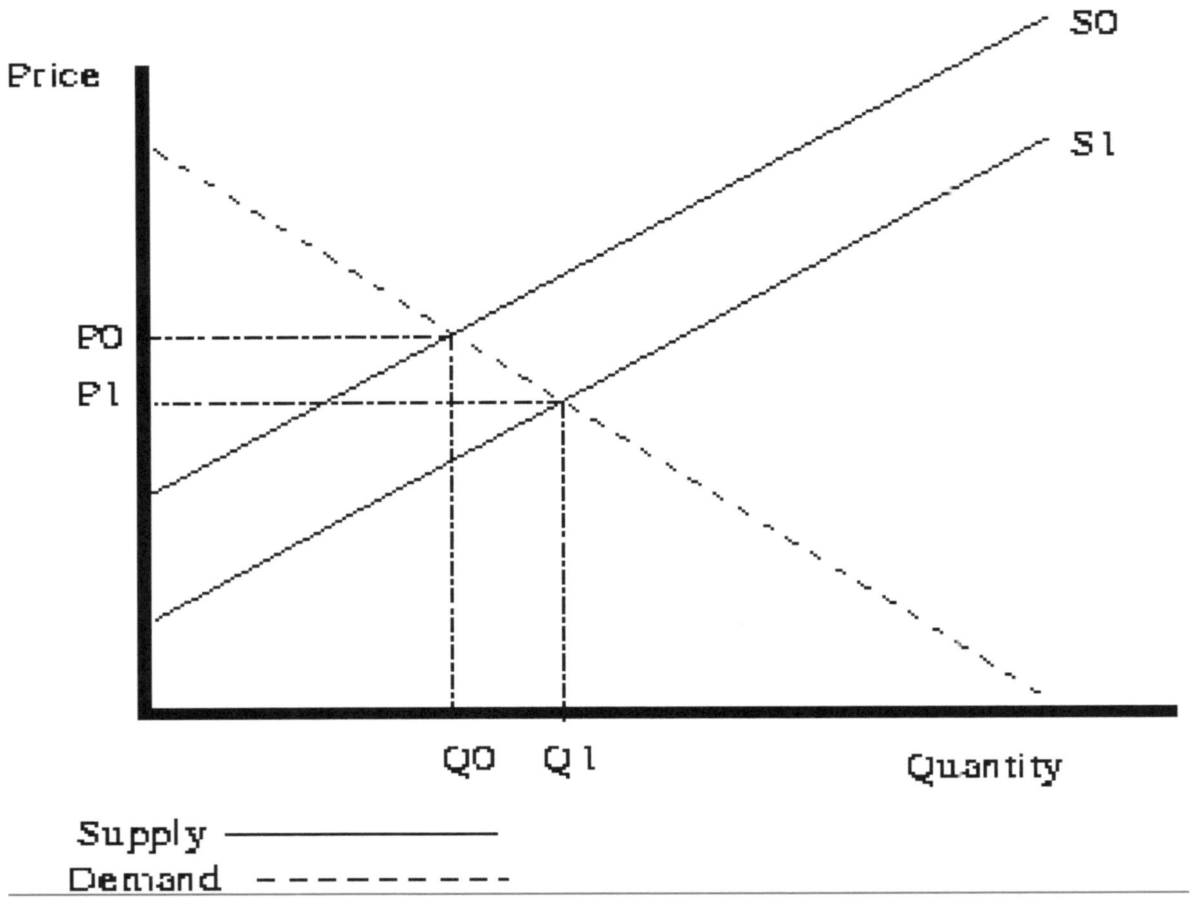

SUPPLY.

Supply is the amount of commodity which sellers offer for sale at a particular price and a given time. Supply is not the same as production. Production is the mere creation of goods and service. Just as it is possible to construct a demand schedule of the amounts of a commodity that will be offered for sale at different prices by individuals as shown below.

Table 43: The market supply schedule of onions in Nanyuki, Kenya

Price of onion offered (Birr)	150	100	70	65	50	40	30
Quantity of onion supplied (kg)	60	70	80	90	100	110	120

From the supply schedule we can plot the supply curve. The supply curve slopes upwards from left to right (this is opposite to the demand curve which slopes downwards from left to right).

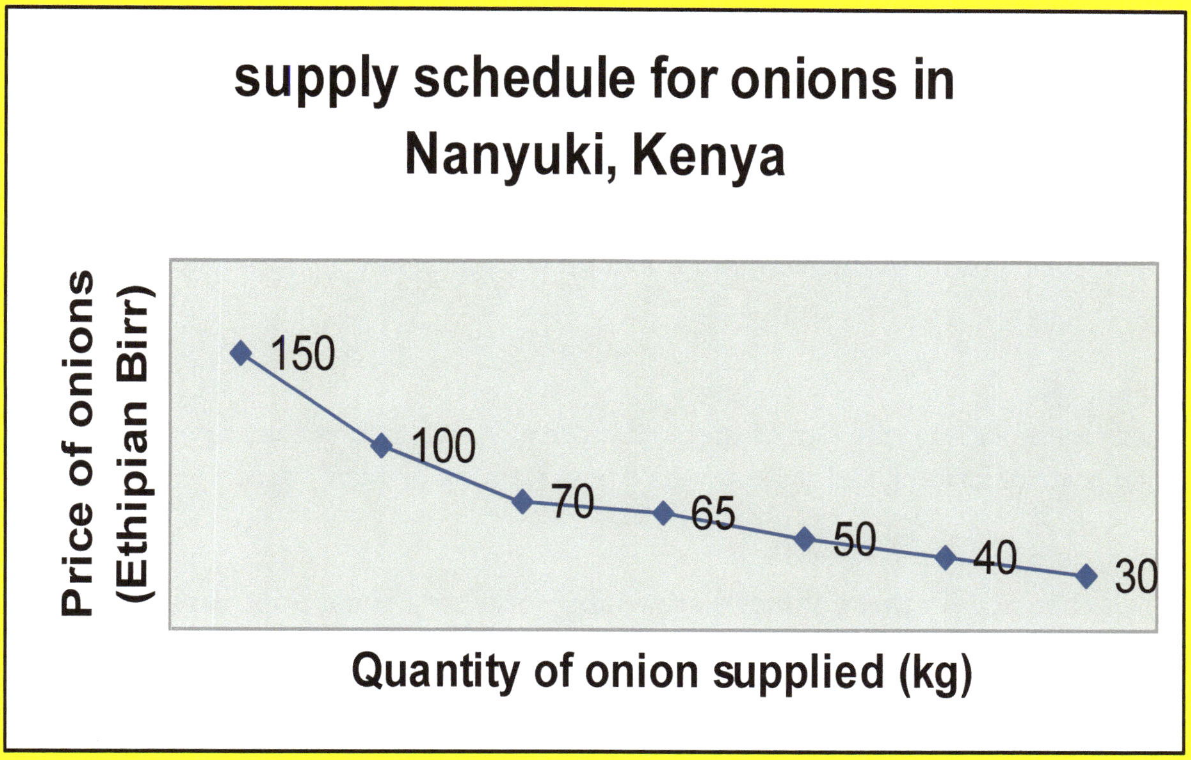

Fgure : The supply curve

From this we have the law of supply, which states that 'the higher the price of a commodity the greater the quantity which will be supplied in the market'.

FACTORS AFFECTING SUPPLY OR DETERMINANTS OF SUPPLY

1. Price of the commodity

Usually the higher the price, the greater the quantity which will be offered for sale since more profit will be made as a results of the higher price. Sellers will normally restrict supply if the price is low.

2. Prices of inputs

Producers can only supply commodity after involving inputs into production. Rise in cost of one input means an increase in the cost of production. Producers may decide to buy less amount of input which is sold at higher price thereby affecting the supply of a given commodity.

3. Taxation

Taxation on agricultural inputs may lead to increase in the cost of production. This may lower the quantity of output to be supplied.

4 Taste or preference of producers

It is quite possible for producers to decide to engage in the production of a particular commodity rather than the other, though both products are likely to yield similar profit / returns e.g. maize and sorghum.

5. The season

Supply of agricultural produce tend to fluctuate with seasons e.g. watermelons, tomatoes etc.

6.Technology

Modern technologies make work faster and more efficient such as the use of tractor leads to larger scale of farming and more produce than farming by hand tools.

Supply

Supply is the quantity that producers are willing to sell at a given price. For example, the bean grower may be willing to sell 1 million kg of potatoes if the price is $0.75 per kg and substantially more if the market price is Kshs 50 per kg. The main determinants of supply will be the market price of the good and the cost of producing it. In fact, supply curves are constructed from the firm's long-run cost schedule. Supply curves are traditionally represented as upward-sloping because of the law of diminishing (marginal) returns.

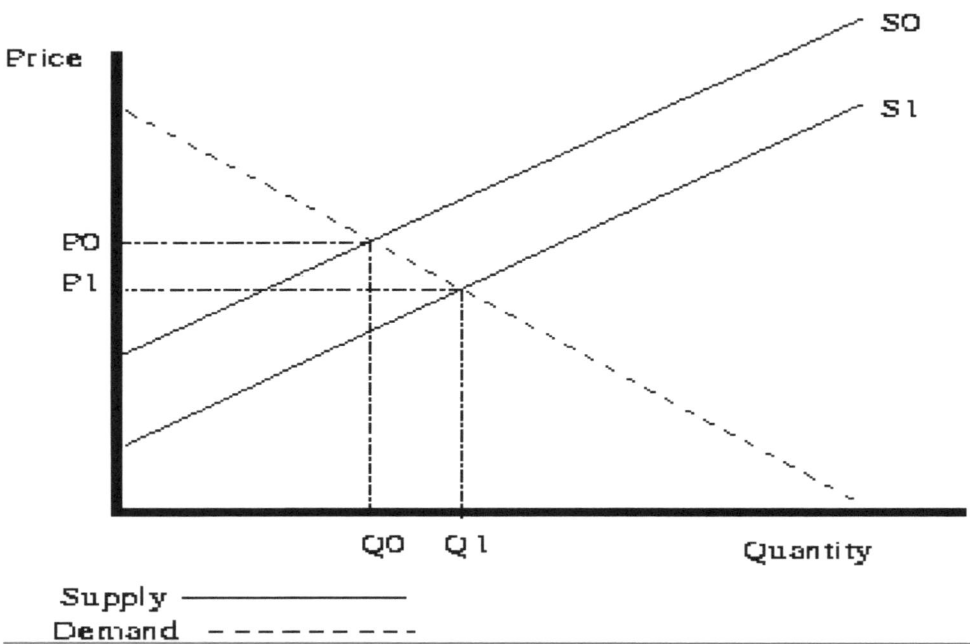

Supply curve shifts Source: http//en.Wikipeadia.org/wiki/image

ELASTICITY OF SUPPLY (ES)

Elasticity of supply shows how prices changes affect quantity of a commodity supplied. It is a rate at which the quantity of a commodity supplied changes as prices changes. Mathematically supply can be stated as:

Elasticity of supply = %change in quantity supplied / %Change in price

The supply of a commodity is elastic when it expands and contracts as a results of a change (fall or rise) in prices. On the other hand supply is inelastic when it is not affected by a small change in price.

PRICE

It is a monetary value or real value of a resource, commodity or service.

EQUILIBRIUM PRICE AND MARKET PRICE

This is the price at which quantity of a commodity demanded equals to the quantity of the same commodity supplied. OR This is the price at which the market will just be cleared of the commodities. Equilibrium price depends both on supply and demand. Consider the supply and demand schedule of beef in Maun as seen in the table.

Table 44: Demand and supply schedules, illustrating the equilibrium price

Price of beef (Pula)	25	20	15	10	5	3
Quantity of beef demanded (kg)	85	95	120	140	164	195
Quantity of beef supplied (kg)	220	180	120	80	50	10

The graph below shows the equilibrium price from the above demand and supply schedule.

Figure 140: The equilibrium price (Pula)

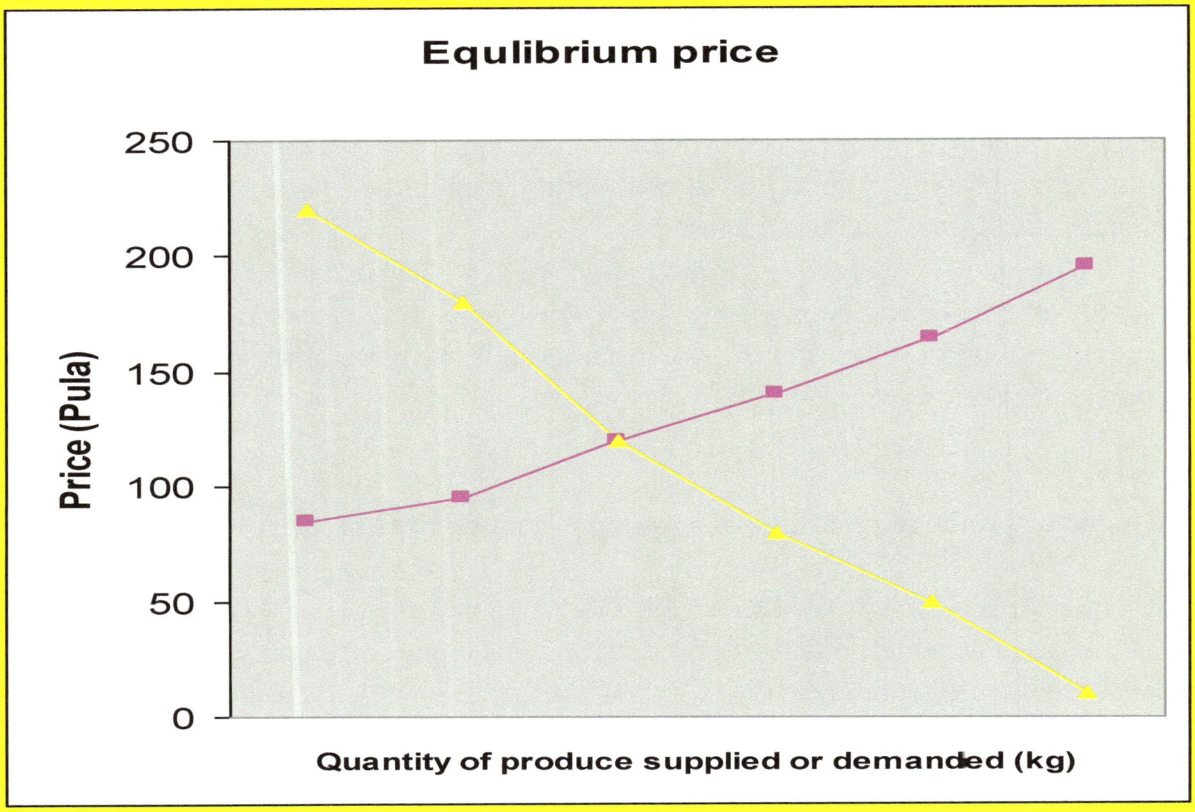

At the price of P 25, 85kg of beef will be demanded while 200kg will be supplied. This means that supply is greater than demand. This causes the price to fall. At the price of P 15, demand and supply are equal. At that price the buyers will purchase all the sellers have offered for sale. The demand of beef is 120kg and the supply is the same. The price at which the market will just be cleared without leaving unsatisfied demands, is known as equilibrium price. This price is 125 pula in this particular case.

If sellers raise the price above equilibrium price there will be more beef offered for sale than buyers will take. The sellers therefore will tend to lower the price in order to sell all the beef. Similarly, if the price falls below the equilibrium price, buyers will be eager to buy much more beef and they will tend to clear the stock, thus raising the price. There is therefore a tendency for market price to move towards the equilibrium price. However, prices will tend to fluctuate around this price level rather than actually stay there, since it is not easy for sellers and buyers to decide the equilibrium price. This is why it is also called market price.

BOX : SUPPLY AND DEMAND

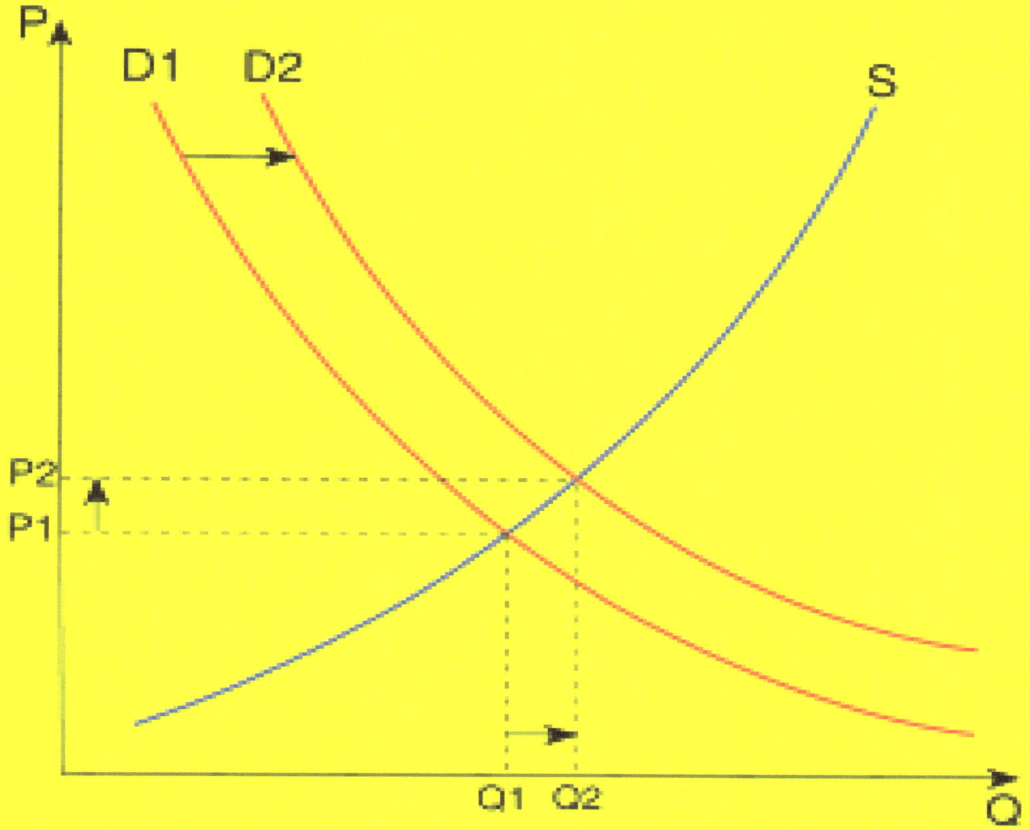

The theory of supply and demand describes how prices vary as a result of a balance between product availability at each price (supply) and the desires of those with purchasing power at each price (demand). The graph depicts an increase in demand from D1 to D2 along with the consequent increase in price and quantity required to reach a new market-clearing equilibrium point on the supply curve (S).

RISK AND UNCERTAINTY

Agricultural production is influenced by many factors, some of which may be beyond the control of the farmer and thus lead to low production. These unforeseeable and unavoidable circumstances which face farmers are called risk and uncertainty.

Risk

A risk is defined as a likelihood of a harmful event or loss to occur. Risk can be estimated from past experience. Examples of risks are Fire, theft, pests and diseases, accidents of employees, floods, frost, storms (weather).

Uncertainty

Uncertainty can be defined as an imperfect knowledge about the future events which may affect the farm business OR event / loss that, though likely to occur, cannot be predicted. Examples are price fluctuation, a farmer may observe that the price of a certain crop is quite attractive and decides to produce it. The production process on the other hand takes a very long time before the crop can be sold. By the time the crop is ready for the market the prices might have fallen and hence the farmer may get lower returns for his crop than that expected. When you plan an enterprise and start it, you cannot know what will happen by the time you come to sell your produce. You cannot be certain what the market will be like, or what price you will get. Factors classified as uncertainty include:

1. Market price- unpredictable;
2. Change in demand - at times a new or cheaper product comes at a time when the farmer is ready to sell his produce; this may affect the demand of his product;
3. Change in government policy - this may be sudden and unexpected and thus affecting farming business;
4. Breach of contracts - a farmer may sign contract to be supplied with farm inputs at times the contract may not be viable. Also the landlord may reclaim leased land from the farmer before the agreed time due to dishonesty.
5. Change in labour supply,
6. Transport reliability etc

WAYS IN WHICH FARMERS ADJUST TO RISK AND UNCERTAINTY

There are steps a farmer can take to play safe and guard against risk and uncertainty. They include:

1. Diversification

This means he can run several different enterprises at once. If one fails, the others may succeed. He does not put all his eggs in one basket

2 Choose the most certain enterprises.

If he knows, for example that there will always be a good market for broiler chickens, this will be a better enterprise for him than rabbit keeping if the local people are not used to eating rabbits.

3. Being flexible

This means he can try to be ready to change his plans. For example, if he puts up a building for keeping pigs, it should be made so that it can be quickly adapted for keeping chickens if necessary. Or he can grow annual crops like maize or potatoes instead of perennial crops like tea or citrus fruits so that he can change quickly if he wants to. e.g. during low price

4. Inputs rationing (if bad weather is predicted)

Here the farmer applies less than optimum quantity of inputs so that should unfavorable production conditions lower yield or prices become reduced, the loss the farmer suffers is less than the maximum he could have suffered. e.g. If 10 units of a compound fertilizer are the most profitable in sorghum production, the farmer may only apply 8 units. If conditions for production become unfavorable, his loss will be the loss of 8 units instead of 10 units.

5 Insurance

Here farmers insure their farms by paying a small charge to insurance company so that should he suffer total loss of a product, the company will reimburse him. Normally only relatively large-scale producers take advantages of insurance.

OTHER CONCEPTS IN AGRICULTURAL ECONOMICS

1. Complementary products

These are products in which an increase in the price of one of the products causes an increase in price of other products e.g. tea and sugar

2. Supplementary products

These are products in which an increase in price of one of the products does not cause an increase or decrease of the price of the other product.

3.Depreciation of assets

Is the decrease in value of assets as they are used in production process. The depreciation of assets can be caused by wear, tear, erosion (land), rust, rot (wood)

OPPORTUNITY COST

Opportunity cost refers to the value of a resource in its alternative use. Opportunity cost is not a real cost e.g. If a farmer is having P40000 and having two alternatives, either to buy a car or to invest in agriculture, but he finally buys a car with P40000, and he does not invest in agriculture. In this case the farmer has lost the opportunity to invest P40000 in agriculture. Therefore the opportunity cost for not investing in agriculture is the P40000.

16.2 FARM MANAGEMENT

FARM RECORDS: Involves keeping details of everything connected with the farm production in suitable book or on a sheet of a paper.

Advantages of keeping records

1. Tells whether a farmer is making profit or loss.
2. Helps a farmer to get bank loans based on the farm performance
3. Helps the farmer to sell his farm-based on the recorded performance
4. Gives a history of the farm
5. To compare the performance of different enterprises.

Inputs

Those factors which have to be applied in a production process in order to produce a product e.g. maize production-seed. Fertilizers, pesticides etc inputs are classified as Fixed and variables inputs.

Fixed inputs

Items /factors whose quantity does not change with the level of production, OR Factors which are not diminished (used up) in the production process but they are affected by depreciation i.e. wearing/decreasing in their value slowly as they are being used e.g. buildings, machinery, tractors and their spares, roads, irrigation furrow, dams, boreholes, breeding stock, Dairy cows, permanent/regular labour etc.

Variable inputs:

Factors whose quantity changes during production, OR Factors which are diminished (used up) during the production process. They include all things like fertilizer, seed, chemicals, foodstuff, chemicals, fuel, causal labour, stock purchase (e.g. broiler chicks).

Fixed costs (F.C)

These are costs incurred in purchasing fixed inputs. Example of fixed costs are cost of purchasing fixed inputs such as tractors and their spares, machinery, paying permanent labour, depreciation etc

Variable costs (V.C)

These are costs incurred in purchasing variable inputs e.g. seed, fertilizer, chemicals causal labour etc.

Total Cost (TC): Is the sum of fixed and variable costs.

Table 45: The total, marginal and average returns from a variable input (fertilizer)

Units of fertilizer	1	2	3	4	5	6	7
Total crop yield	1	1.9	2.6	2.9	3	2.9	2.7
Marginal crop yield		0.9	0.7	0.3	0.1	0.1	0.2
Marginal cost		5	5	5	5	5	5
Average rate of return		15	19	19.5	17.4	15	12.4
Cumulative variable cost		5	10	15	20	25	30
Mrginal returns			27	21	9	3	-3
Profit or loss		15	37	53	57	55	47
Units of fertilizer	1	2	3	4	5	6	7
Average rate of return	0	15	19	19.5	17.4	15	12.4
Cumulative variable cost	0	5	10	15	20	25	30
Marginal return	0	0	27	21	9	3	-3
Total rate of return	0	15	37	53	57	55	47

LAW OF DIMINISHING RETURNS.

Diminishing mens marginal, added or extra. It is a concept used to decide the optial level of input for a maximum output. In ny biological system, such as an agricultural activity, there is a replationship between inputs and outputs. Where there is a fixed level of a resource, such as land, output can be increased only by adding more variable inputs to the fixed resource. Fertilizers, labour, seeds etc are such variable inpus that can be combined as output. Generall, for any of these inputs, as the level of eg fertilizer input increases, the total returns (yield) also increase at first, but ultimately, decline to a point where no extra return is obtained for an extra fertilizer. This is the law of diminishi returns. The law of diminishing returns states that if the inputs of one factor are increased while the use of other factors is held constant then the marginal (and average) return per unit of the variable factor wirll eventually decline. This is illustrated below.

Figure 141: The diminishing returns

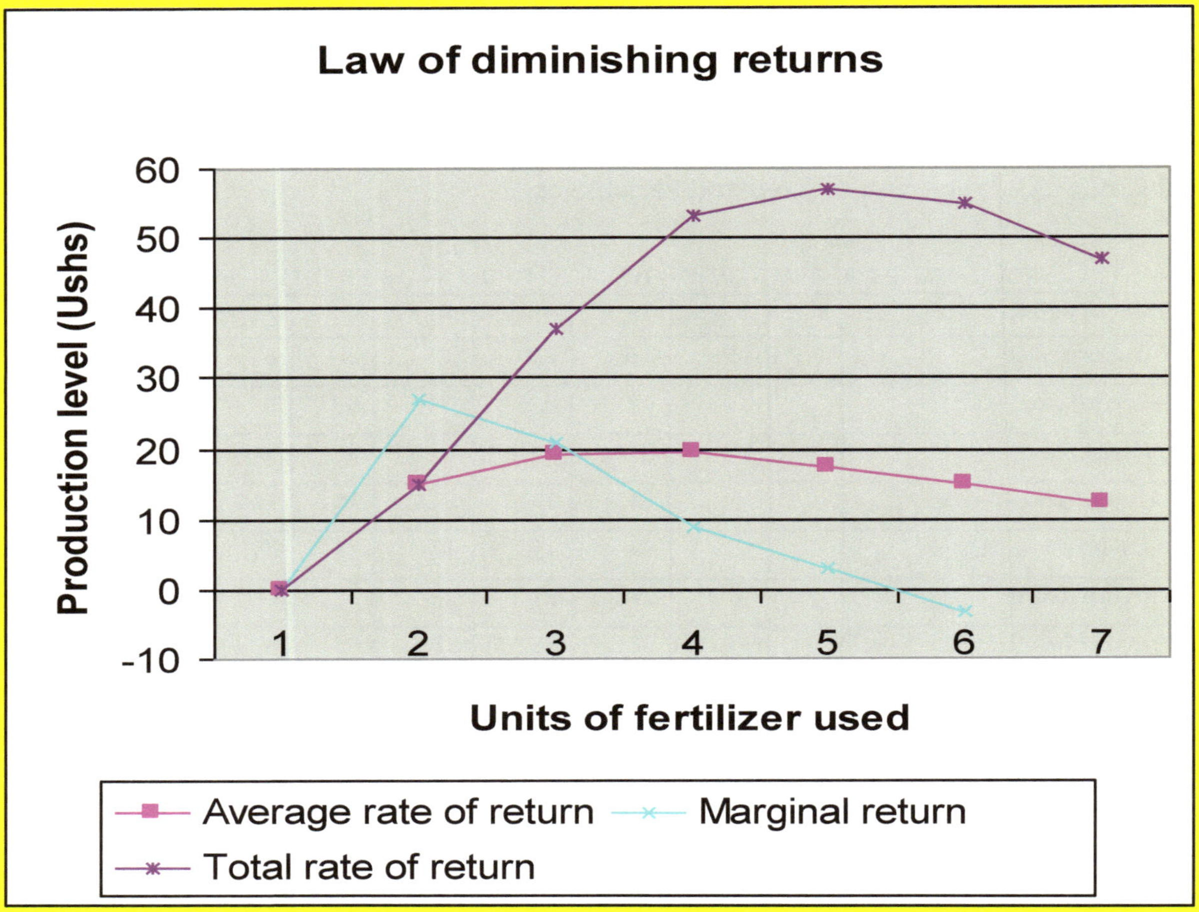

The graph illustrates that at 4 level of fertilizer use, the returns are maximized. However, some extra fertilizer can still be added to give profitable income, albeit not maximum. This extra in between 4 and 4.5 level, register reasonable but reducing profit level. This range of fertilizer application would be reasonable. However, from the graph above, the best level of fertilizer use is 4.5 kg, when the marginal profit is zero. Below this level, the net profit declines (marginal rtuns to fertilizer become negative).

Conclusions:

The point of maximum profit is at 4 units of fertilizer (Ushs 57). If 5 units are used the extra cost is ushs5 but the marginal return is only Ushs, giving a toal profit of only Ushs

55. Thus the point of maximum profit has been passed. Marginal returns are les than the marginal costs at 5 units of fertilizer, with overheads remaining constant.
Because inputs hav a cost, it is only very rarely that the point of maximum physical production (ie at 5 units of fertilizer) is the point of maximum profitable production (4 units), but the tqo may be quite close.
People ususlly mistale or wrongly equate maximum yilds with maximum profits, but the principle of profit maximization shows it is roneous.

TYPES OF FARM RECORDS

1. PRODUCTION RECORDS

Involves keeping all the details of the production of an enterprise of a farm business (i.e. records of crop and livestock production details). They include inputs (seeds, fertilizer, chemicals), yield etc. Examples of production records are farm diary, livestock records and crop records.

CROP RECORD: MAIZE PRODUCTION

2.FINANCIAL RECORDS

BUDGET ACCOUNT

A Budget is a plan that shows expenditure and income on the farm. It estimates the income to be received / generated over a production period and the expenses incurred in producing the income

BALANCE SHEET

It is a statement that shows the financial position of the business at a specific time. It includes assets, liabilities, and the net capital (net worth).

TERMINOLOGIES USED ON THE BALANCE SHEET STATEMENT

Assets:

Assets are valuable properties in the business i.e. properties owned by individual or a business (can be converted into cash)

Long term /fixed assets:

Valuable items which can be used in continuing basis in the production process. Examples are Machinery, land, buildings breeding stock, plantation crops.

Current assets:

These are valuable items which are easily converted into cash e.g. raw materials, finished goods account receivable, cash in hand, cash in bank, also trading livestock and crops which can be sold out within short term

Liabilities:

Amounts of money the business is legally bound to pay i.e. what the business owes to other people

Current liabilities:

Amount payable within a short period of time i.e. within one year e.g. bank draft, account payable, short-term loans, interest, tax payable

Long term liabilities:

Amount payable after more than one year, bonds, mortgage, and long-term bank

Interpretation of the balance sheet

1. If Assets > liabilities, the business is solvent
2. If Liabilities > assets, the business is regarded as bankrupt or insolvent.

OTHER TYPES OF RECORDS:

Main physical records: May include farm map including drainage, electric systems etc

Inventory record: This is a detailed list of assets of the farm with full description and value of each item stated

FARM INVENTORY

GROSS MARGIN

Amount of money that is obtained when all variable costs have been subtracted from the total returns/revenue of the enterprise.

TR -VC = GM.

Profit (P), on the other hand is the value obtained after subtracting all costs i.e. both variable costs and fixed costs.

{TR- (VC+FC)=P} OR {TR-TC=P}

16.4 AGRIBUSINESS AND ENTREPRENEURSHIP

AGRIBUSINESS

Definition:

Skills needed to start an agribusiness:

Resources needed to start an agribusiness:

FORMS OF AGRICULTURAL BUSINESS ORGANIZATION

Various forms of business organizations exist. These are: cooperatives sole proprietorship and partnership businesses.

Sole proprietorship is a business that is owned by one individual. All decisions on farm management are made by the farmer. The owner is responsible for losses, taxes and other business liabilities and he also receives all the profits the business earns. Such business terminates when the owner dies.

Partnership is a business that is owned by two or more people called partners.
All partners make decisions, share profits and they are equally responsible for losses, taxes, and any business liabilities. It is usual to have partnership agreement in a written form and that the management decisions are made according to this written document. Partnership may be terminated by the agreement, liquidation or death of the partners.

Co-operatives are registered organizations of people who decide to work together for mutual economic benefits. Therefore co-operatives consist of groups of people who join together to do something they cannot do successfully as individuals

BENEFITS OF COOPERATIVES

1. Easy to get bank loans
2. All costs and profits are shared among members
3. Low price of agricultural inputs (bought in large quantity)
4. Easy access to agricultural information
5. Storage, processing and transportation can be easily organized

ACTIVITY.

Choose any agricultural business/project and list

i) Skills required to run such a business

ii) Required resources

iii) Design organizational structure of the business

SUPPLEMENTARY GOODS

Are goods that are substitutes, i.e. can be used in place of each other. Eg coffee and tea; magwinya (Mandazi) and bread; beef and mutton; cassava and sweet potatoes; chomolia and spinach; butter and margarine etc. Thus if price of one such a good rises, the demand for the other good (substitute) increases

COMPLEMENTARY GOODS

Goods are complements if people tend to use them together such as camera and film, car and petrol (fuel), petrol and oil, etc. Thus when price of one such a product rises, the demand for its complement decreases. (i.e., their demand increases or decreases in the same direction). These goods go together in that they are jointly demanded (and in some cases, jointly produced as in the case of milk and butter, wheat floor and wheat bran etc.

MODULE 17: AGRICULTURAL ENGINEERING

17.1 INTRODUCTION TO AGRICULTURAL ENGINEERING

Examples of areas of application: Agricultural engineering is applied in the following areas:

Irrigation
Drainage
Fencing
Farm roads and buildings
Animal handling facilities
Farm water supplies
Farm tools, implements and machinery

17.2 IRRIGATION

Irrigation is the artificial application of water to plants to supplement rainfall.

EFFECTS OF IRRIGATION ON CROP YIELD AND QUALITY

It improves yield, as it reduces moisture stress during the life of the crop thus avoiding interruptions at the most needy times of the crop e.g. grain filling stage.
Improves yield, as the quantity applied can be matched with the demands of the crop at different stages
Improves quality as the stresses associated with deficits are overcome;
Keeps the quality of leaves intact (as they are needed in fresh and succulent state)
Lengthens the growing seasons, as it possible to produce even during drought;

IRRIGATION METHODS

These may be surface, subsurface or overhead. Examples of these are:
Surface (a) Flood and (b) Furrow
Subsurface e.g. drip, trickle or weep;
Overhead e.g. sprinkler, mist sprayer,

SURFACE (FLOOD AND FURROW)

Under this method, water is poured onto the surface of a largely flat soil. The water then seeps into the soil (if the soil is very flat), as is the case of rice irrigation. Water loving crops such as paddy rice are irrigated in this way. The method is called flooding.

However, in some cases, the surface is not exactly flat. In such cases, water may be poured from a raised side, and it flows gradually by gravity to the lower sections. This method can cause soil erosion. Therefore the soils should not be too sloppy; it only applies in flat to slightly sloping soils. In some cases, water is let to flow in surface trenches called furrows, laid at intervals. Water will seep into the soil and reach roots of crops as it flows gently through the furrows.

Figure 142: irrigation in West Kano scheme

In some cases, the furrows may be line, and water must be released from them at intervals by use of sluice gates. The flow can be measured, or can be determined from experience.

SUBSURFACE E.G. DRIP, TRICKLE OR WEEP;

Under this method, perforated surface or subsurface pipes are laid and water comes out through the small holes. The holes are put at exactly where there is a crop, thus leading to accurate delivery. Small drops fall onto the soil, thus making it to be called weep, drip or trickle. It is advantageous because it conserves moisture.

OVERHEAD IRRIGATION: SPRINKLER

Water is pumped from a river or a dam and this is supplied to the crop by pipes laid across the land. Jets of water are produced by the sprinkler. The sprinkler turns around gradually so that a full circle is irrigated. The mechanism causing the sprinkler to turn around involves a spring on the top and a swing arm which is bounced off by the jet of water coming out of the nozzle. The watering can is a simple form of sprinkler irrigation.

Figure 143: sprinkler irrigation

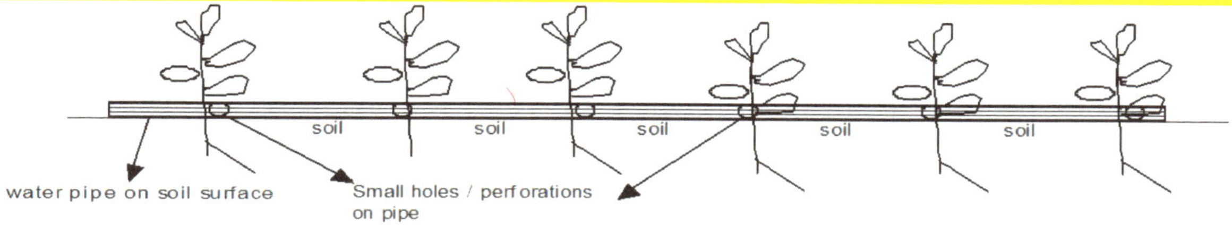

Above: Trickle, Drip or weep irrigation

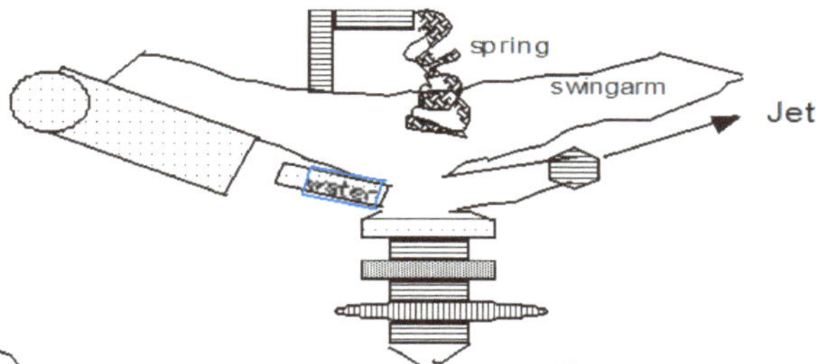

LEFT: SPRINKLER OR OVERHEAD IRRIGATION

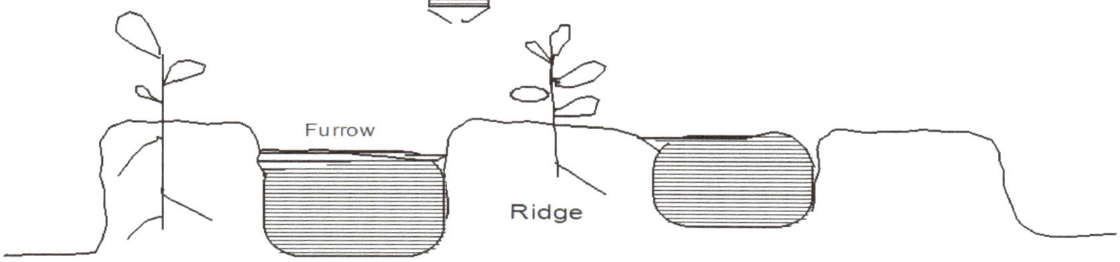

Furrow irrigation

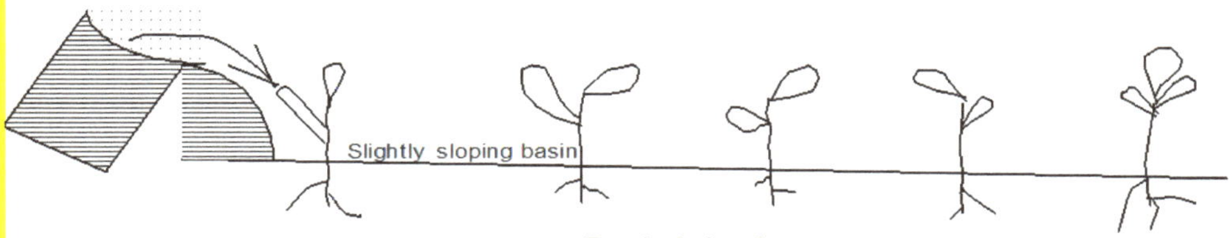

Basin Irrigation

The last 2 diagrams show furrow and basin irrigation. Both are classifies as surface irrigation.

OVERHEAD MIST SPRAYER:

This is a method requiring elaborate capital outlay, and should be used only in very high value crops Such as flowers etc. A pipe is laid horizontally overhead, e.g. 2-4 m above the ground, and at intervals, vertical pipes about 50 cm above the crop height are placed with nozzles at the end. When water is open, it flows into the overhead pipe, then into the vertical pipes, and to the rotating nozzles that then spray the water above the crop in very fine sprays. This creates a misty atmosphere all round. The system can be set so that water comes out either at regular intervals, or after the atmospheric moisture level goes below some level. This method is convenient in ornamental crops especially cut flowers as is used at SANITANS, Botswana. It is expensive, but very effective. It also mimics natural rainfall.

BENEFITS OF OVERHEAD IRRIGATION

The exact amount of water to be used can be calculated and applied;

It can be automated so that it goes on automatically when moisture content goes below some set level. This process can be triggered by a sensor inserted into the soil.

It is very successful on sloping or even flat land.

The water is evenly distributed, especially when well timed so that it is on when there is no wind.

It can be used all year round;

Mimics natural rainfall;

DISADVANTAGES OF OVERHEAD IRRIGATION

It is expensive

Tall growing crops may not be adequately reached by spray.

Some fungal diseases invade due to the moist conditions;

Some weak plants may be uprooted by the water;

Roots of some crops may be exposed.

May displace soil from plant roots

Need skills to lay and manage;

May distribute water unevenly when there is wind.

16.3 DRAINAGE

Definition:

Drainage is a system or process of removing excess water from the soil surface and upper sub-surface by artificial means. It is done while retaining adequate reserves in the soil to carry the crop through dry weather.

ADVANTAGES OF DRAINAGE

Adequate / proper drainage reduces swamps thereby reducing the habitat of livestock parasites such as liver flukes.

Likely to reduce insect vectors of diseases which often inflict man (in this case, farm labourers and management) e.g. malaria, thus enhancing their productivity;

Increases soil aeration

Increases acreage of arable land (e.g. may upgrade land of class V to class I or II)

Reduces incidences of fungal diseases, especially if they are soil borne and reside in stagnant water;
Raises soil temperature - which is essential for proper microbial function; This may enhance the decomposition of organic matter in the soil, thereby improving the humus content and physical property of the soil;
Reduces chances of some water-related livestock diseases such as foot rot, and foot and mouth diseases, etc
Renders the land usable for a wider range of purposes.

DISADVANTAGES OF DRAINAGE
Removes leached soil nutrients with the drained water;
Open ditches occupy extra land that could have been occupied by crops;
Open ditches interfere with farm operations e.g. machinery;
Expensive and add cost of production;
If not well controlled, may lead to formation of gullies or donga resulting from erosion;
May cause ecological imbalance by disturbing some water-loving flora and fauna which reside in the swamps. This is because drainage involves destroying an ecosystem preferred by specific types of flora and fauna, which then may end up being extinct.

EFFECTS OF WATER LOGGED SOILS ON CROPS
Water logged soils generally have the entire void space filled by water. This means that out of the total 25% space normally meant to be under air and soil moisture, the micro pores are also filled with water. This means that:

There is no gaseous exchange by the roots of crops;

Germination of crops is stopped, as oxygen required for the process does not exist in water logged soils;

May increase relative humidity, leading to increase in fungal diseases to crops;

Soil nutrients are in their reduced form, e.g. nitrogen would tend to exist in the ammonium form, which also easily releases ammonia gas- a loss of soil nutrients;

The nutrients in the soil are made less dilute, thus interfering with the crops' ability to absorb them;

Water logging reduces soil temperatures thus reducing the growth rate of crops;

Increases the numbers of anaerobic microorganisms, thereby causing an ecological imbalance in the soil. This can change the soil characteristics, and affect nutrient flow.

Creates anaerobic conditions that make some nutrients exist in insoluble and unavailable form. This can cause nutrient deficiency in crops;

Favours the growth of water loving crops such as rice;

Makes soil difficult to work on, thus suppressing crop production (especially if the soils are the difficult types such as the vertisols common in Kano plains);

Evaporation from the surface waters may reduce atmospheric temperatures due to the moisture laden atmosphere;

Encourages the growth of waterweeds such as Salvinia molesta, and water hyacinth.

Below is a picture of West Kano Irrigation scheme being cultivated under water (rotovated).

Figure 144: Water-logged soil

METHODS OF DRAINING SOIL

The two main methods of draining soils are:

Open ditch surface drainage;

Underground drain pipers or tile methods;

OPEN DITCH (SURFACE DRAINAGE)

This is done using open U or V-shaped ditches or channels laid at a gradient so that water flows freely from the site downslope. These may be lined or unlined. When lined, their maintenance is easier, as they stay longer. Key maintenance requirement for the ditches include:

Weed control;

Removal of soils that may block them;

Removing stagnant water, which can act as a breeding ground for insects such as mosquitoes;

Removing any materials which may find their way into the ditch e.g. grass, plastics, tins etc.

Maintaining depth and slope to ensure uninterrupted flow.

Being based on the soil surface, the drains are likely to interfere with farm operations. However, an advantage of surface drainage is that the structures are visible, so farm work can be planned and done without them interfering- e.g. in use of ploughs. They are cheaper to establish because the do not involve heavy excavation. An illustration is as shown in the diagrams below.

UNDERGROUND DRAIN PIPES

This method involves laying clay or plastic pipes underground. These pipes have small holes or spaces that allow water to enter them and drain away. The pipes must be laid at an angle to ease flow of water through them. This method of drainage does not interfere with mechanized farming. They are more expensive to lay because of the excavation costs.

DRAINAGE METHODS

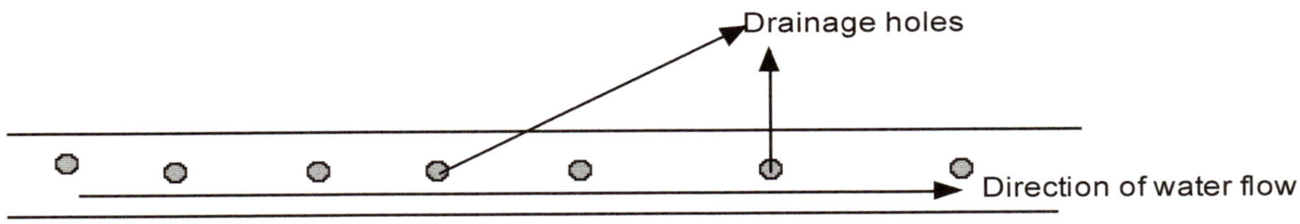

(a) Drainage by use of Underground drain pipes

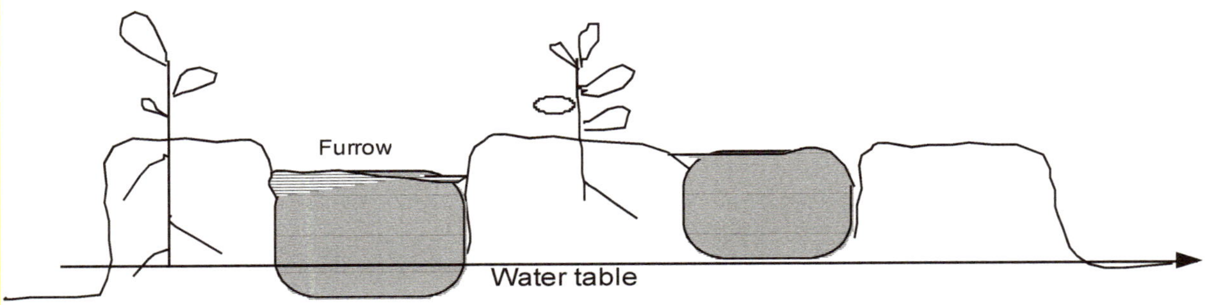

(b) Surface drainage using furrows

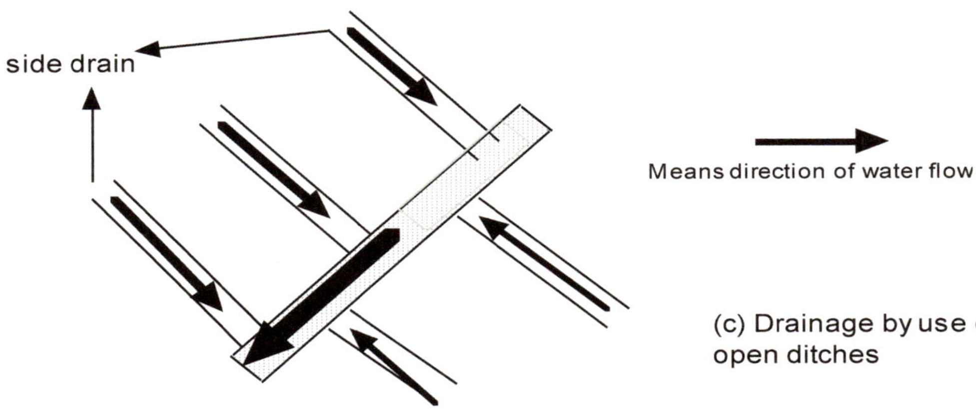

(c) Drainage by use of open ditches

17.4 FENCING

DEFINITION:

Fences are structures for controlling traffic within a farm, e.g. movement of livestock and man, as well as for marking boundaries.

The major functions of fences:

Control livestock movement

Controls man's movement; i.e. avoids trespassing and stepping on delicate cops and special sections of the farm e.g. flowerbeds.

Controls grazing e.g. in paddock systems;

Controls breeding;

Marks plot boundaries with neighbour, thus avoiding conflicts and disputes;

Separates different livestock and crop enterprises;

Improves beauty of the plot (especially brick, block, stone, flower and hedge fences).

MATERIALS FOR CONSTRUCTING FENCES

Generally, different materials are used for making different kinds of fences. Selection criteria depends on proposed use of fence, materials available locally, cost of materials, planned life of fence, available resources, etc. Some of the materials are: Living plants, Bush poles, Posts & rails / sown timber (Off- cuts), Plain wires, Barbed wires, Iron posts, Woven wires, Diamond mesh wires, stones, bricks and Concrete. The table below gives some of the materials, their advantages and disadvantages.

Figure 145: Left: A Corrugated Iron Fence Right: A Brick fence

Stones

Planks of wood

Cement

Fitos

Metals: Y 12's for wall reinforcement and construction of pillars

Table 46: Materials for construction of fences, their advantages and disadvantages

Materials	Advantages	Disadvantages
1. Living plants	Are cheap to maintain	Do not rot Labour required
2. Bush poles	Cheap Easy to replace	Rotting risk Termite attack
3. Posts & rails / sown timber (Off- cuts)	Make stronger fence Do not hurt animals (Smooth)	Timber is expensive Termite attack
4. Plain wires	Strong, last longer Do not hurt animals	Can rust Expensive
5. Barbed wires	Strong, last longer	Can rust Expensive Poles may rot or be eaten by termites Dangerous to animals
6. Iron posts	Strong, last longer	Can rust Expensive
Woven wires (Diamond mesh wires)	Strong, last longer	Can rust Expensive
8. Concrete , blocks or bricks	Durable (see above); very beautiful- god for landscaping	Expensive
9. Corrugated iron sheets	Durable, but can rust in a polluted environment; can be cut by thugs; can be stolen.	Expensive

The main problems associated with some of the fencing materials include:
Rotting (if wooden);
Being eaten by termites (wooden);
Rusting (metallic);
Being expensive (metallic, bricks, stones, and some wood);

To cope with some of these difficulties, the following can be done:
Choose cheaply available material and avoid expensive ones;
Treat material before use e.g. painting against rusting;
Treat with pesticide e.g. creosote or used engine oil to reduce termite attack (see below);
Use only dry timber to reduce rotting;
Use rot-proof wooden material (may be expensive).
Methods of treating fence posts
Soaking posts in creosote: A lot of posts at a time are put into a drum containing creosote and are left for three days;
Then turn the posts upside down and soak again for three days;
Put treated posts into empty drum to drain off the excess creosote.

PAINTING POSTS
Posts are painted with waterproof paint. Iron posts can also be galvanized with Aluminium coat.

Assignment:
Draw the structure of a cattle wire fence, showing all features. You must do this assignment now before you proceed to the next stage.
Describe the construction of a corner in a fence.

TYPES OF FENCES
Fences are divided into 2 major classes: Live and non- live types.
Live fences are made up of either electric wires or hedges (with live plants such as trees, shrubs etc). Below is an illustration of an electric fence.

Figure 146: An electric fence:

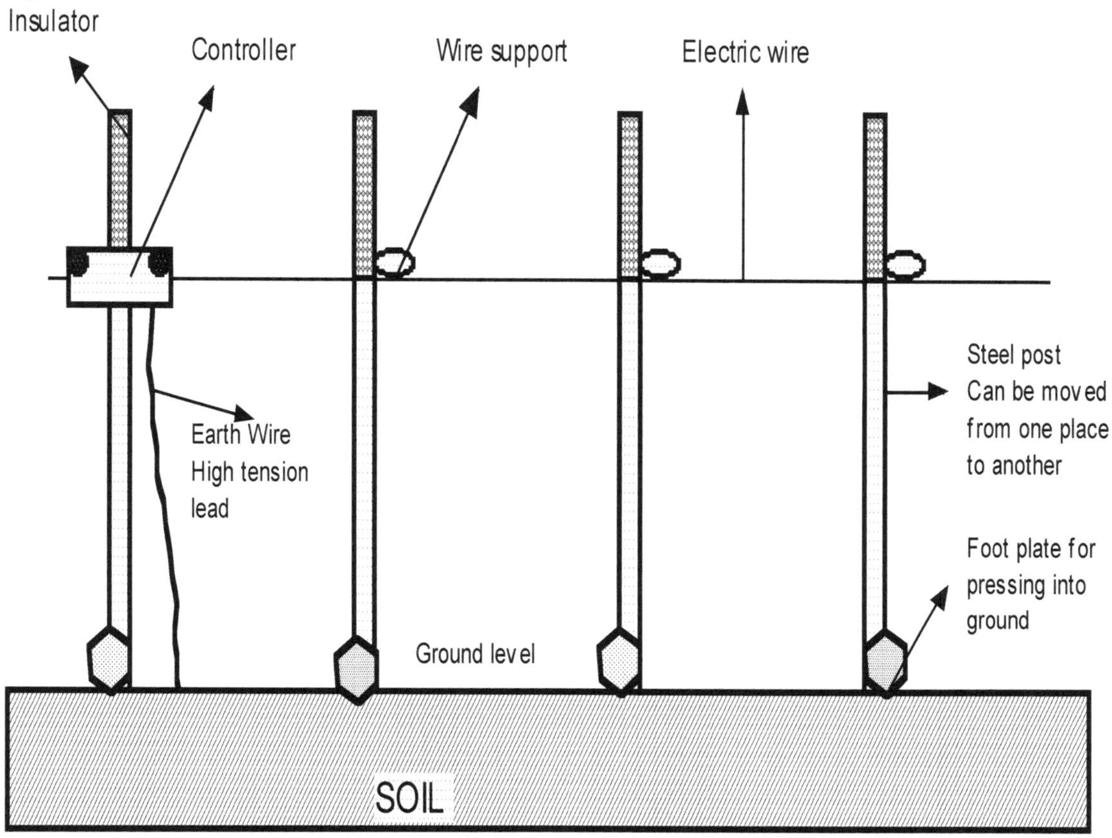

ELECTRIC (WIRE) FENCE

NON- LIVE FENCES
Brick / stone / Block fence;
Barbed wire fence;
Plane wire fence;
Post and rail fence;
Woven wire (cross- link or Chain- link) fence (see photo below);
Reinforced plane wire fence.

Figure 147: Chain link wire fence

Figure 148: Post and rail fence (wooden)

STAGES OF CONSTRUCTION OF A WIRE FENCE

Mark corner and stretch wire between the posts

Dig holes 60cm deep and place the posts

Stretch the wire between comer posts tightly to give straight lines of the fence

Support the comer posts with bracing wire and supporting posts.

Put in posts 4m apart and 50cm deep along the line of the fence. The posts should be straight.

Fix wire to the posts using u-nails (staples). Pull the wire tight using wire strainers and start with top wire and work downward.

SPECIFICATIONS OF THE BARBED WIRE FENCE

Item description	Length (cm)
Depth of fencing holes	60
Depth of post below ground	45-55
Height of post above ground	150
Distance between major posts	45-60
Distance from ground level to the first barbed wire strans	50
Distance beteen first strand and second	40
Distance between second and top strand	45

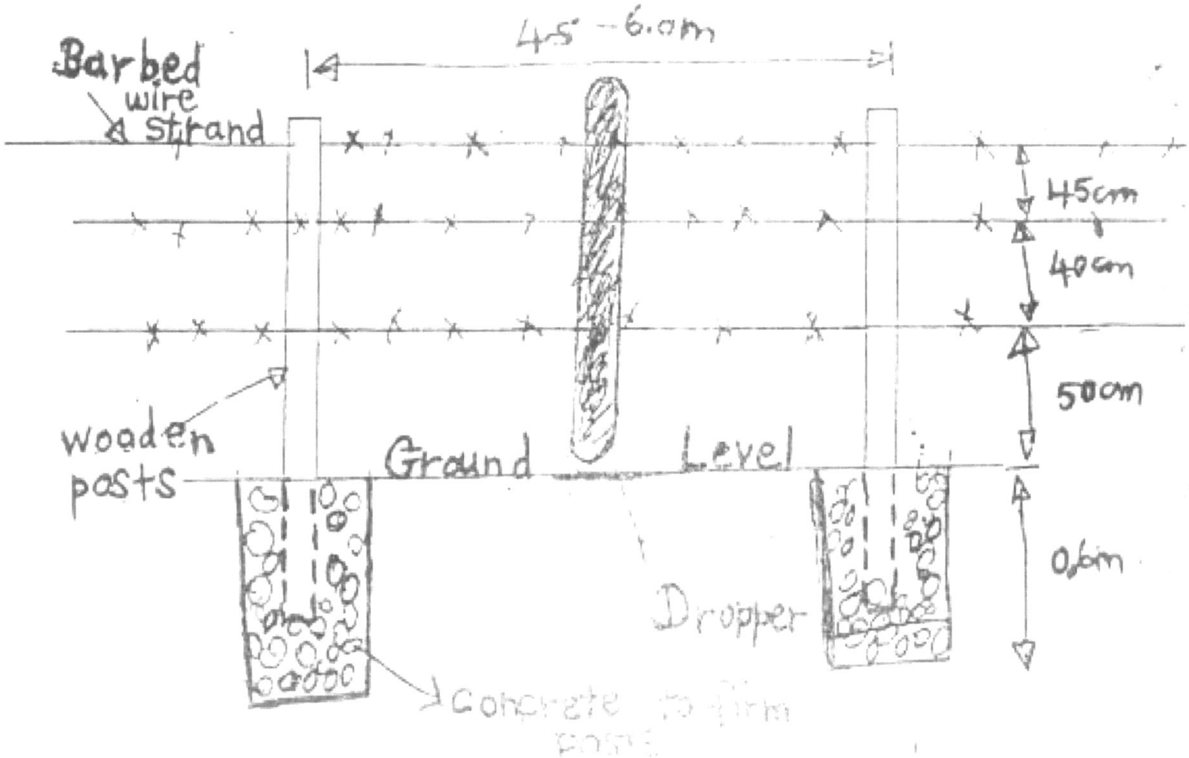

Figure 149: A Barbed wire fence

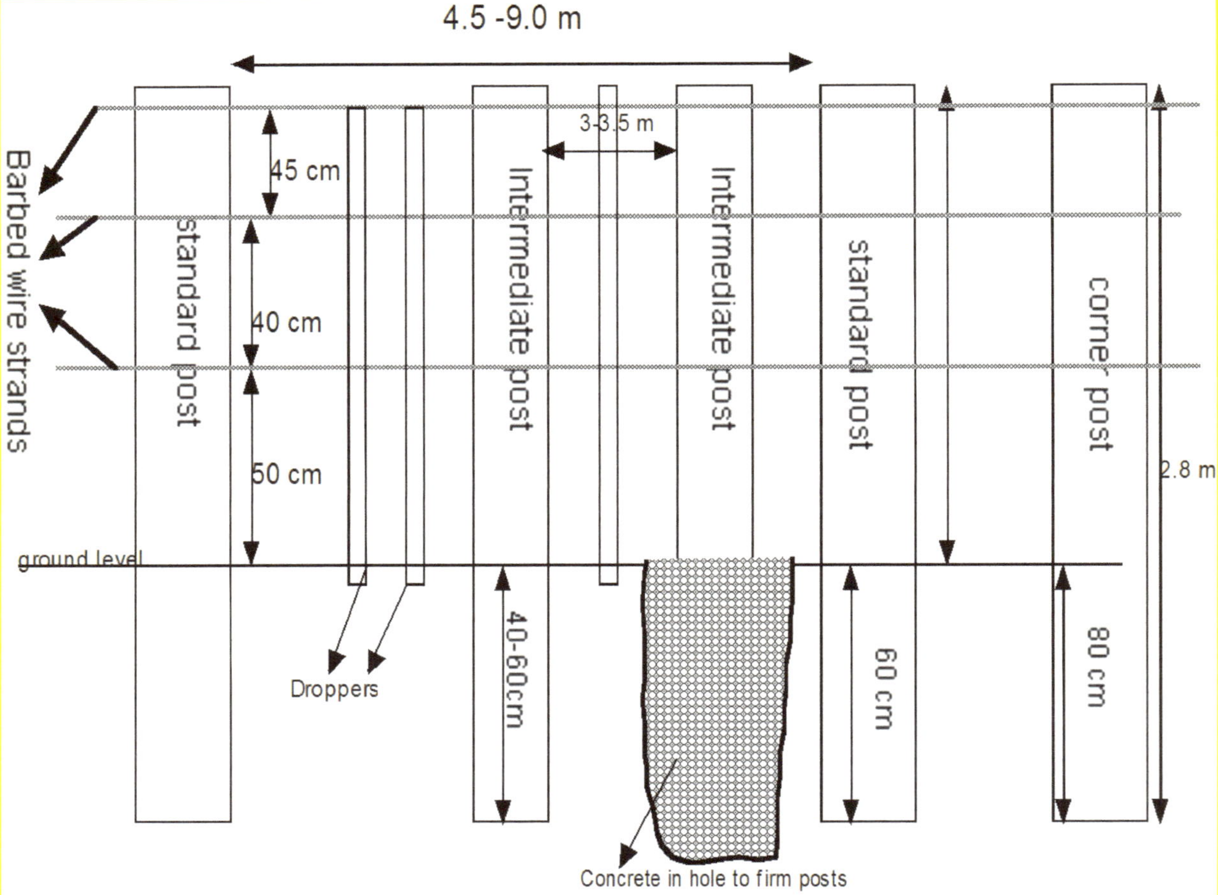

1. The dimensions shown in this type of fence are appropriate for cattle. If 3 strands are used as shown in this figure, the spaces (from down) between strands starting from the ground level should be: 50:40:45 cm (check these in the diagram).
2. If 4 strands are used, the distances should be 45:30:30:35 cm
3. For wooven wire (wirenetting or chain link), 15-30 cm of the wire should be buried in the soil to prevent livestock from passing under it, then 2 m of the wire remains above the ground.
4. When using droppers, the distance apart is not fixed. You use them according to availability and intended purpose of the fence..

GUIDELINES ON CONSTRUCTION OF A FENCE

NB: One can fit one or more middle posts (Droppers) in between the main posts. The droppers need not be fixed to the ground, but need to be firmed by fencing nails on the barbed wire.

17.5 FARM ROADS AND BUILDINGS

FARM ROADS: IMPORTANCE

Easy access of farm by people;

Vehicle access to take farm produce and / or supply farm inputs etc

Eases and controls animals movement;

FACTORS TO CONSIDER WHEN SITING A FARM ROAD:

Spacious: Adequate space for a wide road to allow animals or vehicles to go along it easily;

Soil firmness: soil should be firm to prevent wheels from sinking;

Soil drainage: needs well drained soil to avoid flooding and erosion;

Slope of the site: should be gentle to avoid erosion;

Serves busy part of farm: A road should pass through a busy part of the farm e.g. farmhouse.

STEPS IN CONSTRUCTION OF AN EARTH ROAD

Clear the vegetation and cut the soil with spade and hoes or with mechanical grader;

Remove top/soft soil with the grader and remain with subsoil that is firm enough to make a good road surface;

Add gravel and spread evenly with the grader;

Mix the gravel with loamy soil and press down hard by a roller. This material drains well because it contains some sand, but also it holds together well because it contains clay. Roads sited across hillside should be well leveled with the help of grid measurements.

Lay roadside drains: Roads should slightly have convex shape for proper drainage.

Sand for general construction

ROAD MAINTENANCE

To last longer, a road should constantly be repaired. If erosion damage starts, the road should quickly be repaired to avoid serious damage. Roadside vegetation should be regularly cleared to improve visibility, which renders the road safe from accidents. The side drain should be kept operational so that any surface flow is quickly drained. Any water that accumulates on the road for long is likely to spoil it faster. Where possible, it is useful to place a means of water passage e.g. culverts in places where interruption to water flow is likely e.g. below a road branch.

FARM BUILDINGS

Introduction: Farm buildings are structures made in a farm to serve various functions such as: farmhouse, farm store, milking parlour, tool room, farm workshop, cool house, produce handling and processing, records room etc. Many factors should be taken into account in locating farm structures, which should be part of the broader farm layout. Some of these factors are as shown below.

FACTORS TO CONSIDER WHEN CHOOSING A SITE OF A FARM BUILDING

Direction to face side facing the prevailing wind direction should have solid walls, also consider sun direction etc;

Distance from the farm: should be near enough to be reached easily;

Distance from water supply: should be as near as possible to water supply; It must take into account possibility of water distribution in future;

Drainage: Most structures need non-swampy land;

Firmness of the ground: The base of the site should be firm, avoid swampy areas;

Landscape of the site: avoid rocky, hilly and densely vegetated areas;

Existing legislation: If within a municipality, the structure should be sited in a place that does not render it unusable for the intended purpose. For instance, there may be a need for an occupation license when the construction is over. Before this is issued, the building and its site are thoroughly inspected to allay any fears of a law broken or hazards to user.

ESSENTIAL FEATURES OF A CHICKEN HOUSE

Should be spacious for birds to move about;

Well lit and well ventilated;

Dry and comfortable floor (with litter etc);

Free from predators and thieves;

Free from insect pests and parasites;

Easy to clean;

Fitted with drinkers, feeders or nest boxes (layers);

Well sited to avoid environmental problems such as extreme temperatures, rainfall, draught etc.

Figure 150: A simply built chicken house (Down- bricks and concrete; upper half- chicken wire; roof- corrugated iron sheets)

Figure 151: Roofing a farm house

Figure 152: Use of mud is possible, but it gets washed over time without regular mentenence

MAINTENANCE OF CHICKEN HOUSE

Check for roof leakage;

Check feeder and drinkers for positioning and leakage;

Check floor for holes and wet conditions;

Check doors, windows and wire mesh for any defects;

Check water tanks and water supply system for any blockage;

Check electric system for loose and defective wires;

Check walls for cracks and holes.

Keep the outside of the structure clean and clear from bushes and tall grass to avoid possibility of snakes finding their way inside.

16.6 ANIMAL HANDLING FACILITIES

CRUSHES

Importance: Used to lock up farm animals so that operation such as vaccination, artificial insemination spraying etc can be carried out easily. Below are some examples.

TYPES OF CRUSHES

These include:

Cattle crush e.g. for milking, vaccination, treatment, dehorning/ debudding etc;

Holding crush

CATTLE CRUSH

It is like a small cage that can be opened and closed. The animal is driven into the cage with front of the cage closed. When it is in, poles are put in behind the animal, so that it can be trapped and handled such as spraying without any difficulty. To release the animal, the loose poles at the front are pulled out to let it escape.

CONSTRUCTION OF A CATTLE CRUSH

The cattle crush is 9 m long and the cattle enter through a funnel that is 4m long. The inside width is 60cm, so that the cattle cannot turn around inside the crush.

CATTLE DIP

It is used for treating cattle against ticks. The bottom and walls of the dip are made of concrete to hold the fluid, a concrete footbath is also built along the entrance to reduce amount of mud taken into the main tank. The tank should be 5m long and 1.6 deep. The cattle swim to the other end of the tank where stairs are made to enable the cattle to climb out.

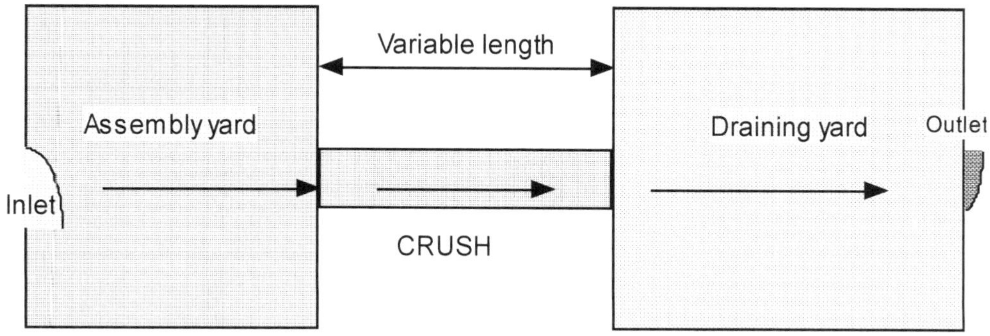

(a) A crush for holding many animals

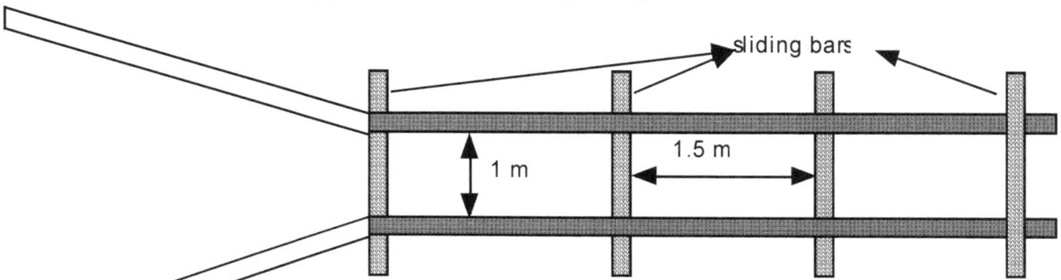

(b) A crush for holding 3 animals

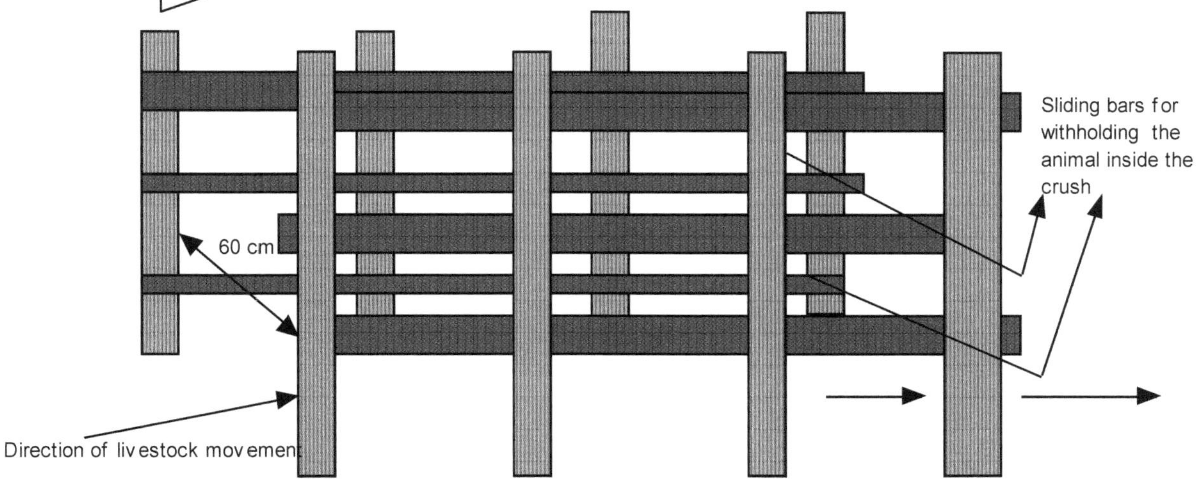

(c) A simple post and rail crush.

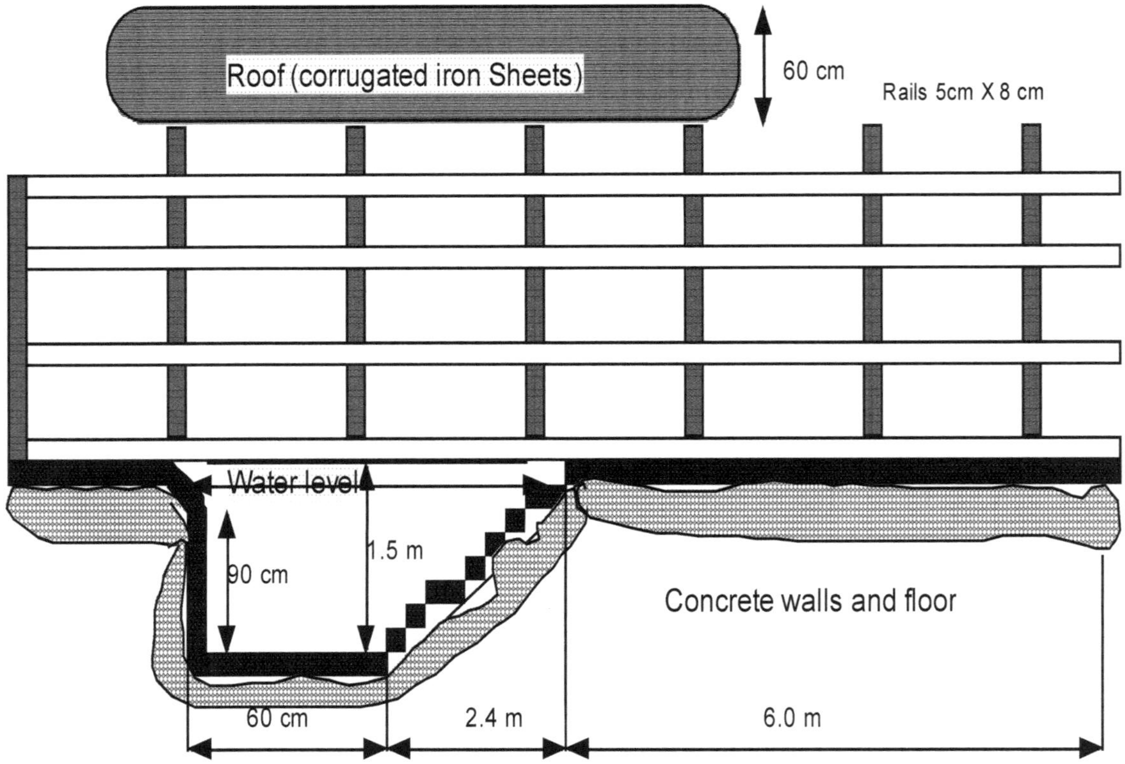

Capacity of small animals' dip = 2 -2.5 kilolitres
Capacity for cattle dip = 13-14 kilolitres of dip wash.(at least 1.6 m deep and 5 m long)

Side view of a dip

Advantages of dip tank

It can be operated by unskilled worker

Does not need pumping machinery

There is a complete chemical coverage on the immersed part of the animal body

Low risk of pasture contamination

Dip wash (chemical) can be used

Disadvantaged

Expensive to build

Small, expectant and sick animals may drown or get injured

Not economical for small scale farmer.

SPRAY RACE

Also controls ticks, it consists of enclosed corridor fitted with spray nozzles. Acaricide (tick poison) is pumped through the nozzles to spray the animal all over. See diagram below.

PARTS OF A KNAPSACK SPRAYER

Nozzle
Cap
Tank
Handle
Delivery hose
Lance handle
Trigger

MAINTENANCE OF A KNAPSACK SPRAYER

Drain the tank;

Wash thoroughly with clean water and detergent e.g. OMO;

Pump through the sprayer at least 5 litres of clean water mixed with detergent to remove traces of chemical / detergent.

Fill the tank with clean water and let it stay overnight

Wash the outer part of the sprayer

Repaint parts which have been corroded

Remove and unblock the filter and nozzle (Do not blow them; it is dangerous; also do not poke nozzles with wire)

Drain the hose and pump, then wash well

Remove hose and store separately in a clean dry place

Check the control valves frequently (Check for wear, tear, damage; replace faulty valves)

Inspect hand lance after detachment of nozzles (replace if worn out);

Grease or oil all moving parts;

Clean all metal parts with warm water and 10% solution of washing soda;

Wash all rubber parts and store in a dry place

Buy and keep spare parts of a sprayer ready before next use.

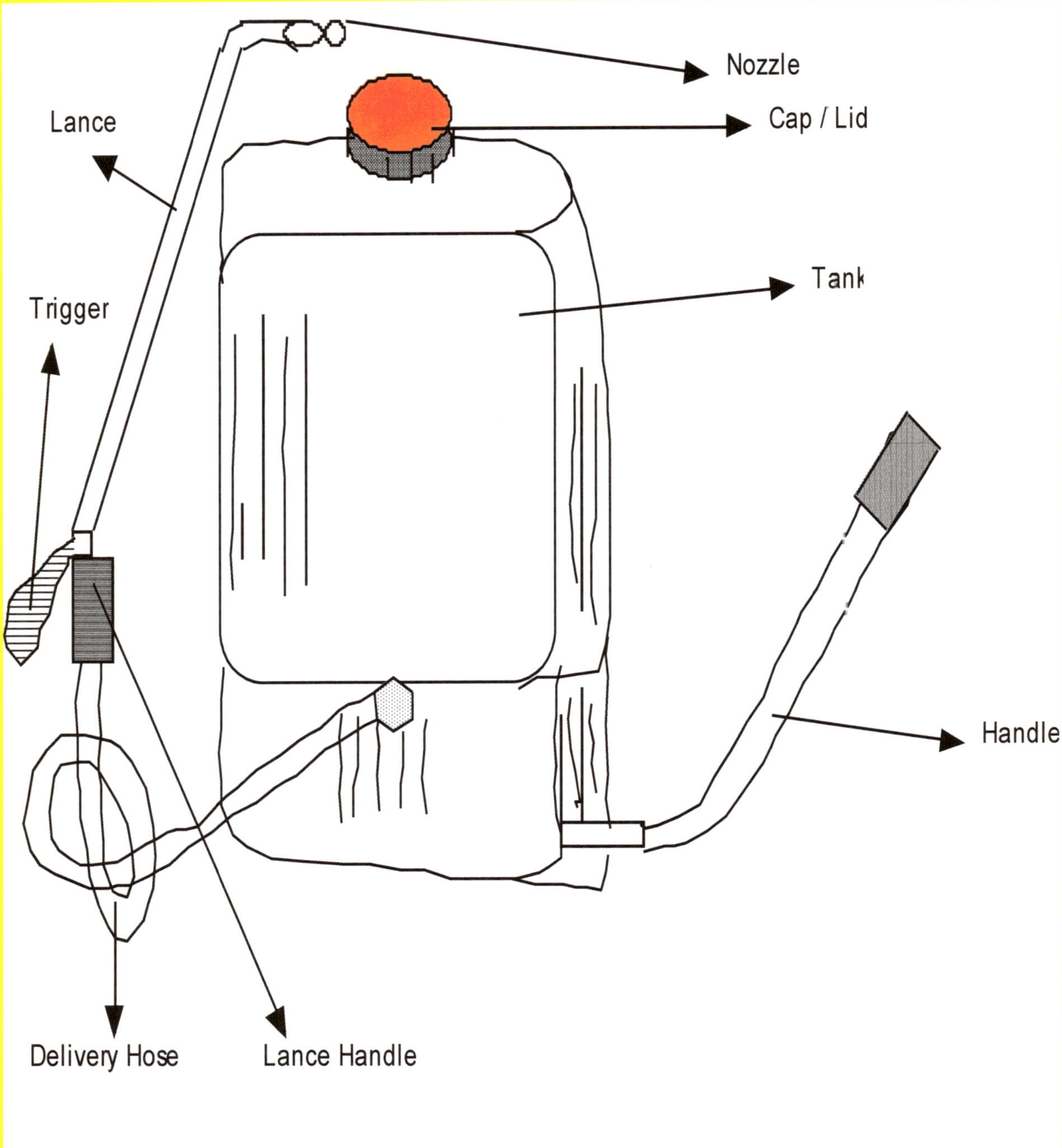

The Knapsack Sprayer.

Activity: to dismantle and reassemble a knapsack sprayer.

Fence

Animals in a fenced area

17.7 FARM WATER SUPPLIES

SOURCES OF WATER IN THE FARM

Water for farm use can be obtained from:

Shallow wells

Deep bore holes (Sediba)

Rivers

Springs

Streams

Lakes

Figure 153: Water of River Nairobi, taken at Naro Moru.
Figure 154: Water from Kajulu Hills (Spring), feeding the Kibos River.

Figure 155: Cattle drinking from a farm pond. Fig 156: Brown, silted water sources only suitable for selected farm uses

Figure 157: Kasau Borehole water (Naikonda) in Bondo District, Kenya

METHOD OF TREATING FARM WATER FOR ANIMAL USE

Store water in tanks and allow settling to become cleaner and freer from sediment and micro-organisms. This is because any suspended material settles as sediment on the bottom. Drinking water for animals can be treated in this way, and then piped to their drinking troughs. Generally, water treatment methods include: chemical disinfection using chemicals e.g. chlorine; Coagulation / clarification using chemicals such as Alum; pH adjustment using calcium compounds; settling in a sedimentation tank; exposure to sunshine (irradiation), etc. These methods can be used singly or in combination to treat water for various uses, which require water of various qualities. Below is a conventional water treatment process useful in producing water suitable for almost all uses.

TREATING WATER FOR FARM PROCESSING PURPOSES

The water is taken in from any of the sources above.

Sieved through a screen to remove bulky solids;

Left to settle in a sedimentation cylindrical tank with a conical base;

Some chemicals may be added to the water to adjust the pH;

Coagulants may be added if the water is turbid, e.g. that from a pond. This facilitates further coagulation and settling, leaving sludge at the bottom and clarified water at the top.

The clarified water is directed to the next chamber where t is treated for microbial elimination (i.e. disinfected) e.g. by use of chlorine.

The water is then pumped to a tank at a raised part o the farm from where I can flow by gravity.

The water is good for all farm activities e.g. drinking, all domestic activities, irrigation, farm produce processing / factory etc. However, it may not be good for mixing of vaccines; this needs chemical-free water that cannot interfere with activities of the microorganisms within. Care should be taken not to use water polluted by dangerous chemicals from industries e.g. heavy metals, pesticides etc. It may be useful to tap river water from upstream of such polluting factories, or else use borehole water.

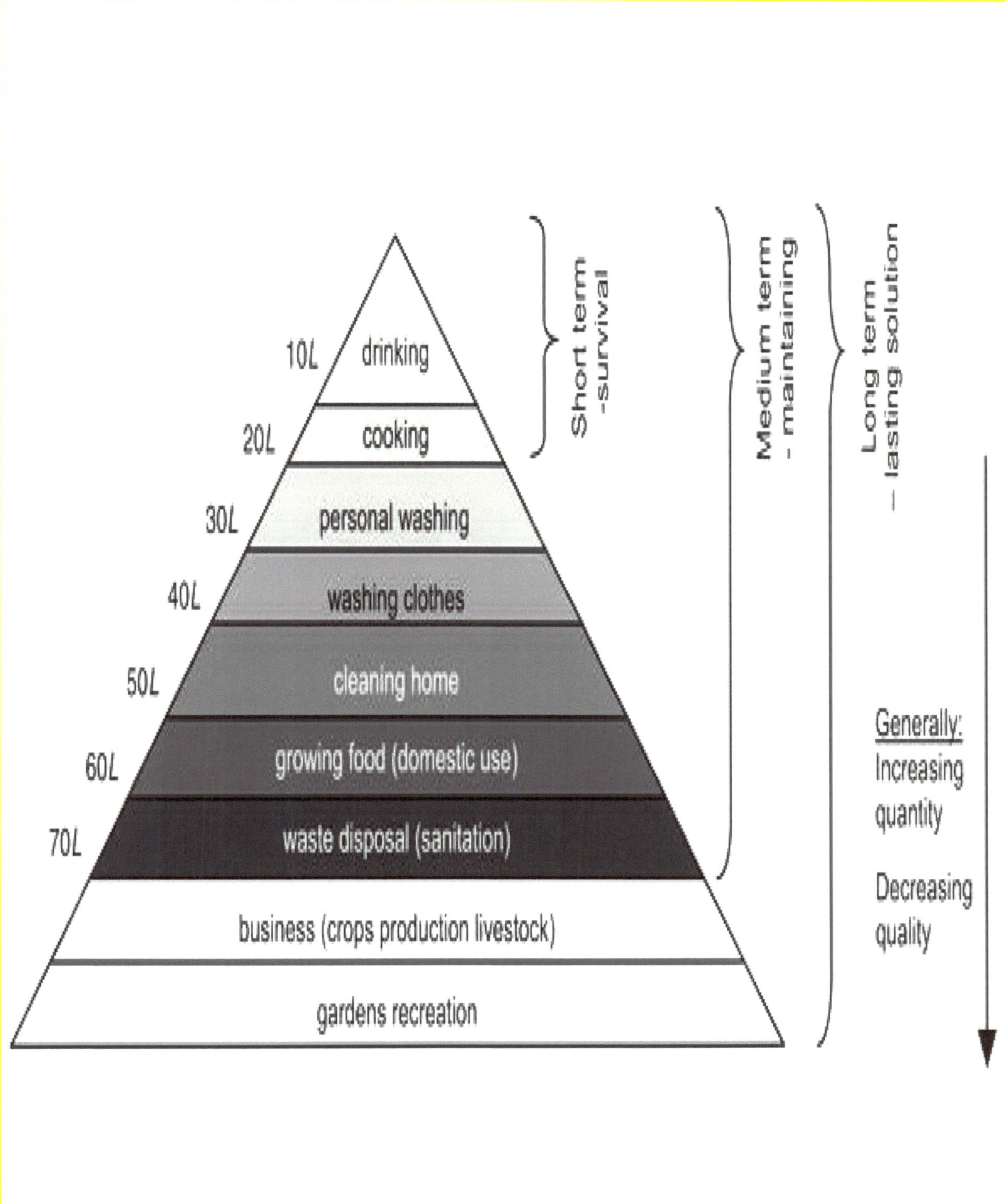

Figure 158: The water need and use hierarchy

Whatever the use, water quality is determined by: Physical quality (colour, taste, suspended solids, smell, texture etc); chemical quality (pH, dissolved chemicals, heavy metals, organic pollutants, Oxygen demand BOD or COD), or biological quality (pathogen/ feacal pollution or load as determined by the Escherichia coli load). These are what the treatment processes strive to remove, or at least to reduce to acceptable levels. Most general water uses do not require very stringent treatment. In such cases, only one selected treatment method e.g. settling may be used. Whatever treatment method is chosen depends on the quality of the water source and the intended use.

METHODS OF FARM WATER STORAGE

Use of large tanks (plastic or metal): Tanks are used to store water for the farm. Water is pumped into the tanks from nearby sources e.g. rives, or waterworks. Some tanks can be put on a raised ground e.g. on top of a roof so that it flows by gravity to where it is needed within a farm. Jojo tanks in Kenya are a common type used in this way.

Above ground water tanks are becoming popular means of water storage

Dam:

Water can be harvested from rain or surface runoff and stored in an above ground tank, below ground tank, or even stored in a dam. Such a dam, which must be leak proof, is as shown below.

Water reservoirs: Water can be piped from its source and kept in reservoirs inside a house. Some of these may include pails, buckets, large tanks such as the 100 or 300 litre oil drums, etc. The problem with these is their low capacity. But they are useful especially in household / domestic chores such a cleaning, washing etc. Some reservoirs and tanks can be underground.

An undergroubnd enclosed water tank An above ground water tank

Underground open tank for water storage Pot for water storage

Special livestock watering troughs: In some cases, a watering trough can be made in the livestock ranch from where they drink any time they are thirsty. The trough can be constantly with water to avoid interrupting man's activities when livestock feel like drinking.

Pots and gourds: these are used to keep clean and cold water for special uses, e.g. in mixing chemicals and medicines such vaccines. The pots are special because in situations where medicines have to be kept cold e.g. NCD vaccine, in hot environments, its cold water can be used and the mixture stored in a normal thermos flask to keep the temperature low. This saves the dangers of the medicine losing its viability as time passes.

STRUCTURE OF AN EARTH DAM

A dam comprises the following: Overflow pipe, concrete splash plate, dam wall, and a dam wall (stonewall). The concrete splash plate is made of stones and cement. It prevents erosion and damage of the dam wall. The overflow pipe or spill way is used to prevent water spilling over the top of the dam. It therefore controls excess water from wearing out the dam. The shape of the dam is constructed wider at the bottom and narrower at the top. This is because water pressure increases with depth. The dam has to hold back a great mass of water and all the pressure it exerts. The inner surface is covered with large stones to prevent erosion by water wave action. The outer surface is covered with grass to prevent rain erosion. Overflow pipe or concrete channel is there to maintain the water level.

STRUCTURE OF AN EARTH DAM

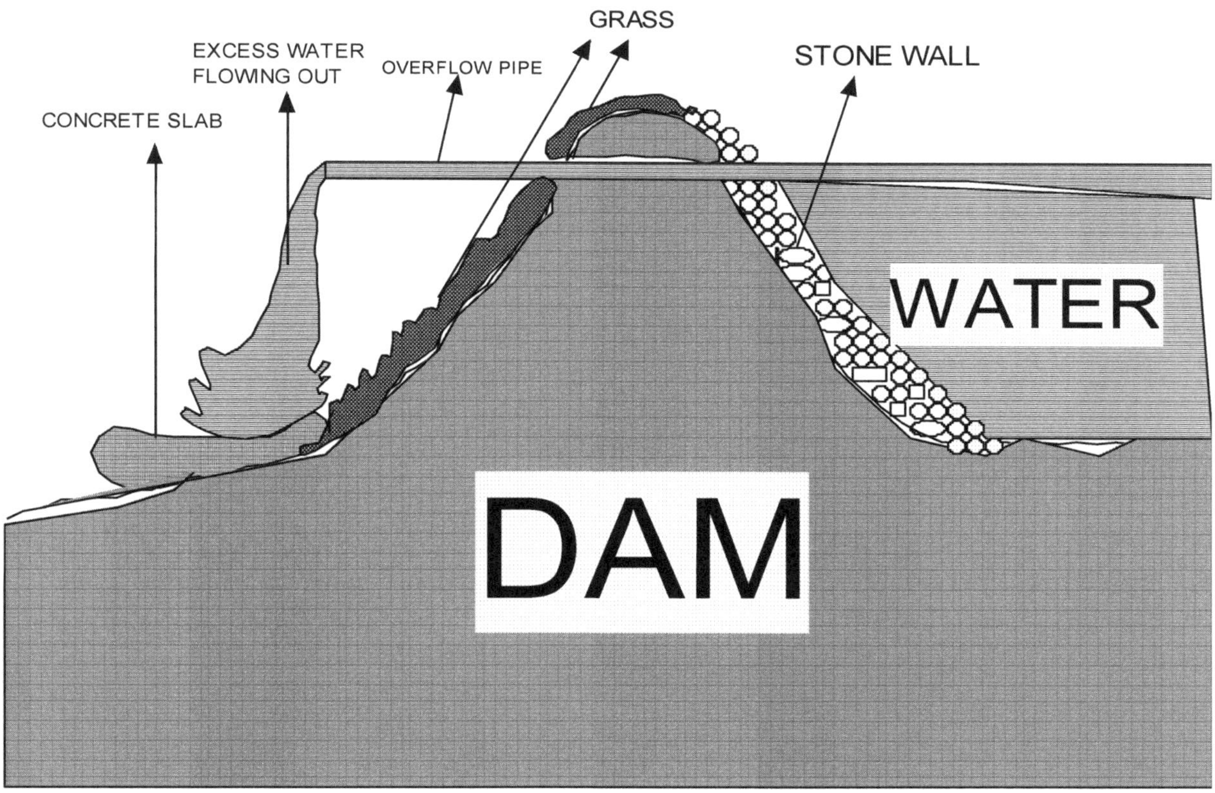

A WATER STORAGE DAM / RESERVOIR.

Figure 159: An earth dam

17.8 FARM TOOLS, IMPLEMENTS AND MACHINERY

FARM TOOLS

Equipment which are mostly hand operated and used to perform small farm work.

Farm implements: which are externally operated by power e.g. plough, harrow etc

Farm machinery: are farm equipment that generate power and used for specialized purposes on the farm e.g. Tractor, water pumps etc.

Figure 161: The farm ower sources: Left- donkey ploughs; Right: The tractor

MODE OF OPERATION OF A 4 STROKE ENGINE

The four-stroke engine operates under the following four stages: induction, compression, combustion and exhaust strokes. These are described below, with an illustration accompanying the explanations.

INDUCTION STROKE

Pistons moving down

Exhaust valve open

Inlet valve open

Mixture of air and fuel drawn into cylinder

COMPRESSION STROKE

Piston moving up and compresses mixture of air and fuel

The compression creates high temperatures due to reduced space in the cylinder

Both valves closed.

POWER STROKE
Spark plug ignites (sets fire) the mixture
Piston moves down again, pushed by expanding and burning gases
Both valves closed.

EXHAUST STROKE
Piston moving up
Inlet valve closed
Exhaust valve open
Burnt gases are driven out of the exhaust Passage.

LUBRICATION SYSTEM OF ENGINE
Lubrication is the application of oil or grease to moving parts of the engine with the aim of reducing overheating and friction. Lubrication system in the petrol and diesel engines consists of the sump, oil filter, oil pump and oil ways.

Sump: This is the reservoir for oil. It collects oil after the oil has gone through the engine.

Oil pump: Pumps the oil from oil sump into parts of engine

Oil filter: Cleans oil as it leaves the oil pump. Only clean oil is therefore circulated round the engine

Oil ways: Are holes drilled in the engine to carry oil to moving parts.
Water is kept in a radiator and supplied to engine through a hosepipe. While in the engine, water surrounds cylinders, cylinder head and absorbs hearts produced in these parts of the engine. When the water gets heated, it becomes less dense and flows to upper part of the radiator and it is replaced by cool water from the bottom of the radiator. The radiator is fitted with a fan to cool hot water. The water is therefore recycled within.

ELECTRIC SYSTEM
The tractor electrical system consists of the battery, generator, ignition coil, distributor, starter motor and the lighting system. These are as described below.
Battery: Supplies the electric power
Generator: Generates electricity for recharging battery

Ignition coil: Changes the low voltage current produced by battery to higher voltage current.

Distributor: Distributes electric spark to each plug.

Starter motor: Uses electricity from battery to start the engine.

Lighting: Headlights, parking lights, break lights etc. All use electric current from battery.

Table 47: DIFFERENCES BETWEEN PETROL AND DIESEL ENGINES

	PETROL ENGINE	DIESEL ENGINE
1	Uses petrol	Uses diesel
2	Has carburetor	Has injector
3	Has spark plug	Has no spark plug
4	Produces less smoke	Produce a lot of smoke
5	Cheaper	Expensive
6	Light in weight	Heavy in weight
7	Less noise	A lot of noise
8	Air and fuel first meet in the carburetor	Air and fuel first meet in the cylinder
9	Air/ fuel mixture is compressed During compression stroke	Only air compressed during Compression stroke
10	Fuel ignited by electric spark	Fuel ignited by compression

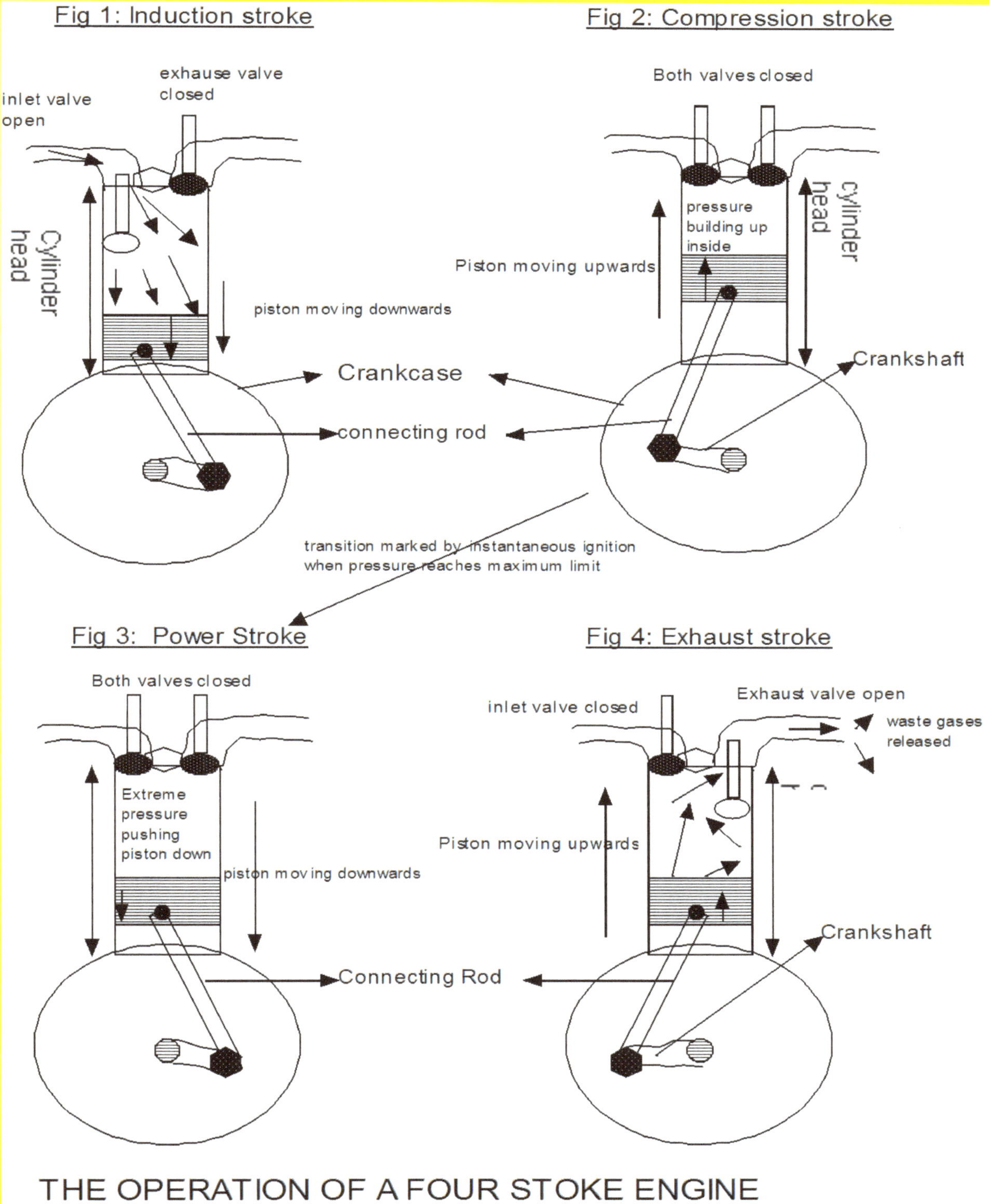

Figure 162: Early Internal-Combustion Engine

WATER COOLING SYSTEM

ROUTINE CHECKS BEFORE AND DURING OPERATION OF A TRACTOR

CHECKS BEFORE OPERATION

Regular checks should be on:

Engine oil using dipstick

Water level in the radiator

Fuel level

Tyre pressure

Loose nuts and bolts

Condition and tension of fan belt

Air cleaner for excessive dust

CHECKS DURING OPERATION

Regular checks should be on:

Brakes

Lights

Clutch

Side mirror

Steering

Horn

MODULE 18: AGRICULTURE FOR SUSTAINABLE LIVELIHOODS

AN OPINION PIECE DEPICTING CASE OF IRONY IN AFRICA'S FOOD INSECURITY: NYANZA PROVINCE JOURNEY TO FOOD INSECURITY 50 YEARS AFTER INDEPENDENCE

INTRODUCTION:

Food Security

Food security is an all-time availability of the right amount and type of food to all individuals regardless of their temporal and spatial circumstances. It is indeed the bottom line of any form of development When achieved by any nation, food security quickly leads to economic empowerment of citizens. This is both direct and indirect-through increased savings and investment, agro-based industrial development, improved nutritional status and health, and increased productivity of the citizens. On the other hand, the resources that could have been utilised for food imports become available for other forms of development such as education , research and infrastructure, thereby enhancing food production further.

Factors Affecting

Food security is an end to a long process it is affected by: seasonality, policy, issues, infrastructure, market efficiency, socio-economic aspects, culture, climate, population structure, distribution and dynamics, resource utilization and planning, technology and edaphic factors. A detailed discussion of each of these follows later.

Seasonality Of Food Availability And Quality

Many factors affect food security (sec 1.2). Seasonality defines a time-based inequity in food availability. Thus for non-relay farming systems, largely rain-fed, harvesting time is usually associated with surpluses, leading to careless handling losses and even generous donations. Gleying is often then left for children and the poor, among other deprived lots, and is inefficiently done. Even the poorest temporarily forget about their misfortune. Food is cheapest, often sold at through-away prices by the poor to enable them have money to purchase other commodities. The beneficiaries of this practice are often the rich who have the money for purchasing the farm produce; the means to transport it to a depot; a store; resources to pay those employed to handle and manage the business, a means of keeping the produce pest and diseases free (i.e. pesticides for store e.g. fumigants) and resources to keep the owners going through the period of plenty without interfering with the stored produce. Of course, not forgetting the security of the produce for the whole period.

Once the poor have depleted their reserves and the money from crop sales, they start purchasing food from the villagers privileged enough to have stored more than their domestic demand. The village reserves soon get depleted, and they resort to the rich depot owners who sell the produce at dictated prices well characterizing monopoly. For survival, the poor villagers have to take the price. With limited, and quickly diminishing capital base, the poor soon start going hungry- experiencing a very abrupt drop from Heaven to Hell. Amidst this hunger, they somehow manage to produce another crop in the new cropping season. This period of hardship is characterised by different types of famines such as 'kech mar amiyi meru?' I.e. the famine of 'Do I give you your mother?' according to the Luo in Kenya.

The Rich Opportunists Or Saviors

The rich thus get very good harvest in both seasons of harvesting and famine. Having accumulated enough resources during famine, they dig some of this onto other businesses while timing the next harvesting season. With time, the little village resources are accumulated by only very few tycoons who may come either from within or outside the village itself. If the persons come from the village, it is better because one is obliged to share the accumulated resources with the poor kins. Matters become more complicated if the rich person originates from outside the village or community. In that case (s), he only sucks the village resources without offering any opportunity to share the profit with the poor farmers. Thus it becomes permanent village poverty with no immediate cure vicious cycle. In this respect, seasonality greatly affects the poor while it benefits the rich. Some benefits however, may trickle to the poor if the rich merchants originate from the villages where they trade. Thus the vicious cycle in this case has weaker boundaries that can be broken with positive will. In Kenya, the major cereal sellers during drought and famine originate from Rift valley and Central Provinces. This is the major cause of income disparity among provinces, with a rather dangerous vicious cycle of poverty hanging in the air in Nyanza, Eastern, North Eastern and Coast provinces. Nyanza is worst placed, but with potentials of breaking the cycle. This is if food production and proper nutrition become the priority agenda in the whole Nyanza.

Food Security And Government Policy

One very important factor upon which food security depends on government policy. This is more-so important for Kenya which is 80% marginal (arid to semi-arid). The

Kenya Government, however, underestimates this % of land that is marginal, since it recognizes Nyanza province as 90-100% high potential. In this regard, the agricultural policies have often been too misplaced for Nyanza province since independence- a period spanning almost 4 decades.

Agro-Ecological Zones Arid Mismatch Of Crop Enterprises.
Thus many urban-based policy makers, unaware of the real ground situation as far as the Nyanza agro-ecological zoning is concerned, have always recommended very wrong agricultural enterprises for the region. Thus the Kisii highlands and Maseno (AEZ UM, LH 1-3), South Nyanza sugarcane zone (UM2-3), Gem (UM1-2,LM1-2), the Nyanza sugar belt (UM1-3, LM1-2), seem to have been considered of the same potential as the arid, semi-arid and transitional Yimbo, Uyoma, Alego, Kano plains, Seme, Karachuonyo, and the marginal parts of south nyanza regions In an effort to be recognized as early adopters, many farmers in these marginal, arid and semi-arid areas have tried to cultivate water loving crops such as coffee and HYVs of maize. The resulting low yields and chronic crop failures have left many residents of Nyanza disillusioned. They have thus lost hope in agriculture and migrate to towns for non-existent jobs.

The Green Revolution: A Blessing Or A Curse?
With the traditional, drought resistant and time tested crops such as millet and sorghum, groundnuts, the grams, among others, neglected or relegated or to the periphery, there have been persistent hunger in the entire Nyanza. The residents thus have had to rely on imported cereals especially maize, and to an extent, rice, for general survival. It is this reliance on imported foods from other Kenyan provinces-especially western, central and rift valley that has worsened the poverty crisis in the whole Nyanza. The little income that the residents earn or receive is spent entirely on purchasing food that, because it has been transported from far, is often very expensive. There are peak months when a 2kg tin of maize (locally known as Gorogoro) is sold at Ksh 60.

A Food -Need Assessment Model
Assuming that an adult who solely relies on maize-meal for energy consumes 3kg per week (ksh100), a one adult family needs 156 kg of dry maize per annum for reasonable feeding. This is equivalent to ksh. 4800 per annum or, to be reasonable, ksh 400 per month.

In a household of 6 the most common in the province, at least ksh. 2400 would be needed for maize purchase only per month. If the other dietary requirements are also taker onto account, and assumed to takes 60% of a family diet, then a total of ksh 6,000 would be needed for food security within the household in Nyanza per month.

However, the average per capita income for Kenya is about US 250 (ksh15,000).Third translates to Ksh 12000 per month. For Nyanza, this is an exaggerated figure because its per capita income is 60-80% of the nation value. Thus for a family in Nyanza, the monthly per capita income could be ksh.900. Often, with only at most half the family working, this would offer the family ksh. 2700 per month at disposal to meet all life demands. If compared with the ksh. 6,000 needed per household per month for food purchase, this ksh.2700 is clearly deficient. Therefore on monthly basis, a deficit of ksh. 3,300 is encountered with respect to food alone. It is clear that food security is clearly hard to achieve with the current structure of families and household, income levels, agricultural trends and feeding habits as they exist in Nyanza.

Malnutrition And Food Insecurity In Kenyan Lake Basin

With food insecurity and malnutrition clearly and deeply entrenched in Nyanza it seems clear that poverty alleviation still remains a tall order. The model set above indicates clearly that with all family household resources pooled together food alone is still deficient; leave alone other basic needs such as housing, water, transport, education etc. The scenarios is worsened by the fact that 80% of this food is imported from other neighboring and far Provinces. Thus all the income (Ksh 2700 per household per month) is again stashed away to other Provinces, with the food importing provinces remaining poorer and poorer, always operating at subsistence and survival level. This worsens the poverty situation. This has been the situation in Nyanza Province and dry land parts of Western Province for the last 3-4 decades.

White Collar Job Syndrome

With no monthly incomes left for non- food needs, many households in Nyanza, disillusioned with agriculture due to cool season and cash crop bias, have had to sell their other property such as land and livestock to be able to take their children to school. This has been their understanding of investment. To an extent, a number of parents have excelled in this respect because their children have attained the highest levels of academic and professional qualifications. These educated sons and daughters who have been lucky enough to acquire white-collar jobs have acted as granaries to their entire immediate families, and at times extend families. Since most such parents sold most of their wealth to use for educating their children, there is often very little, if any, left for family survival nor for educating younger brothers and sisters. Thus in most families in

Nyanza, many have remained un-educated and have had to rely on the income from the only educated brother or kin. This has worsened the dependency ratio beyond manageable levels. The little that trickles from the educated to the next of kin is again used to purchase food imported from other Provinces yet another vicious cycle.

Agriculture As A Profession Of The Last Resort

Most Nyanza residents took to education as a means of avoiding agriculture, farm-work or any form of gardening. This was because right from colonial days, agriculture was like forced on the citizens - who were providing forced labour. The labourers were fed on, according to historical records, maize-meal and salt. This shaped their appetite towards maize, and away from traditional foodstuffs. The seeds of white desired crops were distributed free, while the extension workers were trained to discourage traditional crops and foods. The school-going children were fed on maize and other imported foodstuffs - i.e. those that were not produced within their environment. Agriculture was used as a part-time activity, especially as a punishment for indiscipline. Thus even students and pupils developed a very negative attitude towards the subject and the profession. Thus in Nyanza, the educated preferred other courses such as medicine, law, engineering, education and commerce to agriculture. This is why there is a low percentage of agriculture professionals in (the Luo) Nyanza. The situation was a little different in highland parts of Nyanza such as Kisii. It was regarded a profession of the last resort largely for failures and un-educated.

For the un-educated, agriculture was still inevitable even during the colonial days and thereafter. However, to avoid being called names such as backward, inferior and outdated, they avoided cultivating traditional crops such as millet, sorghum, sweet potatoes, cassava etc. They instead preferred to cultivate maize even if it was clear from experience that there was little to expect from it . This maize -bias in agriculture feeding worsened the food security in dry land parts of Nyanza. In most high potential areas where maize would be appropriate, cash crops such as sugarcane, coffee and tea became too popular. Thus all parts of Nyanza - regardless of agro-ecological zone- experience food insecurity to varying levels- from subtle to extreme.

The Vicious Cycle Of Poverty In Sight

Due to persistent cultivation of wrong crops in the Province, coupled with unsustainable feeding habits it has been too costly availing sufficient food. The population has therefore remained poorly fed, worsening the already low levels of labour productivity, and intensifying urban migration further. As a result, largely low income; physically weak and sickly elderly men and women, helpless dependents and children are left in the Nyanza villages. The educated and/or income earning, able bodied young men and women have all literally moved to urban centres, cities and towns. The village development is thus left at the mercy of the helpless. Armed with only the traditional management skill, the weak elderly villagers can hardly be relied upon to shape the destiny of the villages. With their physical weaknesses, the villagers produce too little for any honorable living. They largely rely on the support from their urban-based relatives who send them money for use in purchasing grains (maize) imported from other provinces.

Yet the urban-based educated elites look upon the rural based colleagues to produce food for them to survive. As a result, there is no food production each party is only hoping that the other party produces it. It is this phenomenon that has worsened food crisis and indeed it is also worsening poverty. This is because the little earned by urban-based Nyanza professionals is sent back to other provinces- and perennially so.

From socio-economic profiles written by Prof. Odada (1990), Nyanza residents are poorest, only compared to Coast and North Eastern Provinces, Eastern, Central and Western Provinces and Kisii highlands in Nyanza have been long entrenched in productive agriculture -with food production well balanced with cash crop production. This effort has led to good production of food and cash to support the communities. Since Kenya's agricultural policies are highland, wetland and cool-season based and biased, these regions have greatly benefited from the economy right from 1930's. It is in these three provinces, besides the cosmopolitan Nairobi, where the best infrastructure to date exists. This has helped them develop their agriculture sector even more. This was strengthened by the establishment of marketing boards for different cash crops by the government. Thus coffee, tea, pyrethrum, maize (NCPB)and related co-operative societies are found in the highland parts of Kenya.

The Dryland Crops And Lands: Some Hopes And Need For Change

Thus it is clear that the neglect of dryland crops and therefore dry land by the government right from pre-independence days to date has worsened food crisis days to date has worsened food crisis and poverty in dryland provinces, in Kenya, Nyanza included. Thus cotton, simsim, groundnuts, cassava, sweet potatoes, sorghum or millet, and dryland fruits have no formalized marketing institutions. They are regarded as subsistence crops, and this has helped to keep their production low, thus worsening food insecurity and poverty in those areas. There is therefore an urgent need to properly classify the existing land resources identify their agricultural potential (AEZ) and policies to ensure farmers are encouraged to appropriately utilise their land resources. It is already clear that regions (e.g provinces) that have strong agricultural base especially in all-year long food production in either monoculture or rotational basis are better placed to curb food insecurity. At the moment, for instance in Kenya, inhabitants of agricultural districts are comparatively richer than their non-farming counterparts.

Link With Colonialism Colonial Hangovers:

In as far as infrastructure is concerned, high-level development is associated with high agricultural potential and activity. This started all the way before independence, a time that the Kavirondo was a neglected region due to its unfavourable climate for both white-man's life and his desired farm enterprises (livestock and crops), as well as being malaria-prone. No machinery was put in place to elevate the economy if this highly populated part of the country. The residents from here used to be popular resident labourers in the white highlands- largely due to their heavy built and work stamina. Thus high from the beginning the majority got used to a salary, however, meagre. They have therefore, to date, internalised the culture of income-earning employment outside their homes and this charactenses the majority of Luos and to an extent Luhyas. It is wrong culture that urgently needs to be changed. Thus it is not uncommon to have school leavers being referred to as 'having finished school' and asked, or ordered to go to towns to look for jobs, or their town-based relatives asked to accommodate and look for them a job. This population dynamism- outmigration- has deprived the rural areas of labour and therefore reduced and/or incapacitated food production-since people live on hopes of getting jobs instead of creating jobs themselves in the villages.

Farm Yields, Food Insecurity And Population Dynamics.

Due to out migration from dry land Nyanza districts, many rural centres have remained lowly/sparsely populated. This has resulted in: low market demand for rural farm products, thereby suppressing further production; uncontrolled land use, with livestock left to roam randomly, increasing in numbers helplessly, with the result that the carrying capacities of lands have been exceeded; soils heavily impoverished; lands seriously degraded and the livestock themselves helplessly emaciated bony and skiny. With the cattle complex inherrent among the luo community, it has been hard controlling wanton increase in livestock/herd sizes. The livestock have thus caused more havoc by trespassing into farmlands (especially off-season crops)and fields with cassava and sweet potatoes. Since most villagers are cousins, it is a taboo to report a neighbour to the authorities (e.g. chief or sub chief) for compensation even in cases of clear crop destruction by livestock. People thus lead such social and communal life that it has been difficult for role models in agriculture to progress, unless they fence their plots using barbed wire and other strong fence types. Since many lack the resources to do strong and durable fencing, people have preferred to keep a low profile. So nobody does anything; the little that is done is destroyed by livestock without the owner having the authority to question. This is how poverty and food insecurity have worked losely to reinforce each other. Until now, there is a poverty cycle-a syndrome that requires a lot of resources and skills to break.

Culture, Taboos And Food Insecurity

The Luo culture has played a very central role in reinforcing poverty and food insecurity in Nyanza province. Initially introduced as a means of enforcing respect and responsibility among extended family members, the taboos have now mostly outlived their usefulness. For instance, all farm activities starting from clearing, to primary cultivation, harrowing, planting, weeding and harvesting were done in order of age, starting from the household of the eldest member of a homestead, or sub-clan. This had clear advantages such as:respect to the elderly; pressure on all families to engage in farming thus ensuring proper labour utilisation; pooling of resources together was possible because of lack of competition for the same within the homestead/sub clan; and lastly, there was co-ordination with all members aware of the state of the others.

The method, however, had a number of disadvantages, some of which are real to date. Some of these demerits include: lack of progress in cases where the elder or senior person or family is sluggish-often leading to either insufficient preparation and/or farming or incomplete non-participation in farming in some years; in cases where one family member died, the whole family chain was affected- since it was a taboo to do any farming upon death of a close kin-especially upto 1 month following the death . In such cases, the crucial timing of farm activities is not done, leading to low or no farm yield.

The Soils Of Nyanza

Edaphic factors refer to soil physical, chemical and biological characteristics. Nyanza soils are all fertile - whether for rain fed or irrigated agriculture. A majority of Nyanza soils located in dry lands are also quite irrigable.Thus with all due respect, Nyanza province is a prime area for dry land and water-loving irrigated. If properly utilised, the land can fetch good returns to the community. The only limiting factor is rainfall, on basis of which 90% of Bondo District, 70% of Siaya district, 70% of Kisumu district, 20% of Migori district, 70% of Suba district, 60%of Rachuonyo district and 70% of Homabay district are classified as either marginal or semi-arid (i.eLM4-6). It, however, happens that the most arid of all these have vast surface water resources in Rivers or Lakes. Thus if the available water resources are fully tapped, Nyanza Province can be a granary for the whole country. This is more so because it has very favorable temperature of 17-26% most of the year and excellent irrigable soils - very good conditions for horticultural development.

Crops Of Hope

In this respect, any heat-loving crop can do excellently in Nyanza. On the current circumstances, studies show that dryland crops can perform reasonably well under rain-fed agriculture. These crops include grams, beans, millet, sorghum, cassava, sweet potatoes , cowpeas, pigeon peas, simsim, groundnuts, sunflower and indigenous vegetables. This combination can efficiently provide a balanced diet to the entire population of Nyanza as long as the right agronomic practises are practised. This can be ever more successful if appropriate cropping system such as relay farming, crop rotation etc. are practised.

The problem of pests, diseases and drought can be skillfully approached by observing the basic substainable intergrated crop management practises such as cultural methods (early planting and other operation; correct land preparation-primary, secondary and tertiary; mulching etc); appropriate and correct technological choice (proper seed quality of the right crop variety suitable for the specific ecological zone use of appropriate fertilizers e.g compost manure prepared from resources such as the water hyacinth -thanks ICIPE for the effort at Kombewa, Seme etc.

Long Term Development Strategies

Under irrigation- that is a priority for long-term food security in Nyanza - all types of warm season crops can do impressively in Nyanza. This is especially true for the wide range of fruits and vegetables. Thus Nyanza has the potential of turning the current food insecurity challenges into riches overnight. The expected increased crop yields will surely give rise to surpluses that will form the basis of cottage industry development. Thus with agriculture intensified using right approaches stated above, food security is sure to be achieved in the whole country with not only surplus for export, but also extra for processing. With the basic tenents of industrial development built, sky will be the limit for Nyanza. Poverty will be sorted out once and for all once the correct agricultural/farming efforts are put in and food security ensured.

MODULE 19: BIBLIOGRAPHY

Abbott J.C. and Makeham . Agricultural Economics and marketing in the tropics. Longman scientific and technical, Hongkong.

Afullo O, Danga B, Adhiambo, H (1994). Background socio-economic survey report of the marginal areas of the Lake Victoria basin, INDICRECE, Kisumu. Pp 1-10.

Alan King (1985) Agriculture: An introduction for Southern Africa Cambridge University Press, Cambridge

Alila, Patrick O. (1979), Higher levels of agricultural production: Role of least developed farmers. IDS, UON.

Anyanzo L. (1988) Farm equipment, Machinery, Structures and buildings. East African Educational Publishers, Nairobi.

Humphreys L R. (1978). Tropical pastures and fodder crops. Longman group Limited, Hongkong.

Areader,1989, Integrated management of Resources in Africa.1st edition,

Arthur, Thomas (1967). Farming in hot climates. 1st edition Faber & Faber Ltd. London pp 28-82.

Atieno Odhiambo, Ouso T I and Williams J F M (1985). A history of East Africa. Longman, London. Pp 1-158.

Brandt Commission (1983), Common crisis-North and South,Cooperation for world recovery.WCED

Botswana government (2000) Botswana general Certificate of Secondary Education Agriculture assessment syllabus. Examinations, research and testing Division, Ministry of education.

Chambers Robert (1983) Rural development: Putting the last first. Longman scientific & Technical, Newyork. Chapters 1-8.

Chengeta J and Matlhare C (1998) Enjoy Agriculture Book 1. Hodder and Stoughton.

Clariana, Roy B. 1982, Teaching Biology in Kenya: A Biology practical manual.

Cone, Wilson and Lipscomb J F (1972). History of Kenyan agriculture. University press of Africa, Nairobi. Pp 13-16.

Daily Nation, 23rd Jan, 4th Feb; 7th July; 9th, 10th, 11th and 13th Sept' 1993.

Elliot I (19977. Elliot I (1997) Agriculture: A practical course for Botswana Book 1 Longman, Botswana.

Elliot RI, Stout GW, Dejardin EJ and Sithole DS (1998) Agriculture for Southern Africa: A practical approach upto Õ Level. Bell & Hyman in Association with Academic Books Zimbabwe, LONDON

Elliot I (1998) Agriculture: A practical course. Book 1 Longman, Botswana.

Elliot I (1997) Agriculture: A practical course for Botswana Book 3. Longman, Botswana.

Elliot I (1997) Agriculture: A practical course for Botswana Book 2. Longman, Botswana.
FAO (1996). (unknown title: Tropical soils biology and fertility unit). FAO, Rome.
FAO of the UN (1991) FAO/Netherlands conference and the environment,
Fernie, John and Spitketh, Alan S. (1985) Resources: Environment and Policy. Herper and Row publishers.
Friedland, J. Andrew, 1989, Readings on Environmental issues in Kenya, Environment and Development. A report on Dartmouth College 1989 Kenya Environmental Studies Foreign Study Program, Dartmouth Foreign study Program, USA.
GOK (1994) Siaya district Development plan. Government printers, Nairobi. Pp 5-9.
GOK (1986). Sessional paper no. 1: Economic management for renewed growth. Government printers, Nairobi. Pp 1-61.
GOK (1994) National Development plan. Government printers, Nairobi. Pp 5-9.
GOK (1994) South Nyanza district Development plan. Government printers, Nairobi. Pp 5-9.
GOK (1989) Siaya district Development plan. Government printers, Nairobi. Pp 1-30.
GOK (1989) South Nyanza district Development plan. Government printers, Nairobi. Pp 1-69.
GOK/ Government of Kenya, 1989/93 National Dev. Plan Ministry of National planning and Development, Government Printers.
GOK (1994). Kenya farm management plan. Government printers, Nairobi.
Henry E. (1987). Soil conservation. College press Ltd, Harare, Zimbabwe.
Hill DS, Waller JM. (1992). Pests and diseases of tropical crops. Volume 1. Principles and methods of control. Longman group Limited., Essesx.
IGAAD (1990) (Unknown title). IGAAD, Nairobi
Jackson R and Hutchinson C (1996). Gardening. Harper Collins Publishers, London.
KIE 1991, Secondary Agriculture: Form 3 pupils Book. Kenya Literature Bureau.
KNAS/ Kenya National Academy of Sciences (1988), Indigenous technologies of Kenya. 1st edition. KNAS.
Kotschi, Ann Waters; Bayer, Reinhard Adelhelm and Johanes, Ulrich Hoelse (1989). Ecofarming in agricultural development: Tropical agroecology 2, Verlagjosef margraf scientific books/GTZ.
Lenilian J and Fletcher W (1977)(eds). Environment and man: Vol 6 in the chemical environment. Blackie, Glasgow and London.
Leslie S Cobley (1976). An introduction to the botany of tropical crops 2nd edition, Longman, Newyork pp 1-59.
Lester R. Brown (1987) State of the world Report on Progress towards sustainable society, Worldwatch Institute.

Madeley, John (1991) When aid is no help: How projects fail and how they could succeed. (1st edition).Intermediate technology Publications, London

Middleton N (1988). Atlas of Environmental issues. Oxford University Press, Oxford.

Migori D.D.P., 1989/93 Migori District Development Plan, Ministry of National planning and Development, Government Printers.

Moran FT. (1992). Success in Vegetable and fruit production. Longman Zimbabwe.

Muchira SG and Ernelholm N. (1989) Agriculture for secondary Schools: A four year course> Heinemann Kenya

Ngugu DN, Karau PK and Nguyo W (1990) East African Agriculture

OAU (1980), Lagos Plan of Action for the economic Development of Africa.

Obara D and Ogonda R (1990). Kenya Secondary School Atlas. McMillan Kenya, Nairobi.

Odada J and OtienoJ (eds)(1990) Socio-economic profiles. GOK and Unicef, Nairobi. Pp 121 – 146.

Owen G (1984) O-Level Agriculture. Longman group Limited, London.

Roger Van (1972) The Agricultural history of Kenya. Historical Association of Kenya, Paper # 1. East African Publishing House, Nairobi. Pp 1-89.

Sakira WA (1981). Ó Level Agriculture. Oxford University Press Nairobi

Siaya D.D.P., 1989/93 Siaya District Development Plan, Ministry of National planning and Development, Government Printers.

Srinivasan, T. N. (1981), Malnutrition: Some measurement and policy issues. World Bank reprint series No. 178.

Srinivasan, T. N. (1977), Development, poverty and Basic human needs:some issues. (1st edition) World Bank reprint series No. 76.

The East African Standard, 23rd, Oct. 1997

The East African Standard, 23rd, Oct. 1993

UNEP (1985) State of the environment (2nd edition): Environmental aspects of emerging agricultural technologies; Population and the environment. UNEP.

USAID 1986,Kenya development strategy statement for 1986.

Watson M J (1983) Agriculture for Botswana. Macmillan Educational and the Ministry of Education, Botswana.London.

WCED / World Commission on environment and Development, 1987. Sustainable Development - a guide our common future.

White H P (1971). Commercial agriculture in tropical Africa. 1st edition. Hichs Smith & Sons Ltd. Wellington. Pp 10 – 35.

Yusuf A. (1981) Mechanisation. UNEP. Nairobi.

WRI (World Resources Institute), UNEP, UNDP, and world bank (1996). World resources: A guide to the global environment. WRI.

Ngugi DN, Nguyo PK and Karau W (1990). East African Agriculture.McMillan, Nairobi
Fish and cereal traders in Kibuye and Kisumu municipal (Chiro Mbero) markets between, Kisumu, between 7th and 11th of June, 2004. Personal interview in June 2004.
Dr Lucas Ngode, Senior Lecturer, TICH, Dfaculty of Environment and food security, Kisumu, Kenya. Personal Communication in June, 2004.
Webster C and Wilson (1996). Agriculture in the tropics: Tropical Agriculture series. Longman. London.
Elliot R I, Stout G W, Dejardin E J, and Sithole D S (1985). Agriculture for southern Africa: A practical approach up to O Level Agriculture.
Bell and Hyman / Academic Books Zimbabwe. Harare.
AMREF (1994). Community nutrition for eastern Africa. Nairobi, kenya. Eds Ann Burgess)

REVISION WORKSHEET

1. Read carefully through your won notes and make summary sheets on the topic of livestock management.

2. Read again your own notes, this time write down the key facts and learn them as you go on the above topic.

3. Without looking at your note try to answer the following questions

   a what do you understand by the term Livestock Management?

   b Outline the reasons for housing livestock.

   c Describe the type of housing needed for any of the following (broiler, Layers rabbit, dairy animal or pigs)

   d Give four qualities of a good livestock house?

   e Explain what you understand by the term "stockmanship"

   f Explain stockmanship in the following areas for one (1) of the following livestock (Broilers, Layers, rabbit diary animals or pigs.)

4 Now check in your notebooks/textbook for the answers. As this information is very important, make sure that you know the answers. Correct any you got wrong. Ask your teacher if you are not sure. Spend some time in re-learning the materials you got wrong.

5. After doing this see if you can answer the questions from memory. Check with correct answers.

www.ingramcontent.com/pod-product-compliance
Lightning Source LLC
LaVergne TN
LVHW070943210926
848482LV00002B/80
* 9 7 8 9 9 6 6 7 2 0 5 6 6 *